SUSTAINABLE LOGISTICS

TRANSPORT AND SUSTAINABILITY

Series Editors: Stephen Ison and Jon Shaw

Recent Volumes:

Volume 1:	Cycling and Sustainability
Volume 2:	Transport and Climate Change
Volume 3:	Sustainable Transport for Chinese Cities
Volume 4:	Sustainable Aviation Futures
Volume 5:	Parking Issues and Policies

TRANSPORT AND SUSTAINABILITY VOLUME 6

SUSTAINABLE LOGISTICS

EDITED BY

CATHY MACHARIS

Research Centre MOBI – Mobility, Logistics and Automotive Technology Research Centre, Vrije Universiteit Brussel, Brussels, Belgium and University of Gothenburg, Gothenburg, Sweden

SANDRA MELO

LAETA, IDMEC-IST, Instituto Superior Técnico, Universidade de Lisboa, Lisboa, Portugal

JOHAN WOXENIUS

Department of Business Administration, University of Gothenburg, Gothenburg, Sweden

TOM VAN LIER

Research Centre MOBI – Mobility, Logistics and Automotive Technology Research Centre, Vrije Universiteit Brussel, Brussels, Belgium

Emerald

United Kingdom – North America – Japan
India – Malaysia – China

Emerald Group Publishing Limited
Howard House, Wagon Lane, Bingley BD16 1WA, UK

First edition 2014

British Library Cataloguing in Publication Data
A catalogue record for this book is available from the British Library

ISBN: 978-1-78441-062-9
ISSN: 2044-9941 (Series)

CONTENTS

AVOIDANCE

ACTING

LIST OF CONTRIBUTORS

Patrícia Baptista	LAETA, IDMEC-IST, Instituto Superior Técnico, Universidade de Lisboa, Lisboa, Portugal
Maria Björklund	Department of Management and Engineering, Linköping University, Linköping, Sweden
Ming Chen	TNO, Delft, Netherlands
Álvaro Costa	Departamento de Engenharia Civil, Faculdade de Engenharia, Universidade do Porto, Porto, Portugal
Sharon Cullinane	Department of Business Administration, University of Gothenburg, Gothenburg, Sweden
Jonas Flodén	Department of Business Administration, University of Gothenburg, Gothenburg, Sweden
Helena Forslund	School of Business and Economics, Linnaeus University, Växjö, Sweden
Jesus Gonzalez-Feliu	Centre National de la Recherche Scientifique, Lyon, France
Anne Goodchild	University of Washington, Seattle, Washington, United States
Egbert Guis	TNO, Delft, Netherlands
Hyun-Chan Kim	Department of Civil and Natural Resources Engineering, University of Canterbury, Christchurch, New Zealand

Diana Kusumastuti	Department of Civil and Natural Resources Engineering, University of Canterbury, Christchurch, New Zealand
Cathy Macharis	Research Centre MOBI – Mobility, Logistics and Automotive Technology Research Centre, Vrije Universiteit Brussel, Brussels, Belgium; University of Gothenburg, Gothenburg, Sweden
Dries Meers	Research Centre MOBI – Mobility, Logistics and Automotive Technology Research Centre, Vrije Universiteit Brussel, Brussels, Belgium
Sandra Melo	LAETA, IDMEC-IST, Instituto Superior Técnico, Universidade de Lisboa, Lisboa, Portugal
Katrin Molina-Besch	Department of Design Sciences, Division of Packaging Logistics, Lund University, Lund, Sweden
Joëlle Morana	Université Lyon 2, Lyon, France
Nina Nesterova	TNO, Delft, Netherlands
Alan Nicholson	Department of Civil and Natural Resources Engineering, University of Canterbury, Christchurch, New Zealand
Henrik Pålsson	Department of Design Sciences, Division of Packaging Logistics, Lund University, Lund, Sweden
Hans Quak	TNO and TU Delft, Delft, Netherlands
Richard Smokers	TNO, Delft, Netherlands
Edith Sorkina	Department of Business Administration, University of Gothenburg, Gothenburg, Sweden

Mathieu Strale	Institut de Gestion de l'Environnement et d'Aménagement du Territoire – IGEAT-ULB, Université Libre de Bruxelles, Bruxelles, Belgium
Lóránt Tavasszy	TNO and TU Delft, Delft, Netherlands
Tom van Lier	Research Centre MOBI – Mobility, Logistics and Automotive Technology Research Centre, Vrije Universiteit Brussel, Brussels, Belgium
Tom Vermeiren	Research Centre MOBI – Mobility, Logistics and Automotive Technology Research Centre, Vrije Universiteit Brussel, Brussels, Belgium
Johan Woxenius	Department of Business Administration, University of Gothenburg, Gothenburg, Sweden
Erica Wygonik	Civil & Environmental Engineering, University of Washington, Seattle, Washington, United States

TRANSPORT AND SUSTAINABILITY EDITORIAL BOARD

THE 4 A'S OF SUSTAINABLE LOGISTICS

SUSTAINABLE LOGISTICS

Mobility and logistics activities have been fundamental to economic development and social well-being for centuries, but it is only over the past 50 years that this has received interest as a major field of academic study and as a key determinant of, for example, business performance (McKinnon, 2010). In the last decades, the evolution towards globalisation and the opportunities presented by technological innovation have greatly increased the importance of mobility and logistics worldwide.

Nevertheless, the growing environmental concern of citizens and governments and the widespread introduction of the concept of sustainability have simultaneously placed increasing pressure on public and private activities to take all effects related to such activities into account, as elaborated on by Macharis and Van Mierlo (2013). Logistics, and especially freight transport representing its most physical component, has accordingly received much attention in the sustainability debate in recent years, due to the numerous external effects and the widespread effects on virtually all individuals (van Lier & Macharis, 2013). This has forced stakeholders involved in logistics processes to address the issue of sustainability, leading to the birth of terms combining adjectives such as sustainable, ecological, green, clean and lean with domain specific nouns such as supply chain management, logistics, freight transport and urban freight. Some specific terms such as logistics shades of green have also been introduced.

These concepts, although inter-related and sometimes with overlapping interests, differ on the focus of their approach. *Green logistics* essentially focuses on ways to reduce the environmental effects of logistics. *Sustainable logistics* is a broader concept, also taking into account the economic and social implications of activities, striving to improve economical, ecological and societal interests simultaneously (McKinnon, 2010). Sustainable logistics attempts to take into account this so-called triple bottom line, indicating that there is some kind of three-way trade-off (Fig. 1). In popular terminology, these three goals are often referred to as the three Ps: people, planet and profit.

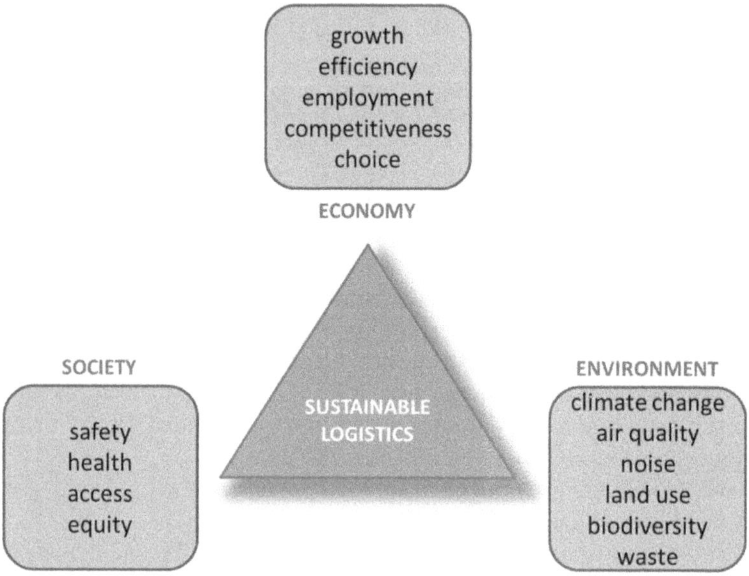

Fig. 1. Framework for Sustainable Logistics. *Source*: Based on www.greenlogistics.org.

Widening logistics from a purely economic concept to include ecological and societal dimensions requires firms to act. In practice however, they will only switch to a more sustainable solution if an economic gain is achievable, even if that is in the long run, or if they are forced to do it. In the next section we look at the positive rate of return from changing logistics into a sustainable direction bit more closely.

HOW TO FOSTER SUSTAINABLE LOGISTICS?

As mentioned, the environmental awareness of the logistics sector has increased considerably over the last years. Ultimately, investments in sustainable logistics should also have a positive return (see Fig. 2). Several studies have shown that within the logistics sector sustainability is a 'nice to have' but does not play a decisive role in most of the tactical and operational decisions of the logistics actors (Lammgård & Andersson, 2014; Vermeiren, 2013).

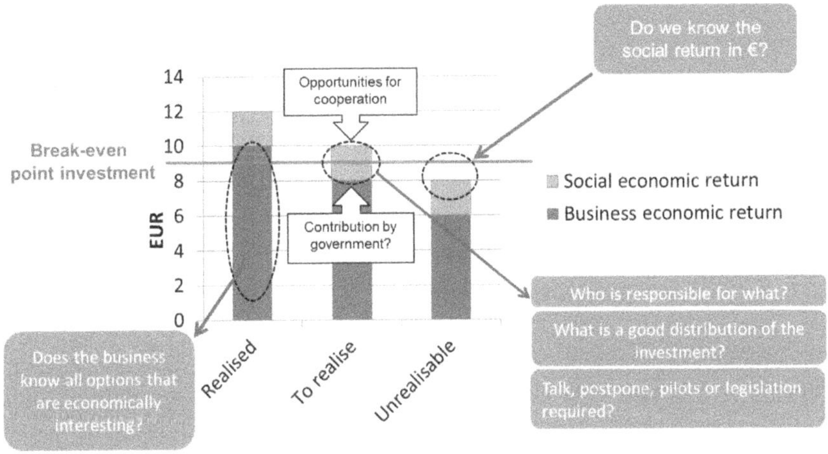

Fig. 2. Selecting Sustainable Logistics Measures.

So the easiest projects are those where the private actors foresee a business economic return clearly higher than the cost to switch to the new way of working, and on top of that the society will also gain from the switch. These are the ripe cherries to pick, and hopefully these measures are already widely adopted.

The second type of projects is the ones where the private actors alone will not get a positive return from the projects although society would gain from them. The question here is how this societal gain can still be reached. Public sector investments and start-up subsidies might be possible answers, or more companies can be brought together so that a critical mass is formed. It is clear that these projects are more difficult to implement and that gain-sharing mechanisms have to be well thought out and put into force.

The third type of projects is where the break-even point cannot be reached even taking the social return into account. Of course these are no projects to implement although one could argue that there is still a lot of scientific work to be done to properly monetarise the external effects of transport. Having a better calculation of these external costs would also allow a better knowledge of the trade-offs in this type of calculations and assist stakeholders when prioritising between measures.

The chapters in this book present several solutions related to all three types of projects and show possibilities for going to a more sustainable organisation and execution of the logistics activities. The chapters are developed from papers presented at the 12th NECTAR International Meeting in the Azores, the 25th Nordic Logistics Research Network (NOFOMA) in Gothenburg and the 13th World Conference of Transport Research (WCTR) in Rio de Janeiro, all held in 2013. The book is an outcome of the activities of NECTAR Cluster 3 Logistics and Freight and of the new WCTRS Special Interest Group B3 Intermodal Freight Transport.

POSSIBLE MEASURES

Possible measures are classified among the four A's of sustainable logistics (Macharis, 2014): *Awareness, Avoidance, Acting and shifting* and *Anticipation*. These A's also form the structure of this book.

Awareness

The first A is about awareness. One possibility to create more awareness is by measuring the effects of the logistics activities. If the effects can also be monetarised into costs (i.e. external costs) this allows for directly incorporating the results in a social cost benefit analysis to assess if a project would be beneficial for the society as a whole. The scientific work on external costs has been substantial over the last years (see, e.g. Fernandez-Barcelo & Campos-Cacheda, 2012; Russo & Comi, 2012) and more consensus has arisen on the way those external effects should be monetarised (IMPACT, 2008; van Lier & Macharis, 2013). Nevertheless, many issues are still open for further exploration, among which the willingness to pay principle (van Wee, 2012), the compensatory effects (De Brucker, Verbeke, & Winkelmans, 1998) and costs incurred over long time frames.

Another way to create awareness is by setting up programmes and certificates in which logistics service providers, transport operators and shippers deliberately show their environmental awareness. Examples of such programmes are:

- SmartWay – a US Environmental Protection Agency programme that reduces transport-related emissions by creating incentives to improve supply chain fuel efficiency,
- Lean and green – a voluntary programme for companies in the Benelux countries committed to a 20% CO_2 reduction in 5 years,

- Green Freight Europe – an independent voluntary programme for reducing CO_2 emissions of road freight in Europe,
- KNEG – a Swedish programme for cutting the CO_2 emissions from a typical road freight transport by half within 10 years,
- Clean air – a programme seeking to improve air quality in cities by bringing together the efforts of leaders from the public, private and NGO sectors, and
- Ecostars – a project establishing a number of fleet recognition schemes in European cities and regions to support energy efficient, cleaner goods and passenger vehicle movements.

These programmes have a positive impact on the daily operations of the stakeholders as they pinpoint the need to bring operational changes into the field.

Within this book, four contributions are presented under the Awareness umbrella. Smokers, Tavasszy, Chen and Guis explore the available options to reduce CO_2 emissions related to freight transport and their reduction potential. The authors propose a framework which, besides efficiency measures with net negative costs, also describes cost-neutral measures and measures with a net positive cost. They argue that the logistics sector cannot achieve ambitious CO_2 emission reductions, by focusing solely on cost-efficient measures. With a lack of incentives to reduce CO_2 emissions, a list of policy measures can push the market to implement measures that go beyond the traditional ones, which combine reductions in emissions with short term cost efficiency.

Cullinane reviews literature for both theoretical and practical mitigation measures from the perspective of long-distance transport providers. The literature review covers both scientific journal articles and reports from various development projects. She finds that firms prioritise energy efficiency measures as these often represent 'green gold', that is, they contribute positively to the private economic outcome for transport providers and their customers. The providers pay less attention to energy transition measures as the gains from these are believed to be more uncertain.

As an example of the first way to create awareness, Wygonik and Goodchild measure the environmental effects of three delivery services – conventional shopping trips with the consumers' own vehicles, local depot deliveries and regional warehouse deliveries. This is of course a very complex and contextual issue, but under certain assumptions, the findings from analysing the three scenarios with a North American data set suggest that the households' last mile transport is actually fairly efficient. The scenario generates the highest number of vehicle kilometres, but the 'professional' scenarios are hampered by much higher emissions per kilometre run by the delivery vans.

On the second way to create awareness, Björklund and Forslund present a classification model and compare four Swedish retail chains and classify their shades of green, evaluating their environmental awareness. Although the retail chains gladly communicate their sustainability awareness to their customers and contracts with logistics service providers stipulate environmental performance, Björklund and Forslund identify a certain lack of measuring and follow-up.

In all, these chapters tell us that there is indeed a wide buffet of measures to take to improve the sustainability of logistics as well as ways of measuring their effects. The public debate has been intense for many years, consumers expect sound commercial activities and it is fair to say that lack of awareness of the need for acting in a sustainable direction is very close to denial. Consequently, firms and policy makers are running out of excuses for passivity and competition is, at least in the long run, likely to favour early moving firms, nations and economic regions.

Avoidance

The second A is for Avoidance. Unlike for passenger transport, where one can think of avoiding travel kilometres by using telework or teleconference, avoidance within the logistics sector is more difficult although products like music are increasingly streamed rather than transported and sold as records. Avoidance can also be done more at the source, for example by 3D printing allowing for dense transport of material in bulk very close to the place of consumption and avoid moving artefacts which will never be sold. Another option is to rethink the design of the packaging in order to avoid wasted volume. A very good example is IKEA, which revolutionised the way furniture is brought to the end consumer.

Ultimately, however, most artefacts are indeed physically needed at the final destination. It should not be neglected that large scale freight transport has allowed the manufacturing industry to build ever-larger and more efficient plants. These are often also more efficient in terms of emissions per produced unit, and often this outweighs the emissions from the additional tonne kilometres. Nevertheless, there are many ways to avoid vehicle kilometres within the freight transport sector.

Improving the load factor by better bundling and balancing flows would have a considerable impact on the amount of vehicle kilometres. During the last 10 years, the load factor for the long haul has been averaging around 50% (European Commission, 2012). Transport for London (2012) found that the load factor was only 38% for the last mile when studying city distribution. By bundling many shipments and utilise larger vehicles, the load factor can be increased considerably. McKinnon and Piecyk (2010) calculated, based on data from Coyle (2007), that carbon emission factors for a 40–44 tonne truck are more than halved from 81.0 to 39.7 (gCO_2/tonne km) when the load increases from 10 to 29 tonnes.

Better bundling is possible by a further internal optimisation or by collaboration within the supply chain (with suppliers or customers) (van Lier, Caris, & Macharis, 2014) or horizontal collaboration with companies in another sector like the cooperation between Procter & Gamble and Tupperware, which allows for better utilisation of the lorry capacity combining Tupperware's voluminous products with Procter & Gamble's heavy products, or – and this is the most difficult one – with competitors (Cruijssen, Cools, & Dullaert, 2007). Horizontal collaboration is difficult, as it requires trust, a good and transparent gain-sharing mechanism, a good information and communication technology (ICT) platform that enables to couple flows and it has to respect anti-trust/competition legislation. Within city distribution, the tolerance for cooperation between competitors is fairly high and bundling might be done by implementing city distribution centres (Olsson & Woxenius, 2014; Verlinde, Macharis, Milan, & Kin, 2014).

In this book, two chapters address Avoidance. Gonzalez-Feliu and Morana add to our understanding by developing a set of performance indicators for pooling of resources in urban logistics. Experts were used for developing and verifying the indicators and structure them into a hierarchic dashboard based on the same four A's that structures this book. To facilitate the comparison between logistics schemes, Gonzalez-Feliu and Morana suggest to use a reference database and they exemplify by using the National Survey Database on Urban Goods Transport in France.

Next, Molina-Besch and Pålsson show how better and more integrated packaging development processes can lead to avoidance of unnecessary vehicle kilometres. In nine Swedish case studies they use interviews to explore what factors affect firms when developing packaging and their approaches to integrating the packaging development processes better with their supply chain partners. Despite the firms' integration aspirations, Molina-Besch and Pålsson find that integration with suppliers and customers is rarely practiced although most include integrative elements in their packaging development processes.

The two chapters show that there are clear opportunities to reduce the ratio between vehicle kilometres and tonne kilometres. Pooling resources has its problems, but there are benefits to gain and reducing packaging by considering needs along the supply chain in an integrative way is a very clear way of avoiding unnecessary transport.

Acting and Shifting

The third A is about acting and shifting, primarily on the operational level. The shift towards inland waterways and rail transport has been advocated for a long time now by the European Commission (see, e.g. European Commission, 2011). Although some expect more from intermodal freight transport, UIC (2009a) estimated that in 2007, about 333,000 trains carried 18.1 million twenty foot equivalent units (TEUs) corresponding to 185 million tonnes in European services combining rail with road and sea. The North American intermodal volumes are even larger; UIC (2009b) estimated them to 28.7 million TEU in 2007 with rather equal distribution between domestic and international traffic. There have been changes since, mainly due to the financial crises, but flows have recovered rather well indicating that intermodal freight transport can compete for base volumes and not only cut the peaks when all trucks are busy.

Within this book, Acting is the focus of four chapters. Kim, Nicholson and Kusumastuti discuss the factors that determine the mode choice in freight transport. The authors perform a revealed preference experiment in New Zealand and find that these modal choice preferences differ among industry groups and business types. In addition they investigate the major constraints for making an actual modal shift to rail or coastal shipping. A rank-ordered logistics model allows dissecting heterogeneity in these modal choice and modal shift preferences. The major constraint for shifting to rail was found to be transport time, while for coastal shipping accessibility and load size seemed to be the main limiting constraints.

Acting in a private economic environment implies that firms need to develop and implement suitable business models to turn ideas into services offered to customers. Intermodal freight transport often aims for a set of customers that in turn offer transport services to shippers. There are, however, examples when the wholesale model is replaced by a retail model where the intermodal operator sells directly to shippers. Yet another option is that shippers develop intermodal services for the sole use of moving its own cargo. Flodén and Sorkina elaborate on such an own-account business model and analyse it in the context of two empirical examples where Swedish shippers have initiated intermodal services without aiming to share the service with other shippers. Nevertheless, few shippers have the equipment, knowledge and permits to start entirely own services but have to rely on a long-term relationship with someone who can operate the trains. The case studies show that implementation can be quicker since the service does not require marketing to several customers to reach an acceptable filling grade in the trains.

In its last white paper, the European Commission (2011, p. 30) stated that by 2050, 50% of road freight transport over 300 km should shift to more energy-efficient transport modes such as rail and waterborne transport. Within their paper, Meers, Vermeiren and Macharis, focusing on intermodal transport, show that this distance may not be the focus of a modal shift policy. They conclude that a list of contextual factors such as the length of the post-haul can influence the actual break-even distance, which might be shorter than 300 km in the case of inland waterways transport. Therefore, they advocate a case-specific approach, based on local, spatial and price-related variables when designing local modal shift policies.

Within city distribution, the interface between long and short distance transport (Behrends, 2011) gains more and more attention and so is the use of trams for freight transport within a city (Arvidsson & Browne, 2013). The successful use of trams for freight transport in Dresden is however not transferable everywhere. Strale looks in his paper if the cargo tram is indeed a good solution for city distribution. He analyses the different cases throughout Europe and derives the limitations and opportunities of an urban cargo tram. Moreover, he applies the knowledge gained to the Brussels metropolitan area in order to assess the possibilities for such a concept. The development of a new cargo tram in Brussels seems very complicated at the moment, due to the complexity of urban freight distribution and operational, political as well as geographical issues. Nevertheless, with a careful approach and preliminary analyses, freight trams seem to have potential to contribute to a more sustainable urban freight distribution.

A second type of shifting is the shift towards night-time deliveries. This enables to avoid traffic congestion during daytime. Several demonstrations have been held in the United States (José Holguín-Veras, Wang, Browne, Darville Hodgec, & Wojtowicza, 2014) and in Europe with low-noise material (in Spain, the Netherlands, Belgium and the United Kingdom) adhering to the Dutch PIEK standard (MDS Transmodal, 2012). The

PIEK label certifies silent vehicles and equipment in order to avoid noise disturbances for residents, making them suitable for deliveries during the early morning, late evening and night (Browne, Allen, Nemoto, Patier, & Visser, 2012).

Action is obviously needed by some actor if the theories, analyses and all creative ideas should have effect on sustainable logistics. The four chapters contribute with examples of actions that contribute to a more sustainable organisation of the supply chain.

Anticipation

The last A is the one of Anticipation of new technologies. Certainly the shift towards new vehicle technologies will be essential in approaching the goal of CO_2 free deliveries. Several types of new drive trains are possible, ranging from gas, electric, hybrids to hydrogen. Messagie, Boureima, Coosemans, Macharis, and Van Mierlo (2014) have analysed the impact of the whole life cycle of the production and use of the car on CO_2 emissions, hence including both well-to-tank (WTT) and tank-to-wheel (TTW). In Fig. 3, one can see that the impact of electric vehicles is clearly lower than the diesel variants.

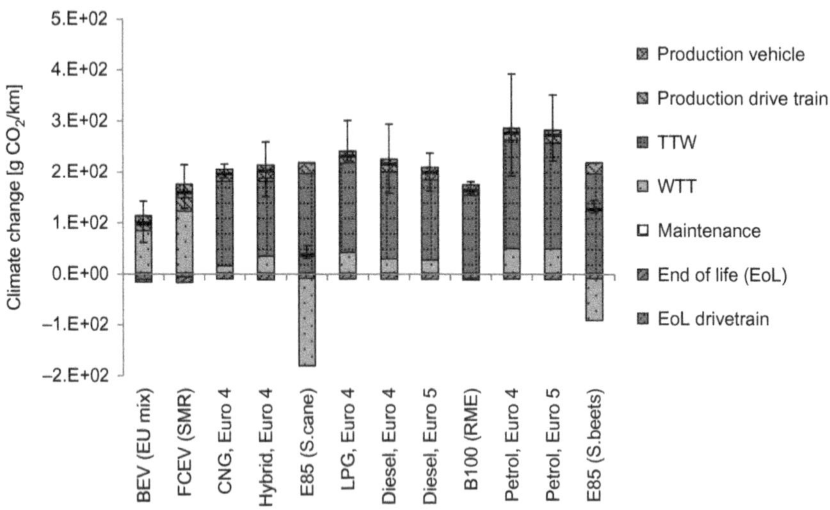

Fig. 3. The Effect of Various Vehicle Technologies on Climate Change. TTW = Tank-to-Wheel, WTT = Well-to-Tank. *Source*: Messagie et al. (2014).

Sensitivity analysis of how the electricity is produced shows that in most reasonable scenarios (e.g. not going all the way back to the coal production), the impact on CO_2 stays lower for electric cars than for diesel cars (Messagie et al., 2014).

Nevertheless, the introduction of the electric vehicles does not come without hurdles. Within the first of the two contributions addressing Anticipation, Quak and Nesterova present the drivers and barriers for the implementation of electric freight vehicles. The analysis strongly relies on results from the state of the art in Europe, in particular from the project FREVUE. The authors conclude that companies could be driven by sustainability awareness, the availability of subsidies to make the business model more attractive, the support from local government and the adaptation/anticipation from logistics operations.

In a last chapter, Melo, Baptista and Costa estimate the cost and effectiveness of using small electric vehicles (SEV) on city logistics operations. The impacts on traffic performance, operational costs and external costs are quantified and compared at different geographical scales, under scenarios of variable market penetration and different replacement rate of SEV's. Results show that the rate should be lower than 5% if the goal is to enhance sustainable mobility, which confirms the use of SEV's as a niche market working complementary to conventional vans. Leaving battery issues of SEV's aside, the comparison of external, purchasing and operating costs supports the replacement of vans by SEV's both on city as on street level. However, if battery costs are taken into account, current uncertainties lead to an unbalanced comparison.

Other technologies can also be used, such as sensors that allow for a better enforcement of parking at loading/unloading bays, which was demonstrated in Lisbon by the Straightsol project (Johansen et al., 2014) and the use of big data and ICT in order to improve the possibilities of bundling and synchronisation of supply chains.

The chapters under Anticipation show that electric vehicles will find gradually their place in the logistic chain and that also other technologies can help to make logistics more sustainable.

CONCLUSIONS

This book gives an overview of recent assessments and new developments in all the four A's: Awareness, Avoidance, Acting and Anticipation. These chapters show that indeed reconciliation between the economic and environmental goals is possible. We hope to give the reader inspiration for bringing these concepts into practice.

Does this classification of the four A's cover all the possibilities of a more sustainable logistics? Probably not. A fifth A should surely be added – the A of Actor involvement. Logistics in general and city distribution in particular, takes place in an environment where several actors interact, influence each other and might even have contradicting goals (Macharis & Melo, 2011; Macharis & Milan, 2014). When implementing these different possible solutions as classified above and further elaborated in this book, actor involvement will be key to enable a successful and lasting implementation of new concepts.

Cathy Macharis
Sandra Melo
Johan Woxenius
Tom van Lier
Editors

REFERENCES

Arvidsson, N., & Browne, M. (2013). A review of the success and failure of tram systems to carry urban freight: The implications for a low emission intermodal solution using electric vehicles on trams. *Trasporti Europei.* [European Transport], *54*, 1–18.

Behrends, S. (2011). *Urban freight transport sustainability – The interaction of urban freight and intermodal transport.* PhD thesis, Chalmers University of Technology, Gothenburg.

Browne, M., Allen, J., Nemoto, T., Patier, D., & Visser, J. (2012). Reducing social and environmental impacts of urban freight transport: A review of some major cities. *Procedia – Social and Behavioral Sciences*, *39*, 19–33.

Coyle, M. (2007). *Effects of payload on the fuel consumption of trucks.* London: Department for Transport.

Cruijssen, F., Cools, M., & Dullaert, W. (2007). Horizontal cooperation in logistics: Opportunities and impediments. *Transportation Research Part E*, *43*(2), 129–142.

De Brucker, K., Verbeke, A., & Winkelmans, W. (1998). *Sociaal-economische evaluatie van overheidsinvesteringen in transportinfrastructuur. Kritische analyse van het bestaande instrumentarium. Ontwikkeling van een eclectisch evaluatie-instrument.* Leuven: Garant (in Flemish).

European Commission. (2011). *White paper – Roadmap to a single European transport area – Towards a competitive and resource efficient transport system.* Luxemburg: European Commission.

European Commission. (2012). *Final report study on urban freight transport.* Brussels, Belgium: DG Move & MDS Transmodal Ltd.

Fernandez-Barcelo, I., & Campos-Cacheda, J. M. (2012). Estimate of social and environmental costs for the Urban distribution of goods. Practical case for the city of Barcelona. *Procedia – Social and Behavioral Sciences*, *39*, 818–830.

IMPACT. (2008). *Handbook on estimation of external costs in the transport sector*. CE Delft. Retrieved from http://ec.europa.eu/transport/themes/sustainable/doc/2008_costs_handbook.pdf

Johansen, B. G., Andersen, J., Eidhammer, O., Verlinde, S., Filipe, L. N., da Rocha, J., … Moolenburgh, E. (2014). *Straightsol deliverable 5.1, demonstration assessments*. Norway: European Commission. FP7-SST-2011-RTD-1.

José Holguín-Veras, J., Wang, C., Browne, M., Darville Hodgec, S., & Wojtowicza, J. (2014). The New York city off-hour delivery project: Lessons for city logistics. *Procedia – Social and Behavioral Sciences*, *125*, 36–48.

Lammgård, C., & Andersson, D. (2014). Environmental considerations and trade-offs in purchasing of transportation services. *Research in Transportation Business & Management*, *10*, 45–52.

Macharis, C. (2014, April 28). Innovative solutions for sustainable logistics. Presentation at Logistics Day 2014 – Cluster for Logistics, Luxembourg, invited speech.

Macharis, C., & Melo, S. (Eds.). (2011). *City distribution and urban freight transport: Multiple perspectives*. Cheltenham: Edward Elgar Publishing.

Macharis, C., & Milan, L. (2014). Transition through dialogue: A stakeholder based decision process for cities: The case of city distribution. In K. Kourtit, P. Nijkamp, & M. Painhi (Eds.), *Habitat international special issue on: Cities, urbanisation and spatial interpretation*. Amsterdam: Elsevier. doi:10.1016/j.habitatint.2014.06.026

Macharis, C., & Van Mierlo, J. (Eds.). (2013). *Sustainable mobility and logistics*. Brussels: VUB Press.

McKinnon, A. (2010). Environmental sustainability – A new priority for logistics managers. In A. McKinnon, S. Cullinane, M. Browne, & A. Whiteing (Eds.), *Green logistics – Improving the environmental sustainability of logistics* (pp. 3–30). London: Kogan Page.

McKinnon, A., & Piecyk, M. (2010). *Measuring and managing CO_2 emissions of European chemical transport*. Edinburgh, UK: Logistics Research Centre, Heriot-Watt University.

MDS Transmodal. (2012). Study on urban freight transport. DG MOVE European Commission, Brussels. Retrieved from http://ec.europa.eu/transport/themes/urban/studies/doc/2012-04-urban-freight-transport.pdf

Messagie, M., Boureima, F.-S., Coosemans, T., Macharis, C., & Van Mierlo, J. (2014). A range-based vehicle life cycle assessment incorporating variability in the environmental assessment of different vehicle technologies and fuels. *Energies*, *7*, 1467–1482.

Olsson, J., & Woxenius, J. (2014). Localisation of freight consolidation centres serving small road hauliers in a wider urban area: Barriers for more efficient freight deliveries in Gothenburg. *Journal of Transport Geography*, *34*, 25–33.

Russo, F., & Comi, A. (2012). City characteristics and urban goods movements: A way to environmental transportation system in a sustainable city. *Procedia – Social and Behavioral Sciences*, *39*, 61–73.

Transport for London. (2012). *What are the main trends and developments affecting van traffic in London?* Roads Task Force – Technical Note 5, London. London: Transport for London.

UIC. (2009a). *2007 Report on intermodal rail/road transport in Europe*. DIOMIS – Developing infrastructure and operating models for intermodal shift. Paris: International Union of Railways.

UIC. (2009b). *Benchmarking intermodal rail transport in the United States and Europe*. DIOMIS – Developing infrastructure and operating models for intermodal shift. International Union of Railways, Paris.

van Lier, T., Caris, A., & Macharis, C. (2014). Bundling of outbound freight flows: Analyzing the potential of internal horizontal collaboration to improve sustainability. *Networks and Spatial Economics*. Article in press, doi:10.1007/s11067-014-9226-x (available online at SpringerLink).

van Lier, T., & Macharis, C. (2013). External costs of transport. In C. Macharis, & J. Van Mierlo (Eds.), *Sustainable mobility and logistics* (pp. 54–88). Brussels: VUB Press.

van Wee, G. P. (2012). How suitable is CBA for the ex-ante evaluation of transport projects and policies? A discussion from the perspective of ethics. *Transport Policy*, *19*, 1–7.

Verlinde, S., Macharis, C., Milan, L., & Kin, B. (2014). Does a mobile depot make urban deliveries faster, more sustainable and more economically viable: Results from a pilot test in Brussels. Mobil. TUM 2014 *international scientific conference on mobility and transport*, Munich, Germany, May 19–20.

Vermeiren, T. (2013). *Intermodal transport: The Delta in the Delta*, PhD thesis, Vrije Universiteit Brussel, Brussels.

Awareness

CHAPTER 1

OPTIONS FOR COMPETITIVE AND SUSTAINABLE LOGISTICS

Richard Smokers, Lóránt Tavasszy, Ming Chen and Egbert Guis

ABSTRACT

Purpose – *Logistics as a sector has a key role to play in reducing greenhouse gas emissions and in reducing the dependency of our economy on non-renewable energy sources. The challenges are enormous: by 2050 the sector needs to have achieved about 50% lower fossil fuel use and* CO_2 *emissions. If freight volumes grow according to expectations, this requires over 70% less* CO_2 *emissions per unit of transport. This chapter explores the options for reducing* CO_2 *emissions from freight transport and their reduction potential, and analyses whether the logistic sector would be likely to achieve the required reduction based on its intrinsic drive for cost reduction alone.*

Methodology/approach – *In this conceptual chapter we identify options for sustainable logistics and discuss the necessary economic conditions for their deployment using a simple cost/benefit analysis framework.*

Sustainable Logistics
Transport and Sustainability, Volume 6, 1–30

ISSN: 2044-9941/doi:10.1108/S2044-994120140000006001

We distinguish between three regimes of measures for improving sustainability: efficiency measures with net negative costs ('low hanging fruit'), cost-neutral measures and measures that allow to reach societal targets at net positive costs. Policy measures are discussed that may help the sector to implement cost-effective greenhouse gas abatement measures that, in the absence of incentives, go beyond the point of lowest cost from an end user perspective.

Social implications – *Sufficient energy saving options are available to be implemented in the short to medium term, which can lead to operational cost savings with a short return on investments period. The potential contribution of the logistics sector to sustainability is larger, however, as logistics can make large steps ahead in sustainability with cost neutrality or with small cost increases. The full potential has been underrated by many stakeholders and should be explored further.*

Originality/value of the chapter – *Efficiency measures are a necessary but insufficient condition for sustainable logistics. The industry will need to go beyond cost saving measures, or even cost-neutral measures to reach the long-term energy saving and emission reduction targets for freight transport. We provide a systematic presentation of these options and discuss the additional necessary measures.*

Keywords: Sustainability; emissions; CO_2; logistics

INTRODUCTION

Logistics as a sector has a key role to play in reducing greenhouse gas emissions and reducing the dependency of our economy on non-renewable energy sources. The potential contribution of the logistics sector to sustainability has so far been focusing on measures that reduce logistics costs (Ruijgrok, 2012). Its potential is larger, however, as logistics can make even larger steps ahead in reducing CO_2 emissions with cost-neutral measures or options that lead to small cost increases (Quak, de Bes, & Leijnse, 2011). As such, we argue that the full potential has been underrated by different stakeholders and should be explored further. This conceptual chapter provides some lines of thinking that motivate to increase the current efforts in green logistics to a level that allows the logistics sector to contribute meaningfully to achieving global sustainability targets for the long-term future up to 2050.

The chapter explains how a sustainable approach to logistics can provide an economic advantage for the logistics sector in the short and in the long term. The focus in this chapter is on climate aspects within the planet-side of sustainability and their interaction with the profit dimension. We argue that sufficient energy saving options are available to be implemented in the short term, which can lead to operational cost savings with a short return on investments period. In addition, however, we also state that aiming only for sustainability levels that can be achieved at a net cost reduction will not prepare the sector for the challenges it will face in the longer term.

The challenge with respect to reducing fossil fuel use and CO_2 emissions is briefly sketched in the section 'The Challenge'. General levers and concrete reduction options for dealing with that challenge are described in the sections 'Levers for reducing CO_2 emissions of the logistics sector' and 'Options for reducing CO_2 emissions in logistics'. Considerations on what it means to strike a balance between the planet and profit side of sustainability are discussed in the section 'Sustainability versus cost reduction'. Sections 'The value of sustainable logistics' and 'The value of objective and reliable information' highlight the value of sustainable logistics and the value of having objective and reliable information in achieving sustainable logistics. Possible policy measures for promoting the logistics sector to move beyond the reduction potential that is achieved in the point of lowest costs are presented and briefly discussed in the section 'Policy instruments for improving sustainability of the logistics sector'.

THE CHALLENGE

Scenario calculations by IPCC show that, to have a likely chance of limiting the increase in global mean temperature to 2°C, global greenhouse gas emissions need to be lowered by 40–70% compared with 2010 by 2050, and to near-zero by the end of this century (IPCC, 2014). To allow developing economies room for growth, the industrialised countries will need to have reduced their greenhouse gas emissions by 80% or more in 2050 relative to 1990. The European Commission has embraced this target. As part of its climate strategy the European Commission, in its 2011 Transport White Paper (European Commission, 2011a), has set a goal for the transport sector of 60% reduction of greenhouse gas emissions from 1990 levels by 2050.

Transport is responsible for a quarter of EU greenhouse gas emissions (Hill et al., 2012). Freight transport accounts for about 40% of the transport sector's greenhouse gas emissions. Between 2000 and 2050 the global population will have increased by 50% (see e.g. United Nations, 2013). If this is accompanied by economic growth figures of the order of 2–3% p.a. it is clear that the demand for products will increase dramatically over the next decades, leading to a strong growth in goods transport worldwide. According to the White Paper EU freight transport activity is projected to increase, relative to 2005, by around 40% in 2030 and by little over 80% by 2050. Without measures to improve the efficiency and to reduce the carbon intensity of the transport sector this will result in a strong increase in the use of fossil fuels as well as in the emissions of CO_2. Realising a net absolute reduction in the CO_2 emissions of goods transport against a trend of growing demand requires very large relative reductions of the CO_2 emissions per unit of transport. To give an example: If the total amount of tonne kilometres transported increases by 80% up to 2050, achieving a 50% reduction of the absolute CO_2 emissions requires a relative reduction of CO_2 emissions per tonne kilometres of 72%.

Specific targets for freight transport have not been set. It is generally recognised (see e.g. Cuelenaere et al., 2014; European Commission, 2011b; Hill & Morris, 2012) that the potential for reducing greenhouse gas emissions is larger in passenger transport than in freight transport. In heavy duty vehicles the potential for further improvement of internal combustion engines is smaller, and also the potential for using vehicles running on renewable electricity or hydrogen is significantly more limited than in passenger transport. Biofuels can be used in trucks and ships, but the potential for sustainable production of biofuels appears limited. Overall in passenger transport emission reductions of significantly more than 60% seem possible. It is therefore reasonable to assume that the reduction required from freight transport as contribution to meeting the overall objective of 60% for the transport sector would be of the order of 50% in 2050 relative to 1990.

Besides climate change mitigation, also the scarcity of energy in the period up to 2050 may become an important driver for reducing the consumption of fossil fuels. Chen and Koppelaar (2010) performed a sensitivity analysis to get insight in impact on economic growth of fossil fuel scarcity, especially of oil, which is still used for 97% of all mobility in the world. This scarcity, and the resulting increase in prices, not only results from depleted reserves but also from investments in production capacity which are not able to keep up with increasing global demand. It was concluded that when current (2010) behaviour in mobility and logistics would be maintained, the world would already be confronted with high oil prices and therefore a restriction on economic growth approximately in the year 2016. In case all measures are implemented as defined to achieve the Kyoto agreement up to 2020, one could have an additional 10 years of unrestricted economic growth. The role of a transition to sustainable logistics in securing economic and welfare growth is further elaborated in Chen, Smokers, Guis, and Tavasszy (2012).

A shortage in oil and oil-derived fuels could be overcome by increasing the use of natural gas and other fuels derived from natural gas (e.g. by using gas-to-liquid conversion). Recent, rapid developments in the field of shale gas and oil production from tar sands could provide time to adjust to energy scarcity. From the climate change perspective, however, this is not good news. It will not only postpone the transition to a low-carbon economy but may also increase emissions, due to the high well-to-tank emissions, associated with producing oil and gas from unconventional resources.

Overall, it is clear that Europe's logistic sector is facing a tremendous challenge to halve its fossil fuel use and greenhouse gas emissions, while catering in a cost-effective way for a still increasing demand for goods transportation in the next four decades.

LEVERS FOR REDUCING CO_2 EMISSIONS OF THE LOGISTICS SECTOR

An absolute reduction of the fossil energy consumption and CO_2 emissions of the logistics sector can be achieved by combining several of the following levers:

1. Less products transported (reduced dispatch);
2. Less tonnes or m^3, same dispatch (reduced weight or volume of product or packaging);
3. Less tonne kilometre or m^3 km, same tonne or m^3 (reduced distance production-consumption, optimal hub location, etc.);
4. Less vehicle kilometre, same tonne kilometre or m^3 km (increased load factor, increased vehicle capacity, modal shift, etc.);
5. Less fossil fuel use, same vehicle kilometre (improved energy efficiency of vehicles, efficient driving style, alternative fuels, shift to more energy efficient modalities, etc.).

The first lever of transporting less goods is not only not in the interest of the logistics sector but also unlikely to happen in view of the growing world population and economy. Still, dematerialisation may result in a less freight transport-intensive world and contribute to a decoupling between economy and transport flows (McKinnon, 2007).

Realising lever 5, and to a lesser extent option 4 (specifically for vehicles with higher transport capacity), requires innovations from the vehicle manufacturing industry, more specifically by vehicle manufacturers (OEMs) and component suppliers, to improve the efficiency of vehicles and enable the use of low-carbon energy carriers. These innovations need to be adopted by the logistics sector as part of the transition process. Fuel efficient driving styles also contribute to this lever, and can be realised by driver training and feedback, to be implemented by employers in the transport sector, with the help of driver feedback tools that can be developed by the vehicle industry. Levers 2 to 4 can be realised by innovations in the logistics sector itself. These innovations include more efficient logistics, for example through optimised routing, and changes in the management and design of supply chains. Horizontal and vertical cooperation schemes are vital to develop cost-effective options for supply chain measures. This division into measures affecting emissions of vehicles and measures related to the operation of vehicles in logistics systems is also illustrated in Fig. 1.

Measures that can be implemented by the logistics sector partly consist of reduction options that can be implemented by individual companies. But improvements also can be achieved by improved supply chain management and design, which requires vertical as well as horizontal cooperation between companies. Timely implementation of such measures, in view of

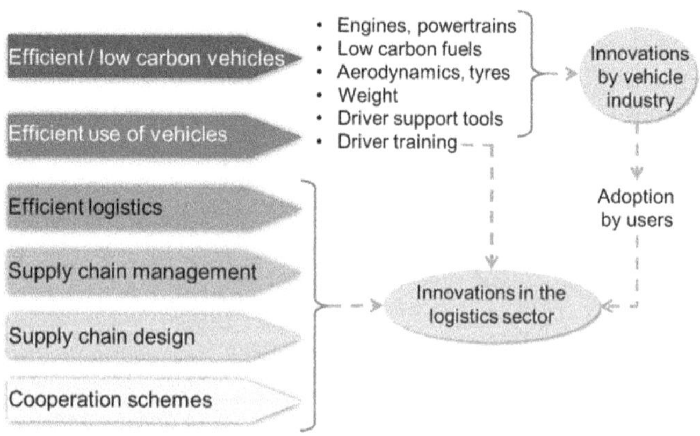

Fig. 1. Categories of Options for Reducing CO_2 Emissions from the Logistics Sector.

achieving long term and intermediate CO_2-reduction goals, may require stimulation of the innovative power of the logistics sector. Eventually, it is the companies (logistic service providers as well as shippers) that have to make the changes in their operation in order to become more sustainable. Of course, companies do not have unlimited resources to make the necessary investments and a sustainable operation should be aligned with other business objectives. Investments that do not lead to additional revenue normally are not considered as feasible. And if they are, the payback period should be sufficiently short. In our view, the difference between conventional logistic innovations and sustainable logistics becomes apparent in this interpretation of measures. We discuss this further in the section 'Sustainability versus cost reduction'.

OPTIONS FOR REDUCING CO_2 EMISSIONS IN LOGISTICS

Fig. 2 lists some examples of options in the technical, operational and logistical sphere that may contribute to reducing the CO_2 emissions of logistics

Technical options	Operational options	Logistical options

• Engine technology
• Alternative fuels / energy carriers
• Advanced powertrains
  • E.g. hybridisation, electric
• Resistance reduction
  • Tyre pressure monitoring
  • Low rolling resistance tyres
  • Aerodynamics
    • Integration truck + trailer
• Weight reduction
	• Driver behaviour
	  • Driver training
	  • Driver feedback tools
	• ITS
	  • Predictive cruise control
	  • Traffic management
	  • Cooperative vehicles
	  • "Electronic tow bar"
		• Network design
		• Networked cooperation
		• Organised consumers
		• Multi-/synchromodal transport
		• Improved planning
		  • Replenishment frequency (supplier)
		  • Delivery frequency and time windows (consumer)
		  • Route planning / management
		• Green warehousing
		• Reversed logistics
		• Cradle-to-cradle

Fig. 2. Three Categories of Options for Reducing CO_2 Emissions in Logistics, with Examples.

operations. For many options the costs are not yet well-known. Most options will require additional investments.

Many of these options have been described in detail (see e.g. McKinnon, Browne, & Whiteing, 2012), but systematic research on the costs and benefits of these options is rare. The World Economic Forum identifies the abatement potential of different measures in the supply chain (WEF, 2009). The potential increases with the scope of the system considered. Interestingly, the abatement impact of broader logistical measures (including sourcing, manufacturing, inventories) is about threefold that of transport oriented measures (reducing congestion, improved driver training and clean vehicles).

Technical Options

Despite the fact that both the logistics sector and truck manufacturers have been striving for high fuel efficiency and low fuel costs over the past decades, there is still a large potential for further reduction of the fuel consumption (and CO_2 emissions) of trucks (for detailed reviews of these options, see e.g. TIAX (2011) and AEA (2011)). Over time, new options are developed or brought to technical and economical maturity. These options relate to technical measures applied at the vehicle level as well as to operational measures that influence driving or that reduce the influence of the driver on the vehicle's fuel consumption. If the vehicle configuration is a given, measures to reduce fossil fuel consumption relate to improvements in:

- Engine technology, ranging from, for example, reduced friction and advanced valve timing and lift to improved combustion concepts;
- Improved transmissions and drivetrains, including start-stop systems or increasing levels of powertrain hybridisation;
- Waste heat recovery from the exhaust to either improve engine efficiency or to generate electricity for powering auxiliaries;
- Improved efficiency of auxiliaries, such as oil pumps and air compressors.

Besides innovations related to the propulsion of the vehicle, also the vehicle body can be modified to reduce the amount of energy needed to propel the vehicle. In urban areas with a lot of stop-and-go traffic weight reduction is an important measure, while for long-distance trucks, which drive most of their mileage at higher speeds on motorways, improvements in aerodynamics offer significant reduction potentials. The latter starts with simple measures such as spoilers, fenders and side skirts and culminates in more drastic redesign of the vehicle through aerodynamic noses and teardrop-shaped trailers. For all applications, dedicated tyres and tyre pressure monitoring systems reduce rolling resistance and thus contribute to lower fuel consumption.

Another way to reduce the CO_2 emissions from transport and to become less dependent on oil is found in the application of alternative fuels and energy carriers. Using advanced dual fuel concepts, which are currently under development, the use of natural gas may offer some CO_2 benefit compared to diesel. Further reductions are possible through the use of biogas. Diesel can be blended with biofuels to reduce the well-to-wheel greenhouse gas emissions. As mentioned, however, the sustainability of various biofuels is currently under debate.

In urban areas electric propulsion not only offers local zero-emission driving but also allows CO_2 emission reduction through the use of electricity from renewable sources, such as wind and solar. For the somewhat longer term, hydrogen offers similar benefits and may expand the implementation of zero-emission vehicles to applications outside urban areas.

Operational Options

Operational measures include measures that improve the driving style of the truck driver. A well-known example is eco-driving courses, the effect of which can be increased and sustained by installing devices that provide feedback to the driver. Devices are also being developed that include gaming aspects in order to stimulate their use, including features such as benchmarking performances among colleagues.

In the future, the impact of the driver on fuel consumption will be more and more reduced by the application of ITS-type measures, such as predictive cruise control, route management and traffic management systems. Schroten, Warringa, and Bles (2012) indicate that by combining a large number of engine, powertrain and vehicle-related CO_2-reduction measures in trucks, reductions in CO_2 emissions per vehicle kilometre of around 40% are achievable in the 2020–2030 timeframe at net negative costs to end-users and society.

A big step forward in city logistics could be regulations that restrict delivery of goods in the inner city to electric or other zero-emission vehicles. This would be in line with the European Commission's ambition, as expressed in the Transport White Paper, to 'achieve essentially CO_2-free city logistics in major urban centres by 2030'. Such a measure would increase the pressure on the degree of loading of vehicles, due to the higher capital costs (van Duin, Tavasszy, & Quak, 2013).

Logistical Options

Many logistical options are available for companies. McKinnon et al. (2012) distinguish between three levers, which relate to logistics organisation, to reduce the carbon intensity of supply chains: freight transport intensity, freight modal split and vehicle utilisation.

'Freight transport intensity' is defined as the ratio of freight transport performance (expressed in tonne-kilometres) to economic output (expressed in monetary terms). Options for making supply chains more sustainable are wide ranging and include the following (WEF, 2009):

- Optimised distribution networks: By taking into account emissions as costs and reducing the speed of delivery, bundling of goods can reduce the unit costs and unit emission intensity of transport.
- Low-carbon sourcing and manufacturing: As the carbon intensity of alternative products may differ, a more environment friendly orientation of sourcing practices can reduce emissions upstream. In the same vein, goods can be produced that carry less water (e.g. milk powder by drying milk), which saves transport volume. Printing on demand at the Centraal Boekhuis (the central book distributor for all publishers in the Netherlands) has eliminated stocks and the associated replenishment deliveries of books for around 80% of the titles.
- Packaging design: By improved packaging, the amount of air that has to be moved is reduced, leading to lower handling and transport costs.
- Reverse logistics/recycling: By optimising the logistics and use of return goods, the use of raw materials is reduced, waste processing becomes more efficient and transport distances are reduced. In extremis, fully circular production (the 'cradle-to-cradle' ideal (McDonough & Baumgart, 2002)) provides zero waste and local-to-local supply networks.

Freight modal split is the lever that has been most addressed by policy. A move towards larger scale modes of transport would reduce emission costs per tonne and could substantially contribute to achieving emission reduction targets (Woodburn & Whiting, 2012). In the past decades, however, the direction of shift between modes within Europe has been unfavourable from a climate change perspective. Due to various causes (poor access to rail and waterways, low flexibility, high prices, low visibility of services), only those firms that are in a captive situation towards rail and waterways have maintained their flows over these modes. This includes typically the heavy industries and intermodal traffic across the Alps. Despite the fact that the share of road transport has steadily been increasing, there is growing awareness in the industry that unit costs of alternative modes are lower. The industry maintains its interest to experiment with redesigning its supply chains, in a way that larger scale flows can be organised, fitting to the scale needed for efficient rail or waterways operations. In addition, the efforts to increase the connections between modes of transport – in order to allow easy interchanges and the building of intermodal transport chains – are professionalising. Most recently, the attention has shifted from the provision of rail and waterways infrastructure to the synchronization of operations between modes of transport, by minimising delays and aligning timetables. This development is known as 'synchromodality' or synchronized multimodality (ALICE, 2013).

Vehicle utilisation in road freight transport in Europe is below 50%, and around 20–40% of vehicles drive empty (EEA, 2010). From a logistical rationale, however, the optimisation potential is limited. In round trips, for example, trucks starting fully loaded will quickly be half-utilised on the average, and drive the final stretch home empty. In point-to-point movements, the return ride cannot always be loaded. Note further, that utilisation is measured in tonnes, while trucks may be full in terms of space (either in terms of cubic metres, or loading area). Optimising for both space and weight would be ideal. A better cooperation between companies could further reduce the share of empty rides of trucks. This process is hindered, however, by traditional habits such as for instance having the name of the company on the truck, which prevents others to make use of the same truck for their cargo. The resulting costs of this marketing tool (empty rides) are mostly not calculated and it is regarded to be free publicity. Further well-known hindrances to optimise vehicle utilisation are regulatory constraints on sizes and weights of vehicles, on driving hours and on cabotage trips, where hauliers are limited in the number of trips that can be made between foreign countries.

The Total Potential

As indicated above, technical measures related to vehicles provide a cost-effective CO_2 emission reduction potential of around 40%. The potential of improvements in logistics operations and supply chain management/design is not yet well known, but through several case studies it seems likely that at least 10% reduction is possible (see, for an early example, Groothedde, Ruijgrok, & Tavasszy, 2005). This already adds up to a well-to-wheel greenhouse gas reduction potential of the order of 50% per unit of transport.

In addition, well-to-wheel CO_2-reduction potentials of 50% or more per unit of energy used are possible for biofuels and other alternative, low CO_2 energy carriers (see e.g. JEC, 2014). Currently the European Commission requires a minimum well-to-wheel reduction percentage of 35% per unit of energy for biofuels used in the EU. This is increased to 60% by 2018 (European Commission, 2009). There may however be limitations to the availability of sustainably produced biofuels or their application potential in heavy goods transport (electricity).

Combining these, it seems possible overall to reduce CO_2 emissions per unit of transport by 70% or more. This is a requirement for achieving an absolute level of CO_2 emissions from freight transport in 2050 that is 50% lower than the 1990 level.

SUSTAINABILITY VERSUS COST REDUCTION

Without going into the details of the different existing definitions of sustainability, it is clear that improving the sustainability of activities requires striking a balance between the impacts of these activities on people, planet and profit at a local/regional as well as global scale in a way that makes them bearable, viable and equitable in the short term (intra-generation) as well as the long term (intra-generation). This is indicatively illustrated in Fig. 3. Ignoring for the moment possible negative impacts that CO_2 reduction in goods transport may have on people (e.g. negative effects of biofuel production on land availability for food production) as well as potential people benefits of these measures, the focus in the remainder of this chapter is on the interaction between the profit dimension and the climate change aspects within the planet-side of sustainability.

The logistics sector is strongly focussed on improving efficiency and reducing costs. As fuel costs are a significant part of many logistics operations, it could be argued that the logistics sector is already intrinsically striving for improved sustainability in the form of lower CO_2 emissions resulting from lower fuel consumption. The question, however, is whether long term CO_2 emission reduction goals, as discussed in the section 'The

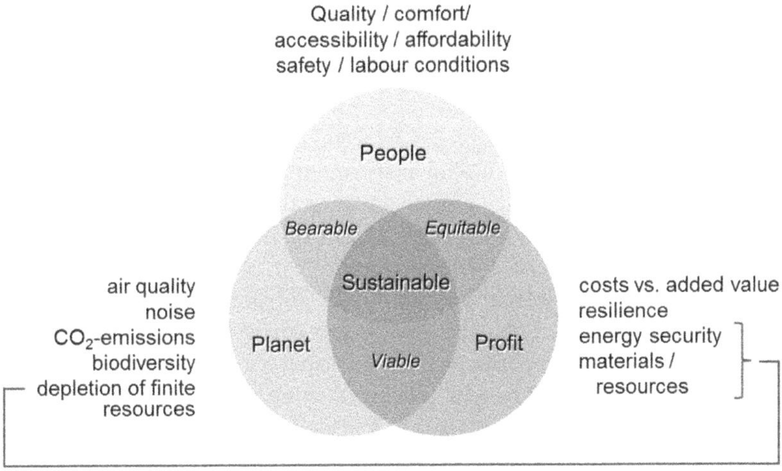

Fig. 3. Aspects of Sustainability.

Challenge', can be achieved while maintaining a 'win-win' situation between planet and profit.

As motivating as the perspective of creating a win-win between planet and profit may be for various actors to engage in improving sustainability of economic activities, it also poses a danger that less reduction is realised than is necessary or even justified from a macro-economic perspective. We will illustrate this with an example, starting from a simple cost-benefit framework.[1]

In general there are three reasons why the natural drive of the logistics sector to reduce costs may not be sufficient to realise the CO_2 emission reductions, required to meet long term climate goals:

(1) As actors in the logistics sector strive for lowest cost it is likely that they will not implement the full reduction potential that is available at zero net costs. If the emission targets require reductions to a point where cost of operations are increased compared to the current situation, they will definitely not be taken up autonomously;
(2) Reduction options that have the potential to be cost effective may not be adopted because of high initial costs. Usually new technologies are expensive when first introduced, but become cheaper with increasing production volumes as a consequence of economies of scale and learning effects;

(3) Rebound effects: reduced costs of transport will lead to increased demand which will counteract part of the reductions achieved per unit of transport.

The last reason will be discussed to some extent in the context of policy instruments (see the section 'Policy instruments for improving sustainability of the logistics sector'). The first relate directly to the issue of creating a 'win-win' between planet and profit and will be further elaborated here.

Lately, the concept of 'green growth' has created expectations that achieving long term sustainability goals can not only go hand in hand with a growing economy (i.e. that economic growth is possible within sustainable limits imposed by people and planet aspects), but that improving sustainability may even lead to new jobs and economic growth. The notion of green growth strongly relates to the 'Porter hypothesis' which states that strict environmental policies provide incentives that lead to process optimisation and a more efficient use of production means. In the economic literature, however, this hypothesis was often claimed to lack empirical proof. Recently, green growth has gained popularity by a study commissioned in 2011 by the German Federal Ministry for the Environment, Nature Conservation and Nuclear Safety. Jaeger et al. (2011) use economic modelling to show that measures are possible that, besides meeting the EU's climate goals for 2020, also lead to growth of GDP and the number of jobs in Europe. Using a similar modelling approach, Cambridge Econometrics and Ricardo-AEA recently estimated that the proposed European CO_2 regulation for passenger cars and vans for 2015–2020 leads to economic growth and more jobs in Europe (Cambridge Econometrics, 2013a, 2013b).

The results of both studies, however, cannot straightforwardly be generalised to a notion that meeting longer term CO_2-reduction goals would not need to go at the expense of economic growth or would even generate additional growth. Many measures available for meeting targets in 2020 can be considered 'low hanging fruit' and deliver high reductions (and associated energy cost savings) at relatively low investment costs. Payback periods are often shorter than the lifetime of the measures. Marginal abatement costs for reducing CO_2 emissions, however, have a tendency to increase non-linearly with increasing levels of CO_2 emission reduction. As fuel costs to first order reduce linearly with the reduction in fuel consumption or CO_2 emissions, beyond a certain level of CO_2 reduction the net impact on costs (from a user and/or societal perspective) will be an increase compared to the reference situation. Below that point simultaneous reduction of costs and emissions is possible. This is illustrated in Fig. 4.

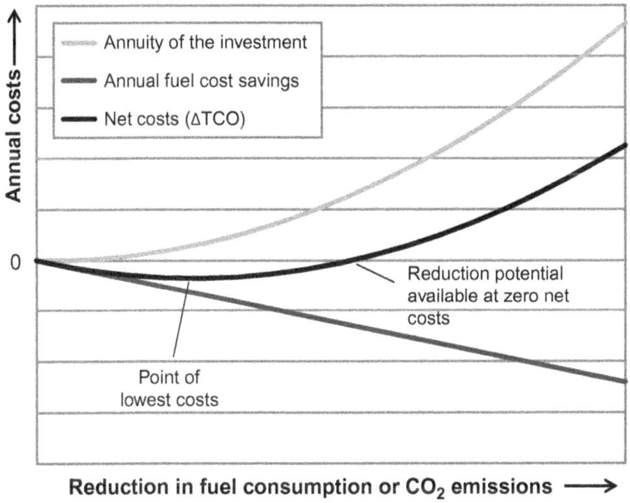

Fig. 4. Generalised Illustration of Impacts on Investment Costs, Fuel Costs and Total Cost of Ownership of Increasing Reductions in Fuel Consumption and CO_2 Emissions in Road Freight Transport.

Fig. 4 shows how investment costs increase non-linearly as a function of the desired fuel (or CO_2 emission) savings, generalised from assessments of the cost implications of improving the energy efficiency of trucks. The sum of the discounted investment costs and the operational benefits (in the figure: fuel cost savings), leads to the net costs curve. For most transport applications there is reduction potential that can be achieved at net negative costs to the user. The net savings, however, depend heavily on important parameters of the calculation, such as the depreciation period that is used for determining the annuity of the investment, the annual mileage and the fuel price.

The figure also shows that we can distinguish three regimes of sustainability improvements. The first steps in improving sustainability (in this case focused on reducing fuel consumption and CO_2 emissions) of a logistics operation generally result in a net cost saving (the 'low hanging fruit'). Companies that focus on cost optimisation, or that interpret the profit/planet balance in such a way that there should always be a win-win situation in which improved sustainability also saves money, would therefore only implement sustainable options up to the point of lowest costs. Beyond this point, however, further significant reductions are generally still possible at net negative costs. Interpreting the balancing of profit and planet as striving for the maximum reduction that can be achieved without additional costs, would lead to implementation of reduction measures up to the point where the net costs are zero. It is to be expected, however, that such measures will not be taken up by the market except by companies that expect additional benefits from being 'greener' than their competitors. These benefits could be based on an increased attractiveness of services for users who have a non-zero willingness to pay for green transport.

Above the point of zero net costs, further reduction of CO_2 emissions will lead to a net increase in costs. As markets are in general profit oriented, market circumstances will need to be different than in the other two regimes, for firms to adopt measures at this level of reductions. In brief, additional revenues will need to exist in the market, that balance the cost increase, or opportunity costs of alternative actions will need to be high enough.

In the Netherlands, the research institute TNO is involved in the organisation of the Lean and Green programme (Quak, de Bes, and Leijnse, 2011). Companies voluntarily join to benchmark their CO_2-reduction achievements. It was observed that for many a 20% reduction in CO_2 emissions turns out to be easily achievable and that therefore the point of lowest costs as indicated in the figure is in many (or most) cases not achieved. For the front-runners, achieving additional reductions is of course more difficult. Although this looks promising, new ways to make the next steps in CO_2 reduction should be found and implemented effectively, so that progress will not stagnate when companies reach the point of lowest costs.

If striving for sustainability is interpreted as 'doing things differently in order to reduce the planet impacts of our operations', applying CO_2-reduction measures up to the point of lowest costs can be interpreted as business as usual for the logistics sector, rather than as sustainable behaviour. A true drive for sustainability would imply a willingness to at least implement all measures that do lead to increased costs. In first order it could be argued that measures that do not lead to additional costs do not go at the expense of profitability. This, however, is not true in practice as a result of the way in which the market operates. The logistics sector is characterised by strong price competition so that companies that implement CO_2 reduction beyond the point of lowest costs would lose market share to competitors that do not strive for the maximum CO_2 emission reduction achievable at zero net costs. This prisoner dilemma could be solved if customers of logistic services would be willing to pay for increased sustainability, but this willingness still seems absent or limited. We return to this in the next section.

Following the above discussion, we can note that cost curves are not fixed or static and may develop over time, due to policies or external economic circumstances. We illustrate this in Fig. 5 where examples (b)–(d) show different plausible developments for the cost curves in Fig. 4.

If energy prices increase, the cost reduction associated with CO_2-reduction measures increases leading to a larger reduction potential that can be achieved at minimal or zero net costs (example (b) in Fig. 5). Costs of technologies generally decrease due to learning effects (economies of scale and innovations in product and production processes), while technological improvements and new innovations may increase the available CO_2-reduction potential. As a consequence, the CO_2-reduction potential that is available at negative or zero net costs increases over time (example (c) in Fig. 5). This development, however, is very uncertain. The combination of

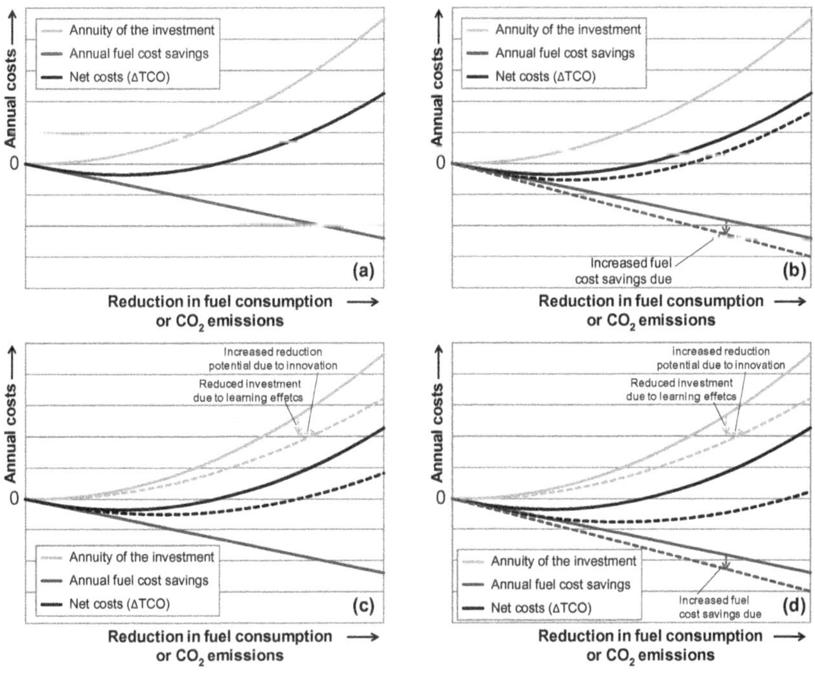

Fig. 5. Illustration of the net costs of CO2 reduction in transport (a) and its evolution due to increased fuel prices (b), reduced costs of existing CO2 reduction measures and increased potential due to new measures (c) and the combination of both (d). Source; TNO. *Source*: TNO.

these two effects (example (d)) may lead to an evolution of the cost curve for CO_2 reduction in freight transport in such a way that over time the potential for cost-effective measures keeps increasing. At the same time, it is difficult to predict to what level of reductions such an evolution can be maintained and whether that level is sufficient for the logistics sector to achieve its long term CO_2-reduction targets.

THE VALUE OF SUSTAINABLE LOGISTICS

Already, we see companies making investments in sustainability that are not fully earned back through direct operational cost savings. They motivate this on the view that the value of sustainability goes beyond direct operational savings. Recent research shows that companies are willing to pay for sustainability (Beltran, Chorus, Tavasszy, & van Wee, 2014), and that this increases as the transport happens more closely to the consumer (Fries, 2009). Companies that work in a more sustainable way often use their green operation in their promotional activities. Clients might also be willing to pay more for more sustainable services. And focus of clients on sustainability may also make that front-runners in this field can gain market share over laggards. The long term positive business impacts of this may justify investments in sustainability that first of all go beyond the point of lowest costs and may even go beyond short term direct cost neutrality. Nevertheless it has become clear from discussions with many transport companies that shippers that put strong pressure on their logistics service providers to lower their carbon footprint still expect this to go hand in hand with lower prices for the contracted services.

The implied value of sustainable logistics can be derived from the measures that regulators are willing to take to protect the environment. In general, modal shift is regarded as a promising measure to reduce emissions of transport (Woodburn & Whiteing, 2012). The European Commission has promoted the idea in its latest White Paper on Transport that 50% of all road freight transport above 300 km should be shifted to rail and waterways. As Tavasszy and van Meijeren (2011) show, forcing a modal shift from road to other modes of transport may imply a very high shadow price for emissions (of over 1500 Euro/tonne CO_2 for high value goods, in their example). This shadow price may be much higher than would ever be considered feasible in an emission trading market (current prices being below 10 Euro/tonne CO_2). Moreover, it is probably higher than necessary to reach the desired reduction of emissions by means of other measures in the supply chain than modal shift.

THE VALUE OF OBJECTIVE AND RELIABLE INFORMATION

An important barrier for logistics operators to invest in CO_2 reducing measures is uncertainty about the fuel cost savings that can be achieved in their specific situation as well as about the reliability of innovative technical measures in daily operations. For the many options available with relatively small reduction potentials (of a few per cent) it is difficult to prove their reductions in actual operation as fuel consumption of trucks varies by a larger amount due to all kinds of factors such as variations in loads, trips and road conditions and ambient weather conditions.

Uncertainty about the benefits of new technologies can be reduced if their potential is assessed using a uniform, standardized assessment procedure. Despite the economic importance of fuel consumption in the logistics sector, CO_2 emissions from heavy duty vehicles are currently neither measured nor reported. The European Commission has developed a strategy that in the short-term focuses on certification, reporting and monitoring heavy duty vehicle CO_2 emissions, as an essential first step towards curbing these emissions (European Commission, 2014). For this purpose an assessment methodology has been developed which combines results of engine, vehicle or component testing with computer simulations, using the VECTO vehicle simulation tool (see TÜ Graz, 2014), to generate CO_2 emission and fuel consumption figures at the level of complete vehicles. The method and the VECTO tool can do this for a wide range of vehicle configurations and a large number of duty cycles (load profiles representative for specific vehicle applications).

With the support of this tool the Commission intends in 2015 to propose legislation which would require CO_2 emissions from new trucks and other heavy duty vehicles to be certified, reported and monitored. Having objective, reliable and specific information available on the fuel consumption of complete vehicles or the fuel saving potential of specific technologies will make it easier for fleet owners to invest in such fuel saving and CO_2 reducing technologies.

Another important type of information for promoting sustainable logistics is the CO_2 footprint of operations. This footprint is expressed in gram

CO_2 per tonne, tonne kilometre, m^3 or m^3 kilometre, depending on the type of transport and whether CO_2 performance is assessed from the perspective of shippers or logistic service providers. Having access to objective and reliable assessments of the CO_2 footprint of (the logistics operations of) a company, of a specific logistics service or even of an individual shipments, allows shippers and transporters to make effective decisions for improving their environmental performance and to monitor their progress.

Currently the widespread application of CO_2-footprinting in the freight transport sector is hindered by the multitude of initiatives that have developed their own specific methodologies. The available international CEN 16258 standard helps to provide some level of harmonisation in the attribution of CO_2 emissions to individual shipments, but does not provide a procedure for how attributed emissions are to be expressed in performance indicators for the CO_2 footprint of companies (CEN, 2012).

POLICY INSTRUMENTS FOR IMPROVING SUSTAINABILITY OF THE LOGISTICS SECTOR

As indicated above, achieving the CO_2 reductions required from the logistics sector as its contribution to meeting long term climate goals, or even achieving the maximum potential of cost-effective reduction options may not be realised on the basis of the sector's intrinsic strive for cost reduction and efficiency improvement alone. Policy instruments may be needed to speed up the transition. This is especially the case when achieving long term reduction targets requires the application of reduction options that will increase the net costs of transporting goods. In general strong policy instruments are necessary to motivate the market to make such investments and to create a level playing field.

Policy instruments may be specifically needed to solve the following issues:

- Increasing the acquaintance with and acceptation of new vehicle technologies and logistic concepts;
- Setting binding targets for CO_2 reductions to be achieved in the logistics sector;
- Market formation for new technologies that may be expensive at first, but become cost effective once costs are reduced due to the economies of scale and learning effects that come with increased production volumes;

- Accelerating the development and application of CO_2 reducing technologies by vehicle manufacturers, including promoting a complete vehicle approach which is hindered by the current division of the sector into truck manufacturers (OEMs) and body and trailer builders;
- Improving the attractiveness at a micro-economic level of options that are cost effective at a macro-economic level. This includes implementing the full reduction potential that is available at zero net costs from a macro-economic perspective as well stimulating the implementation of the most cost-effective solutions if CO_2-reduction targets require application of reduction measures with positive societal abatement costs;
- Creation of a level playing field, regionally such as in the European Union or even globally;
- Reducing rebound effects resulting from increased demand triggered by reduced costs of transport.

Possible policy instruments include:

- Promoting the development and application of standardised carbon footprinting methods;
- Pilots and demonstration projects as well as communication actions;
- Standardized testing and evaluation procedures for (impacts of new technologies on) vehicle CO_2 emissions;
- CO_2 legislation for vehicles;
- Internalising external costs by means of taxation;
- A CO_2-tax on fuels;
- A cap & trade system for CO_2 emissions from the transport sector;
- Incorporating the transport sector in a wider cap & trade system.

This list is not exhaustive. Several options are discussed in more detail later.

Carbon Footprinting

Monitoring of the achievements of the companies in a fair and objective way combined with a labelling or benchmarking system can be an important tool to stimulate companies that already are well ahead of others, to take the next steps in the transition process towards sustainable logistics.

Pilots and Demonstration Projects

Pilots and demonstration projects are not only tools for gaining experience with new technologies, but also help to create visibility and disseminate (hopefully) positive experiences. Besides on proving fuel saving potentials (which can be difficult to measure in practice for a large number of 'smaller' options), such pilots and demonstration projects should focus on testing the reliability of new technologies in daily practice.

Standardized Testing and Evaluation Procedures for (Impacts of New Technologies on) Vehicle CO_2 Emissions

As mentioned above, the availability of objective and application specific information about the fuel saving and CO_2 emission reduction potential of new technologies will help to reduce uncertainties with fleet owners. It is expected that in the heavy duty vehicle sector, where costs play a more important role and are valued more objectively than, for example, in the passenger car market, the availability of objective information may already promote the uptake of CO_2 reducing measures to such an extent that CO_2 legislation may not be necessary.

CO_2 Legislation for Vehicles

European CO_2 standards already exist for passenger cars and light commercial vehicles.[2] For heavy duty vehicles such standards could up to now not be implemented due to the lack of a whole-vehicle type approval test procedure for CO_2. Although harmonised assessment methods have recently been developed (see section 'The value of objective and reliable information'), it is firstly necessary to gain sufficient experience with these procedures and to improve them if necessary, in order to avoid that shortcomings of the assessment procedure create perverse incentives or loopholes in the CO_2 legislation for heavy duty vehicles.

Internalising External Costs by Means of Taxation

In Fig. 5(b) a tax that becomes lower with a higher sustainability performance (specifically a tax on CO_2 emissions) will lead to a larger cost savings potential (tax in addition to fuel costs) for CO_2-reducing technologies and therefore will make the 'point of lowest costs' in the green curve shift towards the right. In other words, more solutions become commercially feasible that previously were not so that the market will by itself implement more measures. The extent to which this happens depends on synergies between reducing CO_2 emissions and reducing other externalities such as pollutant emissions, congestion or accidents as well as on the shadow prices used in the taxation for these externalities.

Internalisation of external costs can and should go beyond the current taxes, if a fair and efficient regime can be implemented. Recent calculations (Tavasszy, Harmsen, Ivanova, & Bulavskaya, 2014) show that the increase of costs and the effect on the economy would be minor, but that the reduction in external effects would be substantial. This would lead to a net positive welfare effect for the Netherlands, even in the scenario that internalisation is not implemented in all countries simultaneously.

A CO_2-Tax on Fuels

A CO_2-tax can be considered a specific example of internalising external costs. But in contrast to a taxation based on external cost estimates, a CO_2-tax should be as high as is necessary to achieve the desired effects, which is basically determined by price elasticities.

A Cap & Trade System for CO_2 Emissions from the Transport Sector

A cap & trade system for CO_2 sets a cap on absolute emissions of the participants covered by the system. These buy (e.g. through auctioning) or receive (for free) CO_2 emission allowances. If a participant emits more than the allowances owned by the participant, the participant has to buy more allowances from other participants that have more allowances than emissions, or that invest in CO_2 mitigation measures. The choice between the first and the latter will depend on the price, so that, at least in theory, all CO_2 mitigation measures with cost (per tonne CO_2 avoided) lower than the cost of the emission allowances will be implemented.

Theoretically economic instruments such as a cap & trade system promote the most cost-effective reduction options. The advantage of a cap & trade system over a CO_2 tax is that the target is set and the CO_2 price follows from the reductions that are necessary to meet the target. With a CO_2 tax, the price incentive is given but the total CO_2 emission reduction is uncertain.

As shown in Schroten et al. (2012) a large reduction potential is available for heavy duty vehicles at negative cost for end-users and society. This illustrates the existence of some market barriers to achieving economically optimal levels of greenhouse gas reduction and fuel efficiency for HDVs which would also inhibit the effective operation of a market instrument. Another shortcoming of a cap & trade system is that it does not automatically stimulate timely action that is required to get longer term, transitional options (such as electric vehicles) implemented. Most new technologies are expensive at first and prices only decrease with increasing production volumes. But under a cap & trade system investments in such initially expensive options may not take place.

Incorporating (Parts of the) Transport Sector in a Wider Cap & Trade System

As explained, for example, in Essen, Blom, Nielsen, and Kampman (2010) the inclusion of the transport sector in the EU Emission Trading System (EU-ETS) can be implemented by means of upstream or downstream trading. In a cap & trade system, an upstream trading system implies that the cap will be put on companies that sell transport fuels. They need emission allowances for the CO_2 emissions caused by the fuels sold by them, and these will be capped. In an upstream trading system, the fuel consumers that actually use the fuels and thus emit the CO_2 will be the trading parties. The latter option seems most appropriate if only freight transport were to be included in the EU-ETS.

Up to recently it was believed that CO_2 reduction in transport would have higher abatement costs than in other sectors. As a consequence inclusion of transport in EU-ETS would lead to the transport sector funding CO_2-reduction measures taken in other sectors, while fuel consumption in transport would not be affected. Even though a significant reduction potential is available at negative costs it could still be the case that no reduction measures are taken in the transport sector. At current CO_2 prices under EU-ETS the impact on fuel prices is very small, so that it will not provide a strong incentive for fleet owners to make the effort to invest in new and initially uncertain technologies.

TRANSITION PROCESS

Achieving sustainable logistics is in essence a transition process. Applying options for sustainable logistics, as mentioned in 'Options for reducing CO_2 emissions in logistics', to the extent that are cost effective is a requirement to make the transition successful. On the other hand it is also the task for sustainable logistics to identify the requirements for new options. It is clear that the options currently on the market or ready for market penetration are not generating sufficient fuel use reduction to meet the long-term reduction objectives. Furthermore the market uptake for the available options is going slowly or is not happening at all, even where it can lead to cost savings for companies. Without a doubt the transition process requires active directing and stimulation in order to come to the required situation.

First of all, a clear strategy should be set out taking into account the long term objective and intermediate steps to be taken. This includes the market uptake of successful options (people, profit planet) in the short term as well as the development of new options for the medium and long term. Important is also to have proper insights in the enablers and disablers of the uptake of sustainability options. What are the bottlenecks, how can they be removed, what is working well, how can this be stimulated, etc.? Objective and reliable advice should be given to companies on which options would work for them and what are the costs savings to be expected. Furthermore the companies should also be informed on the long term process they are in and when next steps can be expected.

The transition is taking place in a situation where although the 'big picture' might be clear, the individual interests often lead to sub-optimal solutions. For instance there is the regular competition between companies that makes cooperation and exchange of information a difficult process even though large logistics efficiency gains can be made by horizontal and vertical cooperation in supply chains. Another example is that where shippers might be interested to have a detailed monitoring of CO_2 emissions in transportation of their commodities, the logistic service providers are less eager to give this information since this might give insight in the costs structure of their company.

Another dilemma is that energy demand is also a result of (global) economic growth. We are facing a situation where we have a policy for economic growth while at the same time energy is becoming more scarce. Higher prices in the coming decade(s) are inevitable. We are now in a situation where one could anticipate to such increasing prices. Given the uncertainty of this phenomenon, companies cannot come to a proper assessment of the required investments. There is a risk of making large investments for which the advantages will come too late to achieve a proper return on investment. It is therefore tempting to react to the changing situation rather than to anticipate. It is important, however, that options are ready for implementation and have been demonstrated by leading companies when time comes for majority to react. Technologies can only be scaled up quickly if they have already gained a certain market penetration and a size of production that leads to economies of scale. This is especially true for CO_2-reduction options that also require investments in new energy infrastructures (such as for electricity or hydrogen).

CONCLUSIONS

In view of the growing volume of freight transport over the next decades, very large reductions in the CO_2 emissions per unit of transport are needed if the logistics sector is to contribute in a meaningful way to the greenhouse gas emission reduction target of 60% in 2050 relative to 1990, as set by the European Commission.

In the short term this requires implementation of energy saving options. Recent assessments have shown that the reduction potential for such options in trucks is larger than expected and can largely be achieved at net negative costs. In the medium to long term a transition to (CO_2 neutral) alternative energy carriers is necessary which should be initiated by field trials and market formation activities in the short term in order to make sure that these technologies are ready to be scaled-up when necessary.

The full potential of sustainable logistics will be reached by adoption and use of more efficient vehicles and low CO_2 fuels on the one hand, and improved logistic efficiency and supply chain innovations on the other hand. Technological options need to be developed by vehicle manufacturers but still require acceptance by end-users in the sector. The potential of improved logistics can only be harvested by mobilisation of the innovative power of the logistical sector itself. The latter can be achieved by giving a higher weight to sustainability objectives in the complex logistical optimisation function applied.

Recent studies show that technical measures related to vehicles provide a cost-effective CO_2 emission reduction potential of around 40%. The total potential of improvements in logistics operations and supply chain management and design is not yet well known, but it seems likely that at least 10% reduction should be possible. Combined with the well-to-wheel

CO_2-reduction potentials of biofuels and other alternative, low CO_2 energy carriers, it seems overall possible to reduce CO_2 emissions per unit of transport by 70% or more allowing for a reduction of total CO_2 emissions of freight transport in 2050 by 50% relative to the 1990 level.

Sustainability will be crucial for the right-to-play of the logistical sector and improves the competitiveness of the companies on the long run. Companies can justify investments that on the short term lead to higher costs in case they anticipate for this situation.

The logistics sector is strongly focussed on improving efficiency and reducing costs. As fuel costs are a significant part of many logistics operations, it could be argued that the logistics sector is already intrinsically striving for improved sustainability in the form of lower CO_2 emissions resulting from lower fuel consumption. Nevertheless it is unlikely that the sector will achieve the necessary reduction by 2050 based on this intrinsic driver alone. As actors in the logistics sector strive for lowest cost it is likely that they will not implement the full reduction potential that is available at zero net costs. If the emission targets require reductions to a point where cost of operations is increased compared to the current situation, they will definitely not be taken up autonomously. Furthermore reduction options that have the potential to be cost effective may not be adopted because of high initial costs. Usually new technologies are expensive when first introduced, but become cheaper with increasing production volumes as a consequence of economies of scale and learning effects. Implementation of policy instruments may be necessary to shift the micro-economically optimal CO_2 emission reduction to higher levels.

The transition has to be made in a difficult economic and social situation which on the one hand complicates the process but on the other hand will stimulate creativity and will force us to let go of old paradigms. Changes will not only be driven by environmental concerns. Scarcity of food, energy and water may be expected to lead to drastic changes in transport flows and supply chains anyhow. The required response of the logistic sector to these challenges offers great opportunities to also improve the sustainability of the sector itself.

NOTES

1. Note that much of our reasoning applies also to other sustainability impacts of transport.
2. See http://ec.europa.eu/clima/policies/transport/vehicles/cars/index_en.htm and http://ec.europa.eu/clima/policies/transport/vehicles/vans/index_en.htm

REFERENCES

AEA. (2011). *Reduction and testing of greenhouse gas (GHG) emissions from heavy duty vehicles – Lot 1: Strategy*. Retrieved from http://ec.europa.eu/clima/policies/transport/vehicles/docs/ec_hdv_ghg_strategy_en.pdf

ALICE. (2013). *European technology platform for logistics innovation*. Vision Document. Retrieved from http://www.etp-logistics.eu/alice/en/publications/. Accessed on June 30, 2014.

Beltran, M., Chorus, C., Tavasszy, L., & van Wee, B. (2014). *Preferences relating to container transportation from the port of Rotterdam to the hinterland: Results from a choice experiment*. Mimeo, Delft University of Technology.

Cambridge Econometrics. (2013a). *Fuelling Europe's future, how auto innovation leads to EU jobs*. Cambridge Econometrics & Ricardo-AEA. Retrieved from http://www.camecon.com/Libraries/Downloadable_Files/Fuelling_Europe_s_Future-_How_auto_innovation_leads_to_EU_jobs.sflb.ashx

Cambridge Econometrics. (2013b). *An economic assessment of low carbon vehicles*. Cambridge Econometrics & Ricardo-AEA. Retrieved from http://www.camecon.com/Libraries/Downloadable_Files/Economic_Assessment_Vehicles_-_FINAL.sflb.ashx

CEN. (2012). *Methodology for calculation and declaration of energy consumption and GHG emissions of transport services (freight and passengers)*. CEN standard EN 16258. Retrieved from www.cen.eu; http://www.en-standard.eu/csn-en-16258-methodology-for-calculation-and-declaration-of-energy-consumption-and-ghg-emissions-of-transport-services-freight-and-passengers/?gclid=CKaz9tzUv78CFWPItAodmxwAJw

Chen, T. M., & Koppelaar, R. (2010). Restricted oil supply in the transport sector and economic growth. *Proceedings of European Transport conference*. Association for European Transport, London.

Chen, T. M., Smokers, R., Guis, E., & Tavasszy, L. (2012). Sustainable logistics securing our welfare: An economic perspective of sustainable logistics. *Proceedings of the Vervoerslogistieke Werkdagen conference 2012*. University Press, Zelzate.

Cuelenaere, R., Koornneef, G., Smokers, R., van Essen, H., van Grinsven, A., 't Hoen, M., … Usmani, O. (2014). *Scenarios for energy carriers in the transport sector*. Retrieved from https://www.tno.nl/content.cfm?context=thema&content=prop_publicatie&laag1=894&laag2=914&laag3=105&item_id=974&Taal=2

EEA. (2010). *Load factors for freight transport* (TERM 030). European Environmental Agency. Retrieved from http://www.eea.europa.eu/data-and-maps/indicators/load-factors-for-freight-transport/load-factors-for-freight-transport-1

Essen, H. v., Blom, M., Nielsen, D., & Kampman, B. (2010). *Economic instruments, paper 7 produced as part of contract ENV.C.3/SER/2008/0053 between European Commission Directorate-General Environment and AEA Technology plc*. Retrieved from www.eutransportghg2050.eu

European Commission. (2009). *Fuel quality directive*. Directive 2009/30/EC. Retrieved from http://eur-lex.europa.eu/legal-content/EN/TXT/?uri=CELEX:32009L0030

European Commission. (2011a). *Roadmap to a single European transport area – Towards a competitive and resource efficient transport system*. Brussels: European Commission. Retrieved from http://ec.europa.eu/transport/strategies/2011_white_paper_en.htm

European Commission. (2011b). IMPACT ASSESSMENT (Accompanying document to the White Paper 'Roadmap to a Single European Transport Area – Towards a competitive and resource efficient transport system'), SEC(2011) 358. Retrieved from http://eur-lex.europa.eu/LexUriServ/LexUriServ.do?uri=SEC:2011:0358:FIN:EN:PDF

European Commission. (2014). *Reducing CO_2 emissions from heavy-duty vehicles*. Retrieved from http://ec.europa.eu/clima/policies/transport/vehicles/heavy/index_en.htm. Accessed on July 11.

Fries, H. T. N. (2009). *Market potential and value of sustainable freight transport chains*. Dissertation, Eidgenössische Technische Hochschule ETH Zürich. Nr. 18685, Zürich, 2009. Retrieved from http://e-collection.library.ethz.ch/eserv/eth:921/eth-921-02.pdf

Groothedde, B., Ruijgrok, C. J., & Tavasszy, L. A. (2005). Towards collaborative, intermodal hub networks: A case study in the fast moving consumer goods market. *Transportation Research Part E: Logistics and Transportation Review*, *41*(6), 567–583.

Hill, N., Brannigan, C., Smokers, R., Schroten, A., van Essen, H., & Skinner, I. (2012). *Developing a better understanding of the secondary impacts and key sensitivities for the decarbonisation of the EU's transport sector by 2050*. Final project report produced as part of a contract between European Commission Directorate-General Climate Action and AEA Technology plc. Retrieved from www.eutransportghg2050.eu

Hill, N., & Morris, M. (2012). *Further development of the SULTAN tool and scenarios for EU transport sector GHG reduction pathways to 2050*. Task 6 paper produced as part of a contract between European Commission Directorate-General Climate Action and AEA Technology plc. Retrieved from www.eutransportghg2050.eu

IPCC. (2014). *Many pathways to substantial emissions reductions are available. Press release*. Geneva: IPCC. Retrieved from http://www.ipcc.ch/pdf/ar5/pr_wg3/20140413_pr_pc_wg3_en.pdf. Accessed on 13 April.

Jaeger, C. C., Paroussos, L., Mangalagiu, D., Kupers, R., Mandel, A., Tàbara, J. D., … Lass, W. (2011). *A new growth path for Europe. Generating prosperity and jobs in the low-carbon economy*. Final report. European Climate Forum, Potsdam, Germany.

JEC. (2014). *JEC – Joint Research Centre-EUCAR-CONCAWE collaboration. WELL-to-WHEELS Report, Version 4.a*. Retrieved from http://iet.jrc.ec.europa.eu/about-jec/downloads

McDonough, W., & Baumgart, M. (2002). *Cradle to cradle: Remaking the way we make things*. North York: New York Press.

McKinnon, A., Browne, M., & Whiteing, A. (2012). *Green logistics: Improving the environmental sustainability of logistics*. London: Kogan Page Publishers.

McKinnon, A. C. (2007). Decoupling of road freight transport and economic growth trends in the UK: An exploratory analysis. *Transport Reviews*, *27*(1), 37–64.

Quak, H., de Bes, J., & Leijnse, M. (2011). Eerste stappen richting een duurzamere logistiek: Korte termijn maatregelen en lange termijn doelstellingen. In S. Weijers, & W. Dullaert (Eds.), *Bijdragen Vervoerslogistieke Werkdagen 2011*. Zelzate: University Press.

Ruijgrok, C. J. (2012). *Innovatie en duurzaamheid in de logistiek: Uitdagingen en ontwikkelingsmogelijkheden voor de periode tot 2040*. Den Haag: Raad van Leefomgeving en Infrastructuur.

Schroten, A., Warringa, G., & Bles, M. (2012). *Marginal abatement cost curves for heavy duty vehicles*. Retrieved from http://ec.europa.eu/clima/policies/transport/vehicles/heavy/docs/hdv_2012_co2_abatement_cost_curves_en.pdf

Tavasszy, L., Harmsen, J., Ivanova, O., & Bulavskaya, T. (2014). Effect of a full internalization of external costs of global supply chains on production, trade and transport. *Proceedings of the 2014 transport research arena conference*. Retrieved from http://www.traconference.eu/papers/pdfs/TRA2014_Fpaper_17858.pdf

Tavasszy, L., & van Meijeren, J. (2011). *Modal shift target for freight transport above 300 km: An assessment*. Brussels: ACEA.

TIAX. (2011). *European Union greenhouse gas reduction potential for heavy-duty vehicles*. Retrieved from http://ec.europa.eu/clima/policies/transport/vehicles/heavy/docs/icct_ghg_reduction_potential_en.pdf

TÜ Graz. (2014). *Development and validation of a methodology for monitoring and certification of greenhouse gas emissions from heavy duty vehicles through vehicle simulation*. Retrieved from http://ec.europa.eu/clima/policies/transport/vehicles/heavy/docs/final_report_co2_hdv_en.pdf

United Nations. (2013). *World population prospects, the 2012 revision*. New York, NY: United Nations. Retrieved from http://esa.un.org/unpd/wpp

van Duin, J. H. R., Tavasszy, L. A., & Quak, H. J. (2013). *Towards E(lectric)-urban freight: First promising steps in the electric vehicle revolution*. European Transport/Trasporti Europei, Report No. 54.

WEF. (2009). *Supply chain decarbonisation*. Geneva: World economic Forum.

Woodburn, A., & Whiteing, A. (2012). Transferring freight to 'greener' transport modes. In *Green logistics: Improving the environmental sustainability of logistics* (pp. 124–139). London: Kogan Page Publishers.

CHAPTER 2

MITIGATING THE NEGATIVE ENVIRONMENTAL IMPACTS OF LONG HAUL FREIGHT TRANSPORT

Sharon Cullinane

ABSTRACT

Purpose – *Long haul freight transport imposes huge negative environmental externalities on society. Although these can never be entirely eliminated, they can be reduced. The purpose of this chapter is to analyse some of the many mitigating measures, or interventions, that can be used.*

Methodology/approach – *The approach used in this chapter is to review the literature and provide an overview of the main theoretical and practical mitigation measures available to transport operators.*

Research limitations – *There are literally thousands of possible mitigation measures and combinations that can be used by operators to reduce their environmental footprint. Each of these measures warrants a separate chapter. This chapter can only present an overview of the principle available measures. Although some mainland European examples are*

Sustainable Logistics
Transport and Sustainability, Volume 6, 31–61

ISSN: 2044-9941/doi:10.1108/S2044-994120140000006002

used, it is acknowledged that the examples used are somewhat skewed towards the United Kingdom.

Originality/value of the chapter – *The value of the chapter is in bringing together some of the many measures and approaches that can be used to reduce the environmental externalities of long haul freight transport. Much of the information on such interventions is based on industrial and EU project sources rather than purely academic research and so is less likely to be found in academic journals.*

Keywords: Long haul logistics; environment; environmental interventions; environmental mitigation; freight transport

INTRODUCTION

The environmental impact of logistics has become an increasingly important issue as global warming and its effects have become internationally recognised and companies have been forced to consider the sustainability of their operations. In this chapter, we look at ways by which the environmental impact of logistics can be reduced. The focus is on the environmental impacts of *long haul* logistics; the subsidiary aspects are only mentioned where they impact on this long haul element.

Logistics is responsible for a variety of negative impacts (negative externalities), including air pollution, noise, accidents, vibration and visual intrusion. This chapter focuses principally on ways of reducing greenhouse gas (GHG) emissions from freight transport. In measuring the environmental effects of logistics it is important to distinguish between different levels of impacts. The International Green House Gas Protocol Initiative (2001) developed jointly by the World Business Council for Sustainable Development and the World Resources Institute, developed the following categorisation:

SCOPE 1 emissions – direct GHG emissions from sources owned or controlled by the entity (e.g. emissions from fossil fuels burned on site and in vehicles).

SCOPE 2 emissions – indirect GHG emissions resulting from the generation of electricity, heating or cooling or steam generated off-site but purchased by the entity and the transmission and distribution losses associated with some purchased utilities (e.g. chilled water, steam).

SCOPE 3 emissions – indirect GHG emissions from sources not owned or directly controlled by the entity but related to the entity's activities (e.g. travel and commuting by employees, solid waste disposal).

This chapter deals primarily with the SCOPE 1 emissions from logistics operations, which could also be termed 'first order' impacts. Additionally, since most of the first-order impacts emanate from the transport of goods, rather than their storage and handling, the attention will focus primarily on this activity. However, it must also be remembered that much long-distance logistics is outsourced to third- or fourth-party logistics operators (3PLs and 4PLs) rather than being supplied in-house. Emissions from these sources could be classed as SCOPE 3 emissions (depending on whether it is the emissions of the consignor or the logistics operator that are being considered). Logistics operators face considerable pressure from their consignors to reduce their environmental footprint.

GHG emissions from freight transport largely depend on the amount and type of fuel used. As discussed later, various combinations of alternative fuel mixes with conventional diesel are possible. However, the main fuel used by long haul goods vehicles (referred to in this chapter as trucks) as well as conventional rail locomotives and ships continues to be diesel. In most countries, a substantial but relatively small amount of freight is moved by electrically powered freight trains and a very small amount by electric road vehicles. In these cases, the pollution arises at the point where the electricity is generated and the nature of that pollution depends on the source of the primary energy (i.e. whether it is generated from sustainable or unsustainable sources). Many, although not all, of the measures discussed in this chapter to reduce the impact of freight transport on the environment depend on reducing fuel usage and energy transition measures.

First, to provide some context, we will take a brief look at the scale and importance of the freight transport sector within Europe.

THE IMPORTANCE OF THE FREIGHT TRANSPORT SECTOR

Logistics has become one of the most important elements of any business and, with the globalisation of trade, its importance, together with its complexity, has increased. In the 27 states of the EU, there are 5 million km of paved roads, including 65,100 km of motorway; 212,800 km of rail lines of which 110,458 km are electrified and 42,709 km of navigable inland waterways. Total inland freight amounted to approximately 2,300,000 million tonne kilometres (mt km) and is expected to grow by around 2.1% a year until 2030 (Eurostat, 2013). Of this, 76.4% was transported by road, 17.1% by rail and 6.5% by inland waterway. This is in contrast to the United States, where rail transport accounted for 41% and road 31% of freight traffic in 2008 (EU, 2012). Expressing the importance of logistics in terms of GDP is notoriously difficult because of the problems of agreeing precisely what is included. Bowersox, Rodrigues, and Calantone (2005) calculated that logistics accounted for around 13.8% of the world's GDP. Rantasila and Ojala (2012), in a paper addressing the problems involved in the calculation, suggested that the percentage of GDP accounted for by logistics ranged from 4.5% in Canada to 20% in Morocco.

In the EU, there is a massive difference in the percentage carried by mode between countries, but overall, road actually increased its share by 2.7% between 2000 and 2010 whilst rail fell by a similar amount. All modes can be used for domestic and international transport of goods. The extent to which rail and inland waterways are used for international transport depends to a large extent on the geographical features and location of the country concerned. The percentage of international transport by rail (ignoring Ireland, Malta and Cyprus which have 0%) ranges from 2% in the United Kingdom to 91% in Latvia (Eurostat, 2013). Considering transport by inland waterway, only 17 EU member states have navigable inland waterways. Table 1 shows the amount of freight carried, by EU state and the modal share.

THE POSSIBILITIES FOR IMPROVEMENT

There are literally hundreds of measures, sometimes called 'interventions', which can be taken to improve the environmental sustainability of logistics operations. For the purposes of this chapter, the interventions are divided into the following (not necessarily mutually exclusive) categories:

1. Modal split,
2. Efficiency of vehicle usage,
3. Efficiency of vehicle routing,
4. Supply chain structure,
5. Technology,
6. Human issues,
7. Specific issues.

Table 1. Inland Freight Transport (mtkm) and Modal Split (%) by Country, 2010.

EU State	Road	Rail	Inland Waterway	Road	Rail	Inland Waterway
	Million tkm			Modal split (%)		
Belgium	33,107	6,268	9,251	69.5	12.5	18
Bulgaria	21,214	3,291	4,311	68.1	10.7	21.2
Czech Republic	54,830	14,316	42	79	21	0.1
Denmark	16,120	2,239	–	87	13	–
Germany	323,833	1,07,317	55,027	64.9	22.2	12.9
Estonia	5,912	6,271	–	45.8	64.2	–
Ireland	10,108	105	–	99.2	0.8	–
Greece	29,815	614	–	98	2	–
Spain	2,08,843	9,748	–	95.8	4.2	–
France	1,85,658	34,202	9,029	82.2	13.5	4.3
Italy	1,42,885	19,787	–	90.4	9.6	0.1
Cyprus	941		–	100	–	–
Latvia	12,131	21,410	–	38.1	61.9	–
Lithuania	21,512	15,088	–	59.1	40.9	–
Luxembourg	8,835	288	305	93.5	2.7	–
Hungary	34,529	9,118	1,840	75.1	19.6	3.9
Malta			–			
Netherlands	73,333	6,378	46,278	62.1	4.9	33
Austria	28,542	20,345	2,123	56.3	39	4.7
Poland	2,07,651	53,746	161	80.6	19.4	0.1
Portugal	36,453	2,322	–	93.9	6.1	–
Romania	26,349	14,719	11,409	49.2	23.5	27.2
Slovenia	16,439	3,752	–	82.3	17.7	–
Slovakia	29,179	8,105	931	74.8	22	3.2
Finland	26,787	9,295	–	75	24.8	0.2
Sweden	36,932	22,864	–	60.7	39.3	–
UK	1,46,685	18,576	–	88.7	11.2	0.1
Liechtenstein	312	10	–	97.8	2.2	–
Norway	19,188	3,496	–	85	15	–
Switzerland	13,828	11,526	–	54.4	45.6	–
Croatia	8,926	2,438	692	71.2	21.2	7.6
Turkey		11,303		94.9	5.1	–

Source: Eurostat (2013) (see Original Tables for Compilation Dates and Notes).

Modal Split

Environmental Credentials

In order to improve the environmental credentials of logistics, probably the first measure that springs to mind is switching the transport of freight from the least sustainable mode (usually road) to the more sustainable modes (usually rail and water). The relative environmental impact of the main modes of transport is shown in Table 2.

Care must be taken in the interpretation of the figures in the above table as emissions will obviously be dependent on many factors, including loading factors, driving practices, geographical terrain, vehicle condition and many other factors which are discussed later in the chapter. A very thorough exposition of the effect of many of these parameters on fuel consumption and emission factors can be found in den Boer, Otten, and van Essen (2011).

According to the DfT (2013), rail freight contributed to 7.59 million avoided lorry journeys in the United Kingdom in the year 2012/2013. An alternative interpretation is that the number of lorry kilometres required to be undertaken to equal the amount of freight moved by rail in the United Kingdom was 1.6 billion km.

Use of Modes

As shown in Table 2, the extent to which different countries in the EU use rail and water to transport freight is very varied. Some countries have actually witnessed considerable shifts of freight *away* from rail and towards road recently. Thus, over the period 2000–2010, in 8 EU member states, and particularly in Austria and Belgium, rail has increased its modal share. However, in many other member states there has been a considerable

Table 2. Average Emission Factors for Freight Transport Modes within Europe.

		Energy Consumption (kj/tkm)	CO_2 (g/tkm)	NO_X (mg/tkm)	SO_2 (mg/tkm)
Aircraft		9,876	656	3,253	864
Truck > 34–40 t	Euro 1	1,086	72	683	
	Euro 2	1,044	69	755	
	Euro 3	1,082	72	553	90
	Euro 4	1,050	70	353	
	Euro 5	996	66	205	
Train	Diesel	530	35	549	44
	Electric	456	18	32	64
Waterway	Upstream	727	49	839	82
	Downstream	438	29	506	49

Source: Table 22 and Table 9 in IFEU (2008).

modal shift away from rail to road (for instance, Poland's road share has increased by 23%, Slovakia's by 22% and considerable increases have also been registered in Bulgaria, Lithuania, Latvia, the Czech Republic and Slovenia) (Eurostat, 2012). This decline in rail has partly occurred because rail has traditionally been more suited to the transport of bulk goods over long distances; so as Europe has become less heavily industrialised, the traditional market for rail has diminished. Containerisation reversed this decline somewhat but other trends, such as just-in-time production and an increasing share of high value, low density goods in the transport mix, have not favoured rail.

The growth in the use of containers, swap bodies, and more recently low-loaders/liners that can accommodate whole semi-trailers has led to the development of inter-modal transport, where rail and/or water is used for the longer distance flows and the road legs are confined to the end of the journey. As Woodburn and Whiteing (2010) state: 'With inter-modal transport, it is the unit in which the goods are transported that is handled at the point of transfer rather than the goods themselves'. For inter-modalism to increase, handling equipment and other network systems have to be standardised across modes and across countries and this has proved to be difficult when rail systems have been developed independently across Europe. Differences in rail gauges, fuel used (i.e. diesel vs. electric) and bridge heights are three obvious examples. Developments such as the European Modular System facilitate inter-modal operations at the end of transport legs.

Major EU Mode Shift Programmes

Effecting a modal shift in order to improve the environmental sustainability of freight transport has been the subject of much EU attention. The EU has a massive programme – the Trans-European Network programme (TEN-T). Initially agreed upon in 1996, this programme has been continuously expanded and in 2013 another €26 billion was allocated to it for the period 2014–2020. TEN-T is 'a core transport network' built on nine major corridors: two North-South corridors, three East-West corridors; and four diagonal corridors. Essentially it seeks to enhance interoperability and connectivity within the European freight transport industry in order to facilitate EU cohesion and development in a sustainable way (EU, 2013). In the 2007 EU Freight Action Plan (EU, 2007), the concept of 'green corridors' for freight transport was established. The aim of these corridors was to enable traffic to flow more smoothly and efficiently and in a more environmentally acceptable manner across the various borders of the EU.

Co-modality and use of advanced technology were fundamental underlying themes for these corridors. An initial 60 corridors were chosen for study, based on the TEN-T priority projects. This was later reduced to 30 corridors. However, even developing a framework of KPIs to measure the theoretical effectiveness of these corridors in terms of the environment has proved to be problematical (see Psaraftis & Panagakos, 2012).

A second EU Programme, entitled MARCO POLO, which also ran from 2007 to 2013, was established to provide funds to help companies shift from road to rail transport efficiently and profitably. Both the TEN-T and MARCO POLO programmes were transferred to the Innovations and Networks Executive Agency in January 2014 (EU, 2014).

Another large European project established in 2006 to disseminate good practice in sustainable logistics was the bestLog programme. Since 2010 it has been operated by the European Logistics Association and is known as ELAbestLog. Its website provides case studies on mode shifting and co-modality. One such case study is Cargo Domizil in Switzerland, which collects less-than-full truck loads (LTL) from customers and takes them to rail-side depots where they are consolidated for sending by rail. At the other end of the journey the goods are then sorted for onward delivery to the final customer. This is an example of the innovative use of rail for less-than-full truck loads.

In the United Kingdom, the Freight Best Practice programme was established, partly to address the environmental externalities imposed by freight transport. It consists of a number of very useful reports on various themes. One (DfT, 2008) contains 35 diverse case studies of companies which have moved all or part of their freight transport from road to either rail or water-based transport.

Developments in Rail-Freight Transport

The use by large retailers of rail in the United Kingdom (sometimes referred to as 'consumer rail freight') has been a big breakthrough which has accelerated the growth in rail freight. This is a sector which has traditionally used road for practically 100% of its logistics, both inbound and outbound. Much of the success of these initiatives can be attributed in the United Kingdom to Freight grants which have been designed to encourage a shift from road to 'greener' modes.

In the shipping sector, a number of container shipping lines such as Feederlink BV, OOCL and K-Line have developed short-sea and coastal services that move containers to ports that are closer to their ultimate destination, rather than relying on land-based onward movement from the major ports (Wang & Monios, 2012).

A recent major development has been the involvement of shipping companies in contracting train space on container services, in some cases even committing to regular full trainloads. Kuehne & Nagel uses a mix of dedicated trains, contracted space on multi-customer trains and spot hire of capacity (Woodburn & Whiteing, 2010). Logistics service providers have also become actively involved in developing successful rail packages to attract retail companies to rail. Such companies act as consolidators, making up viable trainloads. In the United Kingdom, a company called FreightArranger has completed a trial to provide a cloud-based inter-modal brokerage and tracking system for consignors. It takes live container bookings for consignors looking to use rail for loads as small as a single container. This type of brokerage has the potential to improve rail usage in the future.

Rail has historically been used to a greater extent in mainland Europe than in the United Kingdom. For many decades, Austria, Switzerland and Hungary have been using rolling roads (where the whole truck including the tractor and, often, the driver, is loaded onto a train and is carried long distance before finishing the last leg by road). Piggybacking, that is the carrying of semi-trailers on flat rail wagons, has also been happening for 20 years, since the development of the swap-body container.

Barriers to Rail Use

There are still many barriers to the greater use of rail for transporting freight. Loading gauges are still far from synchronised across countries despite the 2002 EU ERA Technical Specifications for Interoperability (TSI) guidelines which seek to harmonise train systems in Europe. In the United Kingdom, where the loading gauge is smaller than most of mainland Europe, low-deck rolling stock can sometimes be used to carry taller (2.9 m) shipping containers on low gauge lines, although the low-deck rolling stock cannot carry as many containers. New 'P400' 'megatrailers' capable of carrying two lorry trailers, each of which can be loaded with 100 tonnes of goods, have just been introduced to the United Kingdom. The aim is to be able to use these vehicles for flows into and out of mainland Europe. However, as yet, their use is very restricted within the United Kingdom. The train is purported to reduce carbon emissions by 20% per trailer. As new types of semi-trailer are introduced into the road logistics market, new demands are placed upon the rail system to be able to deal with them.

Interoperability and, therefore, the potential for modal shift depends not only on the physical and technical attributes of the rail system but also on the IT aspects. Routing and scheduling of international rail freight is currently not easy because of the proprietary nature of the IT systems in use around Europe's rail companies. This too will need to be addressed before there is a smooth flow of goods travelling around Europe on rail.

Flexibility is of key importance in freight mode choice decisions and rail has suffered from being (or as having an image of being) rather inflexible, both in terms of being unable to deliver the door-to-door solutions that road could offer and the unwillingness of the rail industry to respond to modern needs. Flexibility problems include:

- Loads must be booked a long time in advance and cannot be easily changed;
- Last minute orders cannot be accommodated;
- Train timetables do not correspond well with the needs of customer production/distribution patterns.

These flexibility issues occur on top of other perceived service-related problems such as:

- Passenger rail has priority over freight rail meaning it can be quicker to send goods by truck, even over long distances;
- Reliability is not perceived as being good;
- Rail management does not listen to customer requirements;
- A lack of client-oriented focus;
- A lack of understanding of how rail fits in with the whole supply chain of the client.

Without massive investment in rail and water infrastructure, it is difficult to see how all these problems can be addressed.

Of course, although modal switching on its own will produce some improvements in environmental sustainability, it needs to be combined with other measures, discussed below, to maximise the benefits associated with it.

Efficiency of Vehicle Usage

Probably the single biggest measure that can be taken to improve the sustainability of freight transport is to increase the efficiency of the logistics process. McKinnon and Edwards (2010) illustrate this in the following statistics:

- Approximately 25% of truck kilometres in EU countries is run empty, while in Ireland it averages around 37% (Eurostat, 2007).
- In the United Kingdom food supply chain, only about 52% of the available space on laden trips is actually occupied by a load (Freight Best Practice Programme, 2006).
- 44 tonne trucks in the United Kingdom, which can carry a maximum payload of 29 tonnes, transport on average only 17.6 tonnes when laden and 12.7 tonnes if allowance is made for empty running (Knight, Newton, & McKinnon, 2008).

Gucwa and Schafer (2011) found that the greatest variation in energy intensity within the transportation mode is explained by the load carried per vehicle and the vehicle size. Vehicle utilisation has improved dramatically over the past few decades, but as shown by den Boer, Otten, and van Essen (2011), there is still a considerable amount of work to do. It is necessary for logistics operators to engage in 'Smart Logistics' and reduce total tonne-mileage. Of course, not all loads will be weight constrained; many are constrained by volume and not by weight and this can lead to problems in interpreting some efficiency data. Notwithstanding this issue, the major ways of improving vehicle usage efficiency are outlined below.

Improving Vehicle Utilisation Through

(i) Maximising payload – whether for truck, train or vessel, the energy intensity of logistics (fuel/tonne km) is minimised when the vehicle is fully loaded. Maximising payload can be achieved through better use of ICT, load consolidation, proper siting of depots and manufacturing plants and many other factors. In the United Kingdom, statistics from the Freight Transport Association (FTA, 2013) show that the lading factor by weight of HGVs is 59%. Although this has improved over the years, and there will be many vehicles which are volume rather than weight constrained, there is room for further improvement.

(ii) Reducing empty running – is a subset of maximising payload. The FTA suggests that in the United Kingdom, the percentage of HGVs empty running is 29% and that this percentage has not changed over the last 5 years. Finding backloads and collaborating with competitors are key issues here. There exist many load matching companies which will help in providing backloads. However, many companies are loathe to use such companies because their own logistics activities are so finely tuned that any small problem with the backload compromises the whole operation. Finding a particular company or set of companies to work with over a longer time period with specific, appropriate products and locations is a better option.

(iii) Minimising down time – if vehicles are off the road for any reason (whether for technical or staffing reasons), they are not being used efficiently. Keeping the vehicle fleet as new as possible through vehicle replacement schedules in combination with proper computerised vehicle maintenance schedules can prevent breakdowns and maximise engine efficiency (although in terms of a life-cycle assessment perspective, this is not necessarily the most environmentally beneficial policy). Likewise, hiring the right staff and paying attention to staff morale, as well as driver training in vehicle awareness and safe driving practices so as to minimise accidents, will also play a part. As far as possible, pairing particular drivers with particular vehicles will engender more of a sense of responsibility and enable them to pick up on any vehicle peculiarities. Driver involvement in the process and appropriate rewards will result in better vehicle performance.

(iv) Using the correct vehicle size/type – as a general rule, bigger = better in terms of efficiency and sustainability across all modes. Truck sizes and weights have increased over time and there is pressure for them to increase further. Rail trailers and train lengths are also becoming longer, with greater carrying capacity. This is also true of ships. In the case of trucks, high-cube vehicles have been introduced to maximise carrying capacity for low density, high volume loads; double-deckers are being increasingly used for some load types; longer trailers are also being tested. However, larger and heavier vehicles are only efficient if they are fully loaded. A partially loaded 38 tonne truck is less efficient in terms of the environment than a fully loaded 16 tonne truck. Similarly, a partially loaded train is likely to be less efficient than a number of trucks that are full.[1]

Returning to long haul transport, Wincanton Transport has tested the use of longer semi-trailers (up to 15.65 m) which can take two additional rows of pallets. They found that carbon reductions amounted to approximately 15%. However, the costs to them of training the drivers have been considerable. It has been found that two double deck trailers can replace three standard trailers, saving approximately 22% on fuel and CO_2. John Lewis and many other companies are now using double-deckers. They are particularly suited to inter-depot trunking operations of retailers, parcel and pallet operators and manufacturers of high volume, low density products. Some companies, however, have not continued with their use because they have found that the required additional loading and unloading times negate any savings accruing from the trunking operations. The take-up of double-deckers and high-cube vehicles has also been compromised by the EU's attempts to limit the height of HGVs to 4 m, although this policy now seems to have been abandoned.

In an interesting development, John Lewis has designed a vehicle of 31 tGVW. This allows the vehicle to have a smaller engine and only four axles and is more fuel efficient without compromising the payload capacity.

Efficiency of Vehicle Routing

Routing and Scheduling

Routing and scheduling includes the assignment of loads to vehicles so as to maximise payloads and vehicle efficiency. It also includes assigning time windows (for both arrivals (incoming) and departures (outgoing), if appropriate) in terms of both their actual time and their length so as to minimise congestion at depots and vehicle down-time. Similarly, procedures need to be established for the situation when vehicles do not arrive on time and for scheduling backloads. Other routing and scheduling issues include:

- Route planning, taking into account vehicle and time restrictions;
- Scheduling to minimise congestion – including procedures for dealing with unanticipated congestion or road closures;
- Scheduling night hauls where feasible;
- Using green corridors where possible;
- For rail freight, using train-path optimisation.

Backloads

Ensuring that vehicles do not return empty is an obvious element of reducing fuel intensity. Backloads could involve bringing in goods from suppliers (i.e. combining primary and secondary distribution functions); bringing back waste or equipment (such as pallets); organising routes such that trips are circular rather than uni-directional ensuring that vehicles are continuously laden. Maximising backloads could also involve collaboration with other companies, even competitors, although there is always the issues of 'ownership of the product' and fair division of costs and benefits to be overcome. In recent years, many such collaborations have taken place.

Obviously backload products have to be compatible with the main product being carried; if vehicle interiors need to be washed down in between journeys, for instance, it may negate the benefits. Use of non-dedicated 3PLs or 4PLs may facilitate back-haulage.

Companies providing freight or backload exchanges have existed for many years and are now mostly internet based. Such exchanges seek to match loads suitable for backloads with companies looking for backloads. Examples of companies include Returnloads.net, a UK based company which estimates it has saved 750,000 HGV empty runs in the year June 2012–2013; truckspace.co.uk and freightfinder.com.

Freight Best Practice (Freightbestpractice.org.uk, 2009a) have published a document entitled 'Make back-loading work for you' aimed at small and medium sized companies, giving advice on the benefits of getting involved and how to overcome the constraints. It suggests the following services, systems and mechanisms for obtaining backloads:

- Return load specialists,
- Load matching services,
- Freight forwarders,
- Partnerships,
- Reverse logistics,
- Pallet networks,
- Supply chain initiatives,
- Subcontracting.

Use of ICT (Information and Computer Technology)

ICT can be used in a myriad of ways to increase vehicle efficiency and improve environmental sustainability. The most obvious include vehicle and driver data; paperless manifests; asset tracking; satellite navigation; safety and security systems; digital tachographs; traffic information systems and in-cab communications. The selection and purchasing of ICT can be a daunting and expensive experience. There are many off-the-shelf products, but choosing between them can be difficult. Many larger companies opt to have their own bespoke products designed, but much money can be wasted and many companies have paid dearly for products which do not function in the way they were intended. Choice of the wrong ICT products has been blamed for the near failure of some companies (e.g. the large UK supermarket chain Sainsburys (Clark, 2004)). In addition, buying the product is not the end of the matter. The products are only as good as the use to which they are put; the competence of the user and the quality of the data input to start with. Good ICT systems need good management and good information dissemination. It is pointless having digital tachographs, for instance, if the information retrieved is not used effectively.

Supply Chain Structure

Manufacturing Location

It is essentially the internationalisation and globalisation of world trade that has effectively dictated the contemporary pattern of logistics. Manufacturers can choose to manufacture goods anywhere in the world and many have moved to locations which minimise their major cost – that is labour, letting the logistics aspect of the business take the strain. Many goods are now manufactured in Asia and shipped to Europe by container ship, often berthing in a different country from where the goods are demanded, therefore requiring onward transport by another mode.

Warehouse/Depot Location

The issue of how many warehouse/depots to have and where they are best located is a difficult one. In theory, given information on customers and suppliers, the optimal location and number of warehouses can be calculated using mathematical optimisation techniques. Some companies prefer hub and spoke operations and others prefer different structures. However, many companies evolve by merger and acquisition or have developed over decades or longer and find themselves inheriting warehouses and depots in non-optimal locations – perhaps even in the centre of towns or cities – with depots that were built for horses and carts rather than large trucks. When IKEA investigated their supplier locations in Poland, they ended up completely restructuring their depot locations as they found that the original depot locations were organised for the situation where they were supplying the domestic market rather than a large global company. It takes careful monitoring of flows and costs to design an effective configuration.

Load Consolidation/Deconsolidation

Load consolidation and/or deconsolidation at many stages of the supply chain and can occur within organisations or with other companies. A good example of consolidation is that of fish being brought from many different fisheries and ports around Scotland, by different logistics operators often using small vehicles, and then being consolidated in a warehouse in Glasgow for onward shipment to major customers. At an even more local level, fish are brought in small vans and boats to quaysides in Northern Scotland, where they are loaded onto large articulated vehicles to be taken 300 kms south to Glasgow – or further afield. There is the thorny issue of 'ownership' of the product as well as cost apportionment issues if consolidation of goods from several companies is involved.

Logistics Industry Structure. Is Bigger Greener?
Although the logistics industry is still fragmented, with many small haulage companies, each having a very small proportion of the market, the industry has witnessed a considerable degree of concentration over the last 20 years and is now dominated by a few very large logistics companies. Examples of such companies include DB Schenker, Kuehne and Nagel, Exel, Panalpina, UPS, DHL and TDG. Contract logistics, carried out by 3PLs (which provide the logistics services for part or all of the supply chain) or 4PLs (which were originally non-asset based companies but are now often logistics providers which claim to manage the whole supply chain) are now the norm in most industries. Most recently, the so-called 5PLs have emerged. These companies aggregate the demands of the 3PLs into bulk volume in order to negotiate cheaper rates and better resource utilisation. Most large retail/manufacturing companies have contracts with several (sometimes up to 20) 3PLs and/or 4PLs to spread their risk or focus on particular elements of the supply chain.

The greater the level of integration within the supply chain, the higher should be the utilisation of assets and the more likely the company to have the funds to invest in up-to-date equipment and ICT. One of the major functions of a 4PL is to manage the electronic interfaces between different companies in the supply chain. This should result in more seamless, effective supply chains with less waste and duplication and which thus, should be more sustainable. Most large 3PLs and 4PLs make a feature of their 'green' credentials. On the flip side of the coin, however, it could be argued that it is the development of these companies that has facilitated global trade and the global supply chains which have resulted. Their overall impact on the environment is therefore debatable.

Technology

Vehicle Technology
Trucks. Improvements in truck technology have enhanced the sustainability of logistics considerably over the years. Such improvements include

reductions in tare weights, the use of turbo-charging and the design of new types of vehicle which reduce the fuel intensity of goods transport. With the introduction of each Euro Standard (i.e. EU standards regulating the exhaust emissions of trucks >3.5 tonnes), truck engine fuel efficiency has been reduced. In order to maintain fuel efficiency levels, therefore, much effort has gone into improvements in vehicle body and transmission-related measures. Aerodynamics, for instance, has been of key importance. It has been claimed that aerodynamic intervention can reduce fuel usage by between 6% and 20% and is highest when the vehicle is being driven at higher speeds. Cab streamlining includes over-cab spoilers, rounded edges and air dams and trailer streamlining includes the design of the trailer itself, side skirts, curved trailer edges and spats over the wheels. Marks and Spencer have piloted the use of teardrop design trailers (see Fig. 1) and calculate that they save 10% on fuel and increase load capacity by 10%. Integration of the tractor and trailer to reduce the turbulence created between them is also important. The Freight Best Practice programme has produced a 58 page document entitled 'Aerodynamics for Efficient Road Freight Operations' (Freightbestpractice.org.uk, 2009b) which gives guidelines for various types of vehicles as well as explaining the principles behind them and providing the payback periods associated with them.

Fig. 1. Example of a Donbur Teardrop Design Trailer.

In order to improve fuel consumption, fuel additives, lubricating oils, catalysts and magnets can also be used. Their overall effectiveness, however, seems to be somewhat under debate.

CLECAT (2013), which is a very useful document about the environmental impacts of logistics compiled by the European Organisation for Forwarding and Logistics, also discusses the issue of tyres. They suggest that new radial tyres, with proper maintenance can run over 100,000 kms on their original tread and can be re-treaded two or three times if they are carefully looked after. Again, a proper computerised tyre maintenance programme can help with this. Once the tyres have come to the end of their useful life, they should then be recycled properly. Correct tyre pressures are also important. It is claimed that automatic tyre inflation systems can result in fuel savings of around 5%.

Use of lighter weight metals in the construction of vehicles has enabled reductions in tare weight – also called 'lightweighting'. The use of vehicles with lower tare weights will enable the carrying of greater payload, so where the goods being carried are weight constrained, this can be important. The average tare weight of a vehicle varies substantially between manufacturers – and is also influenced by the amount of aerodynamic accessories it has.

The use of speed limiters which reduce the maximum speed of a HGV from 65 to 60 mph produces considerable fuel savings of around 6.8% (AEA consultancy, 2011). It is, therefore, worth considering this option as it is more productive than any other of the fuel improvement measures individually and can be combined with other measures.

Trains. As with trucks, aerodynamics is important for sustainability in trains, particularly at speeds above 100 kph. In the case of inter-modal trains, aerodynamic drag can be reduced by optimising train loading so as to minimise the gaps between wagons. Both the engines and wagons can also be streamlined to improve aerodynamics.

Use of electric traction reduces pollutants at the rail side, but as with any electric vehicle, the overall savings depend on the original source of the electricity. Reduced operating speeds, as with all modes, can improve fuel efficiency as can engine shutdowns in diesel trains when the locomotive is idle for 15 minutes or more. More modern diesel locomotives can much more readily restart once they have been switched off. Again, as with other modes, stop-starting, fast acceleration and breaking and frequent speed changes are bad for the environment. Providing paths through the rail network that enable trains to make steady progress rather than being constantly interrupted, helps the environment considerably (although it is difficult to assess the exact improvement numerically).

Concentration on design is a major area for fuel intensity improvements. The use of high-cube containers, swap bodies and low-loaders as well as increases in the length of trains, have all contributed considerably. These improvements have often been led by the consumer freight sector. Additionally, improvements in handling equipment at rail terminals have helped to make rail more efficient and attractive to potential users. In the case of inter-modal transport, for instance, benefits have accrued from the fact that terminal and transport equipment has become much more standardised. The problem is that not all of the rail network is capable of being used by the more modern trains and not all trains have been adapted/constructed with the more up-to-date features. Also, as new, longer containers are introduced, the technology for handling them needs to be introduced wherever they are being loaded and unloaded, and this is very expensive. Fragmentation of the rail industry across Europe also causes immense problems of coordination.

Ships and Barges. As with other modes, a ship's fuel usage increases with speed. Some shipping companies have implemented policies of 'slow steaming' to take advantage of this, but there is a question of the effectiveness of this policy as slow speeds may in the end mean more ships, which is not very environmentally sustainable (see Cullinane & Cullinane, 2013). Other technical measures include:

- *Sails or kites* – Some companies have experimented with these in order to improve fuel efficiency, but they have not met with a great deal of success.
- *Greater engine efficiency* – Over the past 30 years, more efficient marine engines with lower Specific Fuel Oil Consumption (SFOC) have been developed.
- *Waste heat recovery* – The exhaust gas and cooling water from ships contain substantial energy that could be harnessed, thereby improving the overall thermal efficiency of the engine system by between 5% and 10%. Maersk Line is the first shipping company to install waste heat recovery systems as standard on all its new ships. Maersk estimates that, using this system, its Triple-E fleet has reduced its fuel consumption and emissions by about 9% when operating within a speed range of 18–23 knots. At an installation cost of $10 million per ship, this represents a payback period of somewhere between 5 and 10 years depending on fuel

price and service speed (Maersk Line, 2011). Similarly, Wärtsila, a leading marine engine manufacturer, has developed a total heat recovery system which it claims saves 12% in fuel consumption and emissions.

- *Improved hull design and performance* – To facilitate propulsion, it is necessary to either reduce the weight of a ship by selecting appropriate lighter design materials which do not compromise hull strength or to reduce the resistance of the ship in the water. Innovative research has found that the pumping of compressed air bubbles over a ship's hull surface while in motion can significantly reduce vessel drag, particularly where the hull is polymer-coated to reduce surface tension.
- *More efficient propellers and rudders.* In general, the larger the propeller diameter, the higher the propeller efficiency and the lower the optimum propeller speed. Thus, achieving a propeller speed which is as low as possible (within the design restrictions of a ship) is best.
- *Scrubbers and Filters.* Scrubbers come in two varieties; wet (or seawater scrubbers) and dry (or closed scrubbers). Both types of scrubber practically eliminate both sulphur emissions and particulate matter. Although they appear at this juncture to be very promising, both are still in their infancy and the technology is not yet proven.
- *Shore-based power ('Cold Ironing').* Cold ironing is the use of shore-based power to provide electrical energy to a ship while at berth rather than using its auxiliary engines. This means that all engines can be shut down.

Alternative Fuels
The major alternative fuels available are as follows:

Biofuels (Including Biodiesel and Bioethanol). These are fuels produced from renewable plant material and oils and are, therefore, more sustainable than fossil fuels. They are usually used in combination with diesel. Their use reduces pollutant emissions, but there has been considerable debate about the land-take problems caused by the growth of the biological crops required for their production. Use of biofuels practically eliminates emissions of sulphur and considerably reduces emissions of hydrocarbons and carbon monoxide, although emissions of nitrous oxides increase slightly. Its effect on CO_2 is more debatable. It has been calculated that there are minimal tail pipe emission differences, but that the benefits arise from the renewability of the biofuels themselves (DfT, 2007). Because of the land-take issues and the low efficiency of biofuels, second-generation biofuels are now being developed. These are fuels that can be made from waste materials (such as municipal waste) and cellulosic crops (dedicated energy crops) that can be grown on wasteland. Unfortunately, it is debatable whether there are sufficient quantities of these new sources to make any major impact on the environment.

The UK's Renewable Transport Fuels Obligation (RFTO), introduced in 2008, places an obligation on owners of liquid fossil fuel intended for road transport use to ensure that a certain amount of biofuel is supplied or that a suitable amount of money is paid to support the production of these fuels. Only organisations that supply more than 450,000 litres of fossil fuel per year are affected by this.

Hydrogen. Hydrogen was recently seen as the panacea for the future, but now the outlook is more cautious. The problem for freight vehicles is the weight of the fuel cells required and as yet, this problem has so far proved to be insurmountable.

Gas Filled Vehicles Natural Gas (NG), Compressed NG (CNG) and Liquid Petroleum Gas (LPG). NG vehicles are methane powered vehicles which can be derived from either fossil sources or bio-methane. Some manufacturers produce HGVs that can run purely on NG, or dual fuel/NG vehicles are also available. NG produces no nitrous oxides or particulate matter and about the same level of CO_2 as normal diesel.

DME or Dimethyl Ether. DME is a relatively new addition to the fuel market in Europe, although it has been used in Asia for some time. It is produced primarily from coal or gas or bio-stocks and has massive environmental benefits as it is virtually pollutant free at the point of use. Volvo are piloting it in Europe, but as yet, it is not in general use. It is also being pioneered by Stena in their ferry fleet. It is a gas under ambient conditions but can be stored as a liquid under moderate pressure and requires very little engine adaptation to use it.

Electricity. Not yet suitable for HGVs on long-distance trips. Use of electrically propelled trains is much more common. The sustainability of electric trains depends to a great extent on the source of the electricity used.

The Use of Alternative Energy in Logistics. One problem with all the alternative energy sources is that the supply infrastructure is still not well developed. It is therefore up to the logistics operators themselves to provide the infrastructure. Tesco owns a 25% share of biofuel company

Greenenergy, which buys rapeseed from 1,500 farmers in the United Kingdom to make biodiesel. It runs a 50:50 biodiesel mix in its own vehicles (Tesco, 2008). This was once a very prominent part of its website, but with the debate over the sustainability of biofuels, it has disappeared. Similarly, the companies which belong to the FTA's Logistics Carbon Reduction Scheme (LCRS) are also shying away from the use of biofuels because of the land-take issue. They are looking more towards the use of gas, but are concerned about the lack of infrastructure. According to the Natural and Biogas Vehicle Association (NGVA, 2011), there were 220,000 medium or HGVs worldwide using natural gas.

In the Freight Best Practice Guide, the example is given of United Kingdom logistics operator Howard Tenens who has converted 26 of its 110 vehicles to dual fuel (either CNG or NG). The company has also invested in three refuelling stations. The company estimates that the conversion costs up to 50% of the original cost of the engine, but the conversion kit can be taken out of the vehicle on sale, so can be used multiple times. The company has calculated that the conversion kit has paid its way on its use within the first vehicle, so the cost of its use in subsequent vehicles should be negligible.

DB Schenker uses some hybrid engines and second-generation biofuels in its massive HGV fleet. However, together with many other major logistics companies, their corporate environmental strategy focuses much more on gains made through optimisation of capacity and supply chain structures which enable the 'bundling' of transport, than on the use of alternative fuels. Many companies are using the purchase of up-to-date vehicles which comply with higher Euro standards as their focus for vehicle emission reductions. It is possible that the volte-face over the sustainability of biofuels has deterred logistics companies from spending too much time and money on the use of alternative fuels. This is likely to be even more pronounced in the case of operators of long-distance transport, where tail-pipe emissions are less important than overall cost and environmental standards issues.

In the shipping industry where, new, much tighter sulphur emission standards have been introduced, including the SECAs (sulphur emission control areas) where ships are able to emit practically zero sulphur, huge efforts are now being put into the search for less polluting fuels. One alternative is the use of very low sulphur distillate fuel, but this is very expensive and there are fears that there may be insufficient supplies of it. The evolution and phasing-in of the various regulatory regimes is summarised in Fig. 2.

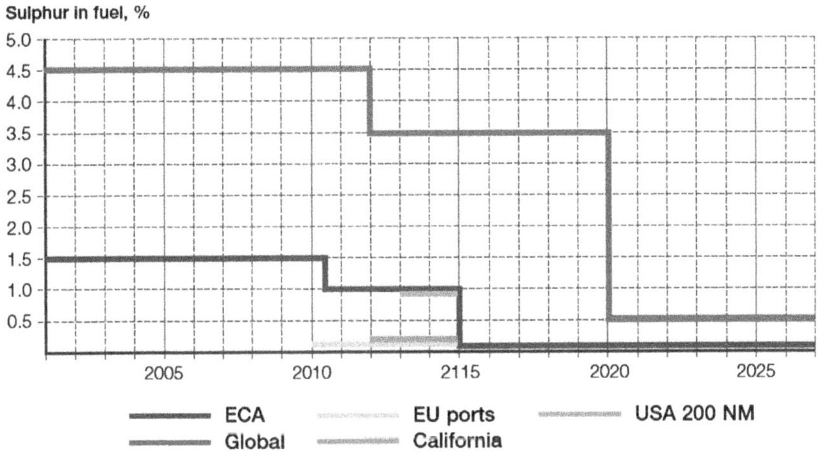

Fig. 2. The Evolution of Global and Local Sulphur Legislation in Shipping. *Source*: Derived from data contained in Buhaug et al. (2009).

In the case of rail, EU legislation under the Fuel Quality Directive means that fuel from rail locomotives must be sulphur free from January 1, 2012. This is being achieved through a combination of the use of ultra-low sulphur diesel and sophisticated fuel injection equipment. Some companies are using biofuel mixtures, but the problems here are the same as those relating to biofuels in general that is their true sustainability is being increasingly questioned. Many of Europe's trains, of course, have electric traction, which, depending on its source, is often far less environmentally damaging. DB Schenker's website relating to its use of rail makes the use of diesel sound positively archaic. In fact they state that their only use of them is for shunting. Whilst the United Kingdom has some freight trains with electric traction, their use on the United Kingdom rail network is very limited. When trains run on electricity, it enables the use of other technologies such as regenerative braking.

Human/Management Issues

Driver Training

One of the easiest ways of improving fuel intensity and reducing other negative logistics externalities (particularly accidents) is through appropriate driver training and particularly the promotion of eco-driving (i.e. driving in a manner which produces the least environmental dis-externalities). Eco-driving basically involves driving more steadily – trying to reduce fast braking and acceleration; correct use of gears; appropriate tyre pressures and maintenance; driving at optimal speeds; reducing idling time; choosing routes carefully and avoiding congestion and generally understanding the relationship between the vehicle and the environment. Training can involve the use of simulators, which give immediate feedback on the driver's impact on the environment through their driving style. Driver training needs to be carried out regularly to reinforce lessons learned. Studies have shown that eco-driving can reduce fuel usage by up to 14% (DfT, 2006). Although training is not cheap, it is usually cost-efficient and also gives the drivers more 'buy-in' to the company.

Many companies have introduced eco-driving training. DB Schenker, for example, trains all its drivers in eco-driving, including those of its subcontractors. Many companies have introduced 'league tables' of drivers who perform well in reducing fuel usage and have 'eco champion' awards. Of course, the first step towards minimising fuel usage is an efficient monitoring system.

Eco-driving is just as important for rail locomotive drivers as for truck drivers.

Environmental Reporting

In order to be able to improve anything, there needs to be accurate benchmark figures. A key principle of operations management is that 'Measurement leads to Improvement'. Requests for information alerts management to where possible improvements can be made. Environmental reporting is a fairly new phenomenon which involves reporting key environmental statistics – possibly to an internal department, an external consultant or further.

In September 2013 the UK government introduced mandatory Green House Gas reporting for all UK quoted companies for Scope I and Scope II emissions and this might be extended to all companies by 2015.

The FTA runs a LCRS through which logistics operators report fuel usage to them and the FTA converts this into CO_2 emissions using government sponsored conversion factors. The FTA then aggregates the fuel usage figures to track emissions over time. Part of the reason that they are doing this is because they recognise how difficult it will be for smaller companies to engage in environmental reporting to government and they wish to persuade the government that reporting does not need to be mandatory. In modern vehicles, sophisticated on-board monitoring systems enable fuel usage to be measured accurately, on a per-vehicle and/or a per-driver basis. Green Freight Europe is another body seeking to promote more sustainable freight transport. It aims to be 'the leading independent voluntary programme for improving environmental performance in road freight transport in Europe' (Green Freight Europe, 2014). Since 2004, the United States has run a SmartWay programme, which is 'a market driven partnership aimed at helping businesses move goods in the cleanest most efficient way possible' (EPA, 2014). This includes benchmarking and environmental reporting.

Many large logistics operators are proud of their environmental credentials and make a great deal of marketing them. Tesco markets its multi-modal operations through the slogan 'Tesco lessCo$_2$,' which is very eye-catching. Most large company reports and websites have a section on the environment or sustainability and how they are seeking to address it through policies, often with measurable targets. Companies have become very sensitive about consumer criticism of their environmental policies and procedures and have sought to influence the environmental credentials of all the companies in their complete supply chains.

Management Issues

The role of good management cannot be overstated when it comes to sustainability. Good management means establishing the right priorities within the company, purchasing the right vehicles and ICT, signing contracts with the right companies and people, accounting for risk and uncertainty in the business and much more. It is difficult managing all the different functions, particularly in times of recession, as well as keeping the environment at the forefront of these decisions.

Implementing sustainable logistics policies is fraught with difficulties. Many are what can be described as 'green gold'; that is the policies are good for the environment whilst being simultaneously good for business. The obvious example here is fuel efficiency gains from better fuel usage monitoring. However, some sustainability interventions are not directly cost-effective and may require substantial investment with little financial benefit. Part of the problem is that interventions are not additive – that is if you implement an intervention such as tyre pressure monitoring which individually might improve fuel intensity by 5%, plus a driving training programme which might improve it by 13%, plus fit aerodynamic fairings to your vehicles which might yield a 4% improvement, the total increase in fuel intensity will not be the sum of these, that is 22%. A manager, or the management team will have to decide which is best for the company, taking into account the other company objectives. As discussed in CLECAT, each company needs to decide for itself what is 'green' and needs to analyse the company's operations in relation to this by measuring and benchmarking.

Freight Best Practice has produced a booklet for operators on how to test different fuel interventions, entitled 'Fuel Efficiency Intervention Trials; How to Test and Save' (Freightbestpractice.org.uk, 2009c). It outlines various fuel interventions and gives examples of companies which have tested some of them and the savings they have made. Freight Best Practice has also produced a 'fuel ready reckoner' (Freightbestpractice.org.uk, 2008) which helps a company to estimate how much fuel they could save by adopting various different fuel saving techniques, either individually or in combination. It lists 28 types of intervention, which are:

Increase vehicle fill	Decrease empty running
First driver training	Repeat training
CVRS	Satellite navigation
Under run air dam	Cab roof fairing
Cab side fairing	Body/trailer side panels
Body/trailer front fairing	Tractor side panels
Reducing cab gap	Tipper sheeting system
Sloped roof trailer	Teardrop trailer
Louvred spray suspension flap	Energy efficient tyres
Super single tyres	Replace steering super single tyres
Tyre pressure management	Regrooving tyres
Regular wheel alignment	Synthetic engine oil
Anti idling campaign	Speed limiter at 54 mph
Speed limiter at 52 mph	

Between 2005 and 2010, Walmart improved the fuel efficiency of its fleet by 60%, using a combination of monitoring, driver training and other measures. It reckoned that just by scrutinising and measuring the fuel usage, it was able to improve the average fuel efficiency of its logistics network by 25% (Plambeck, 2012).

Specific Issues

Reverse Logistics

Reverse logistics, that is, the logistics associated with the return of damaged, unsold or returned raw materials, in-process inventory and finished products back up the supply chain as well as the consolidation, handling and disposal of the resulting waste products.

With the introduction of the EC Directives on Waste Electrical and Electronic Equipment (WEEE) (2002/96/EC), the Restriction of the Use of Certain Hazardous Substances (RoHS) in Electrical and Electronic Equipment (2002/95/EC), Packaging and Packaging Waste (94/62/EC) and Distance Contract (97/7/EC), reverse logistics has become very important. Considerable planning is required to make reverse logistics as environmentally less-damaging as possible. However, in many ways, the reverse logistics process is more complex than the forward logistics because the flows of goods in the reverse direction cannot be so easily planned or forecast. This is probably worse in the case of city logistics, where individual consumers can (for whatever reason) return their unwanted internet purchases, but is nevertheless still important in terms of long haul transport. Often specialist recycling plants, for instance for electrical products, are located hundreds of miles away and trips need to be organised efficiently.

E-commerce and Logistics
Both B2B (business to business) and B2C (business to consumer) e-commerce have expanded at a phenomenal rate over the past 10 years and promise to continue increasing into the future. Much of the business of the large parcel delivery companies (such as UPS, Yodel, DHL) is now concerned with delivering goods purchased on the internet. These goods are often bought with little knowledge of, or regard for, their company of origin or despatch and are required quickly – often the next day. As a result, whole national and international networks of logistics hubs and depots capable of moving these goods around the world as fast as possible, have been established.

Interfaces
This chapter has focused on long haul logistics, but when considering environmental sustainability, the long haul element cannot always be considered in isolation. The origins and destinations of the long haul loads are often within cities and built-up areas. Goods are often manufactured/assembled on the outskirts of large cities, but if, for instance, rail transport is to be used, the goods will need to be taken to the railheads for loading/unloading and will often, therefore, have to pass through cities. City logistics is, therefore, very important. The chapter has also ignored the design of warehouses and depots, although there are many examples of how to improve their design to make them less environmentally damaging (green roofs, waste water use, etc.). Even something as seemingly innocuous as the water used for vehicle cleaning can make a considerable difference to sustainability.

EU/Government Policies
Logistics companies must work within the framework of both domestic and EU law. There are hundreds if not thousands of laws that have an influence on the sustainability of logistics, including those relating to the emissions of pollutants, vehicle standards and dimensions, health and safety, driving regulations, etc. Government (in its widest sense) also has an influence through taxation policy, for example, through vehicle taxation and the subsidies available for implementing measures to encourage modal shift. There is talk of carbon-emissions trading schemes and carbon taxes on location of facilities. EU funds a great deal of research on environmental research and seeks to disseminate good advice through programmes such as: BESTUFS, http://www.bestufs.net; ECOSTARS, www.ecostars-europe.eu; ELAbestlog (European Logistics Association), Supply Chain Management best practice. http://www.elabestlog.org/; GIFTS (Global Intermodal Freight Transport System), http://gifts.newapplication.it/gifts/; START (Short Term Actions to Reorganize Transport of Goods), http://www.start-project.org/; TEN-T (transport Infrastructure) http://ec.europa.eu/transport/themes/infrastructure/index_en.htm

SUMMARY AND CONCLUSIONS

This chapter has sought to provide an overview of the environmental impacts of long haul logistics operations and to describe some of the many ways in which these can be mitigated. The negative environmental externalities will never be entirely eliminated, but with considerable thought as to the overall pattern of logistics operations, as well as to the systems and environments within which they operate, the individual vehicles and drivers and the management of all these resources, the environmental impacts can be considerably reduced. In many cases, the resulting beneficiary is not just the environment but can also be the financial bottom line as environmental accounting can often lead to efficiency gains to the logistics operators and/or the consignors themselves. It could be argued that considerable progress is being made in developing and implementing energy efficiency measures because these are more likely to represent 'green gold' to both operators and consignors. Similarly, less attention is paid by operators to energy transition measures, because the cost is so much higher and the benefits much less tangible. Such measures may require appropriate support from governments, either in the form of carrots (e.g. financial incentives) or sticks (e.g. regulatory requirements).

NOTE

1. This chapter concerns long haul logistics, but it is perhaps important to note here that in the case of city logistics, the rule of bigger = better does not necessarily hold. In this case, dividing loads into smaller units that can, for instance, be transported by electric vehicles, could be less environmentally damaging.

ACKNOWLEDGEMENTS

The author wishes to acknowledge the financial contribution of The Sustainable Transport Initiative (STi) of the University of Gothenburg in writing this chapter. Also, the South East Scotland Transport Department (SESTRAN) for giving permission to use an earlier report as the basis of this chapter and the very helpful comments of two anonymous reviewers of an earlier draft.

REFERENCES

AEA Consultancy. (2011). *Reduction and testing of greenhouse gas emissions from heavy duty vehicles*. Retrieved from http://ec.europa.eu/clima/policies/transport/vehicles/docs/ec_hdv_ghg_strategy_en.pdf. Accessed on January 13, 2014.

Bowersox, D., Rodrigues, A., & Calantone, R. (2005). Estimation of global and national logistics expenditures. 2002 data update. *Journal of Business Logistics*, *26*(2), 1–16.

Buhaug, Ø., Corbett, J. J., Endresen, Ø., Eyring, V., Faber, J., Hanayama, S., … Yoshida, K. (2009). *Second IMO GHG study 2009; Prevention of air pollution from ships*. London: International Maritime Organization (IMO). Retrieved from http://www.imo.org/includes/blastDataOnly.asp/data_id%3D26047/INF-10.pdf

Clark, L. (2004). Sainsburys writes off £260m as supply chain IT trouble hits profits. *Computer weekly*, October 25. Retrieved from http://www.computerweekly.com/news/2240058411/Sainsburys-writes-off-260m-as-supply-chain-IT-trouble-hits-profit. Accessed on October 24, 2013.

CLECAT. (2013). *Logistics best practice guide. A guide to implement best practices in logistics in order to save energy and reduce the environmental impact of logistics*. European Organisation for forwarding and logistics. Retrieved from http://www.clecat.org/dmdocuments/sr004osust091104clecatbpgv.1.0.pdf

Cullinane, K. P. B., & Cullinane, S. L. (2013). Atmospheric emissions from shipping: The need for regulation and approaches to compliance. *Transport Reviews*, *13*(4), 377–402.
den Boer, E., Otten, M., & van Essen, H. (2011). *STREAM International freight 2011.* Comparison of various transport modes on a EU scale with the STREAM database. CE Delft.
DfT. (2006). *Companies and drivers benefit from SAFED for HGVs.* Retrieved from Freightbestpractice.org.uk
DfT. (2007). *Biofuels risks and opportunities.* London: Department for Transport.
DfT. (2008). *Choosing and developing a multi-modal solution.* London: Department for Transport. Retrieved from Freightbestpractice.org.uk
DfT. (2013). *Transport energy and the environment statistics 2011.* London: Department for Transport.
EPA. (2014). Retrieved from http://www.epa.gov/smartway/
EU. (2007, October 18). *Freight transport action plan.* COM (2007) 607, Brussels.
EU. (2012). *EU transport in figures. Statistical pocketbook.* Retrieved from http://ec.europa.eu/transport/facts-fundings/statistics/doc/2012/pocketbook2012.pdf
EU. (2013). Retrieved from http://ec.europa.eu/transport/themes/infrastructure/index_en.htm
EU. (2014). Retrieved from http://ec.europa.eu/transport/marcopolo/about/index_en.htm
Eurostat. (2007). *Average loads, distances and empty running in road freight transport – 2005.* Statistics in Focus. Transport 227/2007. Luxembourg: Eurostat.
Eurostat. (2012). *Energy, transport and environment indicators.* Retrieved from www.eurostat.ec.europa.eu
Eurostat. (2013). *Freight transport statistics.* Retrieved from www.eurostat.ec.europa.eu
Freightbestpractice.org.uk. (2008). Saving costs in tough times.
Freightbestpractice.org.uk. (2009a). Make backloading work for you.
Freightbestpractice.org.uk. (2009b). Aerodynamics for efficient road freight operations.
Freightbestpractice.org.uk. (2009c). Fuel efficiency intervention trials.
Freight Best Practice Programme. (2006). *Key performance indicators for the food supply chain.* London: DfT.
Freight Transport Association. (2013). *Logistics carbon reduction scheme.* Retrieved from http://www.fta.co.uk/policy_and_compliance/environment/logistics_carbon_reduction_scheme/index.html
Green Freight Europe. (2014). Retrieved from http://www.greenfreighteurope.eu/
Gucwa, M., & Schafer, A. (2011). *The impact of scale on energy intensity in freight transportation.* Stanford University, USA.
IFEU. (2008). *EcoTransIT ecological transport information tool, environmental methodology and data.* IFEU Heidelberg.
Knight, I., Newton, W., & McKinnon, A. (2008). *Longer and/or longer and heavier goods vehicles.* TRL Project Report 285, TRL, Berkshire.
Maersk Line. (2011). *Technology harnesses waste heat for energy.* Retrieved from http://www.maerskline.com/link/?page=news&path=/news/story_page/11/Technology_harnesses
McKinnon, A., & Edwards, J. (2010). Opportunities for improving vehicle utilisation. In A. McKinnon, S. L. Cullinane, M. Browne, & A. Whiteing (Eds.), *Green logistics.* London: Kogan Page.
NGVA. (2011). *European NGV statistics.* Retrieved from http://www.ngvaeurope.eu/european-ngv-statistics

Plambeck, E. (2012). Reducing greenhouse gas emissions through operations and supply chain management. *Energy Economics*, *23*(1), S64–S74.

Psaraftis, H. N., & Panagakos, G. (2012). Green corridors in European surface freight logistics and the SuperGreen project. *Procedia – Social and Behavioural Sciences*, *48*, 1723–1732.

Rantasila, K., & Ojala, L. (2012). *Measurement of national logistics costs and performance.* International Transport Forum. Discussion paper 2012–4. OECD.

Tesco. (2008). Retrieved from www.tescoplc.com. Accessed on September 8, 2009.

Wang, Y., & Monios, J. (2012). Synthesis Report of Freight Flow Mapping – Scotland Part. Report for SESTRAN. May.

Woodburn, A., & Whiteing, A. (2010). Transferring freight to greener transport modes. In A. McKinnon, S. L. Cullinane, M. Browne, & A. Whiteing (Eds.), *Green Logistics.* Kogan Page.

World Business Council for Sustainable Development and World Resources Institute. (2001). Greenhouse gas protocol initiative.

CHAPTER 3

COMPARISON OF VEHICLE MILES TRAVELED AND POLLUTION FROM THREE GOODS MOVEMENT STRATEGIES

Erica Wygonik and Anne Goodchild

ABSTRACT

Purpose – *To provide insight into the role and design of delivery services to address CO_2, NO_x, and PM_{10} emissions from passenger travel.*

Methodology/approach – *A simulated North American data sample is served with three transportation structures: last-mile personal vehicles, local-depot-based truck delivery, and regional-warehouse-based truck delivery. CO_2, NO_x, and PM_{10} emissions are modeled using values from the US EPA's MOVES model and are added to an ArcGIS optimization scheme.*

Findings – *Local-depot-based truck delivery requires the lowest amount of vehicle miles traveled (VMT), and last-mile passenger travel generates the lowest levels of CO_2, NO_x, and PM_{10}. While last-mile passenger travel requires the highest amount of VMT, the efficiency gains of the*

Sustainable Logistics
Transport and Sustainability, Volume 6, 63–82

ISSN: 2044-9941/doi:10.1108/S2044-994120140000006003

delivery services are not large enough to offset the higher pollution rate of the delivery vehicle as compared to personal vehicles.

Practical implications – *This research illustrates the clear role delivery structure and logistics have in impacting the CO_2, NO_x, and PM_{10} emissions of goods transportation in North America.*

Social implications – *This research illustrates tension between goals to reduce congestion (via VMT reduction) and CO_2, NO_x, and PM_{10} emissions.*

Originality/value – *This chapter provides additional insight into the role of warehouse location in achieving sustainability targets and provides a novel comparison between delivery and personal travel for criteria pollutants.*

Keywords: Warehouse location; criteria pollutants; greenhouse gas emissions; passenger travel; goods movement; city logistics

INTRODUCTION

Worldwide, awareness has been raised about the dangers of growing greenhouse gas emissions. In the United States, transportation is a key contributor to greenhouse gas emissions (United States. Environmental Production Agency [US EPA], 2008). American and European researchers have identified a potential to reduce greenhouse gas emissions by replacing passenger vehicle travel with delivery service (see Siikavirta, Punakivi, Karkkainen, & Linnanen, 2002; Wygonik & Goodchild, 2012). These reductions are possible because, while delivery vehicles have higher rates of greenhouse gas emissions than private light-duty vehicles, the routing of delivery vehicles to customers is far more efficient than those customers traveling independently. In addition to lowering travel-associated greenhouse gas emissions, because of their more efficient routing and tendency to occur during off-peak hours, delivery services have the potential to reduce congestion. Thus, replacing passenger vehicle travel with delivery service provides opportunity to address global concerns – greenhouse gas emissions and congestion.

While addressing the impact of transportation on greenhouse gas emissions is critical, transportation also produces significant levels of criteria pollutants, which impact the health of those in the immediate area

(US EPA, 2013b, 2013c). These impacts are of particular concern in urban areas, which due to their constrained land availability increase proximity of residents to the roadway network. In the United States, heavy vehicles (those typically used for deliveries) produce a disproportionate amount of NO_x and particulate matter – heavy vehicles represent roughly 9% of vehicle miles traveled (VMT) but produce nearly 50% of the NO_x and PM_{10} from transportation (Davis, Diegle, & Boundy, 2013; US EPA, 2008) (see Fig. 1).

Researchers have noted that urban policies designed to address local concerns including air quality impacts and noise pollution – like time and size restrictions – have a tendency to increase global impacts, by increasing the number of vehicles on the road, by increasing the total VMT required, or by increasing the amount of CO_2 generated (Allen et al., 2003; Holguín-Veras, Cruz, & Ban, 2013; Quak & de Koster, 2007, 2009; Siikavirta et al., 2002; Van Rooijen, Groothedde, & Gerdessen, 2008; Wygonik & Goodchild, 2011). The work presented here is designed to determine whether replacing passenger vehicle travel with delivery service can address both concerns simultaneously. In other words, can replacing passenger travel with delivery service reduce congestion and CO_2 emissions as well as selected criteria pollutants? Further, does the design of the delivery service impact on the results?

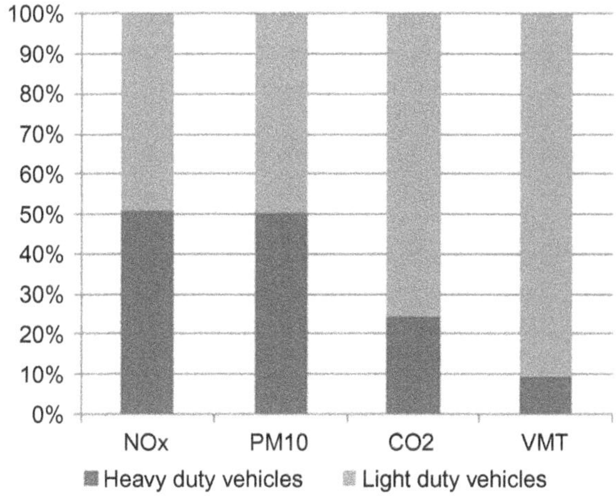

Fig. 1. Emissions and Vehicle Miles Traveled by Source Type.

This work models the amount of CO_2, NO_x, and PM_{10} generated by personal travel and delivery vehicles in a number of different scenarios, including various warehouse locations. The results allow for a comparison of the impacts of greenhouse gas emissions in the form of CO_2 to local criteria pollutants (NO_x and PM_{10}) for each scenario. These efforts will contribute to increased integration of goods movement in urban planning, inform policies designed to mitigate the impacts of goods movement vehicles, and provide insights into achieving sustainability targets, especially as online shopping and goods delivery becomes more prevalent.

LITERATURE REVIEW

Reductions in Externalities with Delivery Systems

Available research has indicated replacement of personal travel to grocery stores with grocery delivery services has significant potential to reduce VMT. Cairns (1997, 1998, 2005) observed reductions in VMT between 60 and 80%when delivery systems replaced personal travel. The Punakivi team found reductions in VMT as high as 50–93% (Punakivi & Saranen, 2001; Punakivi & Tanskanen, 2002; Punakivi, Yrjola, & Holmstrom, 2001; Siikavirta et al., 2002). Wygonik and Goodchild (2012) saw reductions of 70 – 95%.

Both Siikavirta et al. (2002) and Wygonik and Goodchild (2012) examined the impact on CO_2 emissions for passenger travel replacement for grocery shopping. Wygonik and Goodchild observed reductions in CO_2 emissions between 20 and 75% when delivery systems served randomly selected customers and reductions 80–90% when deliver systems served clustered customers. These are comparable to the results observed by Siikavirta et al. (2002).

Hesse (2002) points out limitations in evaluations that directly replace passenger travel with delivery service as other changes to the logistics system are likely. He further comments on the likelihood for e-commerce to encourage more distal warehouse locations. The evaluation presented here attempts to address some of these concerns by incorporating the entire supply chain from regional warehouse to end consumer. Recent growth by Amazon (Wenger, 2013) shows at least some retailers are not moving their warehouses further away, but instead are moving them closer to population centers.

While some research has indicated replacement of personal travel to grocery stores with grocery delivery services has significant potential to reduce VMT, these articles have not addressed criteria pollutants, which are associated with significant health impacts (USEPA, 2013b, 2013c).

Warehouse Locations

Since warehouses (including storage and distribution centers) are frequently an end point for commercial trips, their location can significantly influence the distances traveled by goods movement vehicles. Research about the optimal locations for warehouses is common. Crainic, Ricciardi, and Storchi (2004) found that the use of "satellite" warehouses to coordinate movements of multiple shippers and carriers into smaller vehicles reduced the VMT of heavy trucks in the urban center but increased the total mileage and number of vehicles moving goods within the urban center. This research illustrates the close relationship between warehouse location and the vehicle choice. Likewise Dablanc and Rakotonarivo (2010) found terminal locations have moved further from the city center over the past 30 years resulting in an estimated increase in CO_2 of 15,000 tons per year. They compare this with estimated gains from smaller consolidation centers located close to city centers and found the increase in CO_2 from the relocated terminals was 30 times greater than the savings from the smaller consolidation centers. Filippi, Nuzzolo, Comi, and Delle Site (2010) found greater potential environmental savings through urban distribution centers than through changes to the vehicle fleet, though both were successful.

In contrast, Allen and Browne (2010) found that locating distribution facilities closer to urban centers would reduce the average length of haul and total vehicle kilometers traveled by freight vehicles in and to urban centers, and Andreoli, Goodchild, and Vitasek (2010) found that mega-distribution centers, located to serve multiple regions, increased the distance traveled between the distribution center and the final outlet.

While this area of the literature is well-studied, clear consensus about the CO_2 impacts of warehouse location has not been reached and little research exists on the impacts of warehouse location on criteria pollutants. This research examines the results of shifting shopping behavior from personal travel to delivery service and examines the influence on warehouse structure on those results. It also provides insight into the trade-offs between local impacts (criteria pollutants) and global ones (VMT and CO_2).

METHODS

The methodology used in this analysis is adapted from that used in Wygonik and Goodchild (2012). This section provides a brief summary of the technique and notes modifications to the earlier methodology.

Scenarios

Three scenarios were considered in this evaluation:

1. The baseline scenario, Last-Mile Passenger Vehicles, represents a common form of travel for grocery shopping. A large, combination truck stocks the 49 grocery stores in the city from the regional warehouse. Individual households use passenger vehicles to complete roundtrips from their residences to their closest grocery store and back.
2. The second scenario, Local Depot Truck Delivery, provides delivery service from 5 selected grocery retail locations distributed throughout the city. In this scenario, the local depots are stocked from the regional warehouse using large, combination trucks. Then smaller box trucks complete delivery via a milk-run starting and ending at the select stores and stopping at the sampled customers along the way.
3. The third scenario, Regional Warehouse Truck Delivery, provides delivery service directly from the regional warehouse using small, box trucks. The routes start and end at the regional warehouse and stop at the sampled customers along the way.

These scenarios are illustrated in Fig. 2.

Network Data Set

The base network is pulled from the ESRI StreetMap North America data set (Environmental Systems Research Institute [ESRI], 2006) and was modified in a number of ways. First, the data set was trimmed to only include road segments in King County, Washington to reduce processing time. Next, the length in feet of each road segment was calculated and appended to the data table. Travel time was calculated using the segment length and the speed limit information and appended to the data table. Finally, information regarding the CO_2, NO_x, and PM_{10} emissions associated with each road segment for each vehicle type was also appended to the data table,

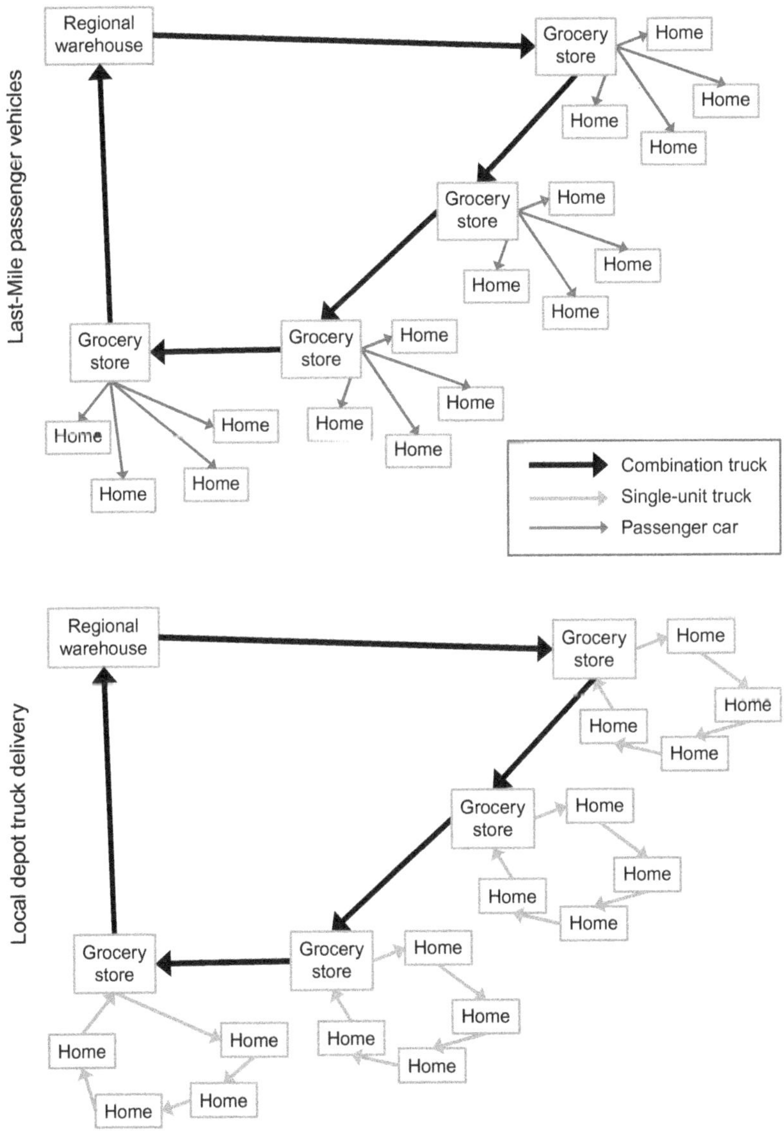

Fig. 2. System Bounds and Vehicle Types for Three Scenarios.

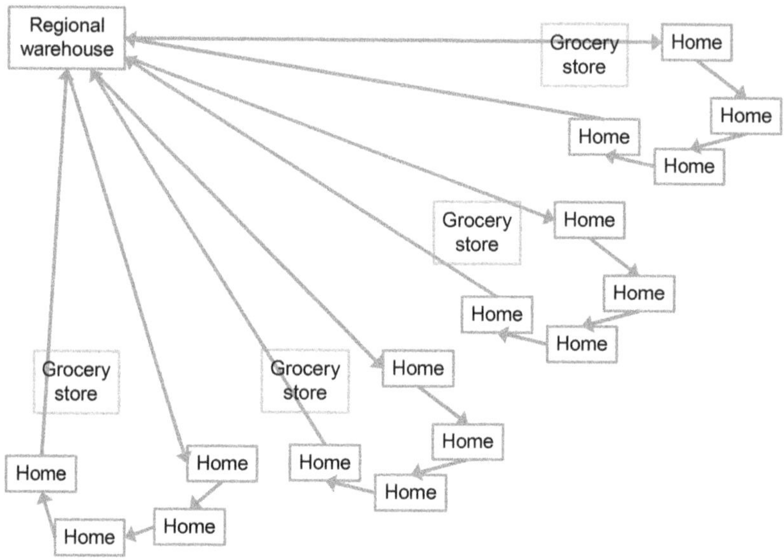

Fig. 2. (*Continued*)

based on the MOVES emissions factors, the roadway speed limit, the roadway functional class, the roadway length, and the vehicle type.

Once the data were added to the StreetMap layer, it was built as a Network for use in the Network Analyst tool set in ArcGIS.

While this evaluation considers link-level travel speeds, it does not include various real-time travel components, including congestion and queuing. These factors may affect the results but are outside the scope of this analysis.

Emissions Factors

Emissions factors were obtained from the 2010b MOVES model (USEPA, 2013a). EPA's MOVES model was used to identify emissions rates as it is the most current emissions model supported by the United States government. The factors in MOVES are sensitive to a number of different parameters considered within this analysis, including speed and vehicle type. This analysis assumed uncongested conditions, so speed limit data from the StreetMap North America data set was used as the default flow speed for each road segment. Running exhaust emissions are tracked.

Personal travel is represented by the emissions factors for personal cars using gasoline. The home delivery vehicle travel uses emissions factors for single-unit short-haul trucks with diesel fuel, and the emissions rates for the vehicles used to move goods from the warehouse to stores relies on data for combination short-haul trucks and diesel fuel. A weighted average of the previous 15 years of data was used according to the distribution reported in the Transportation Energy Data Book (Davis et al., 2013) for passenger cars and trucks, respectively. Because of data restrictions, the distribution of the previous 15 years' data is only released as of 2001. This distribution is applied to 2014.

Emission factors were selected for an analysis year of 2014. Hourly kilograms per mile of CO_2 equivalents, NO_x, and PM_{10} were extracted and averaged over each hour of the day, for weekdays, throughout the year for the King County, Washington region. Roadways with speeds of 5, 20, 25, and 35 miles per hour used urban unrestricted roadtype emissions factors, and roadways with speeds of 45 and 55 miles per hour used urban restricted roadtype emissions factors (see Table 1). Since the trucks work with hot engines due to their short stopping time, only running exhaust emissions are tracked.

Depot Locations

Delivery services are generally clustered into two primary types – ones that rely on existing brick-and-mortar retail locations for depots and those that use warehouses as depots. While other models exist, this research compares these two main types: a brick-and-mortar storefront depot with a warehouse-based model. This analysis considers replacing one roundtrip by a household to its nearest grocery store with delivery from a local store-based delivery service or service from a regional warehouse. Earlier work by the authors (Wygonik & Goodchild, 2012) used one service area for personal travel and delivery service, and this work is designed to develop a more realistic model of the delivery service. For companies operating a delivery service out of a store-front, they are unlikely to operate that company out of every store front. Rather, they would likely pick a small subset of available options that would serve as depots for different quadrants of the city.

Puget Sound Regional Council provided a shapefile with the locations of the major grocery stores within King, Kitsap, and Snohomish counties. The service areas of the Seattle stores were calculated (using the Service Area tool within ArcGIS Network Analyst) and households were assigned

Table 1. Emissions Factors (Kilograms Per Mile of CO_2 Equivalents, NO_x, and PM_{10}) from EPA's MOVES Model (USEPA, 2013a) by Travel Speed (in Miles Per Hour).

		Urban Unrestricted				Urban Restricted	
		5 mph	20 mph	25 mph	35 mph	45 mph	55 mph
Passenger cars	CO_2 (kg/mi)	1.05917	0.41817	0.37320	0.33967	0.30813	0.29773
	NO_x(kg/mi)	0.0004980	0.0002969	0.0002943	0.0003189	0.0003020	0.0003128
	PM_{10} (kg/mi)	0.00002615	0.00000865	0.00000842	0.00001183	0.00000736	0.00000720
Single-unit short haul	CO_2 (kg/mi)	3.8027	1.4837	1.3319	1.1308	0.8667	0.7403
	NO_x (kg/mi)	0.016566	0.005898	0.005196	0.004357	0.003390	0.002950
	PM_{10} (kg/mi)	0.0007268	0.0002548	0.0002240	0.0001876	0.0001566	0.0001448
Combination short haul	CO_2 (kg/mi)	4.8386	2.5148	2.3542	1.9788	1.9175	1.7228
	NO_x (kg/mi)	0.023531	0.010781	0.009821	0.008475	0.008198	0.007719
	PM_{10} (kg/mi)	0.0010048	0.0005433	0.0005058	0.0003797	0.0003296	0.0002410

to their closest store's service area for the personal travel calculations. For this analysis, assigning customers to their nearest store is reasonable and provides a baseline for comparisons between personal travel and delivery vehicles (see Wygonik and Goodchild, 2012 for a complete discussion supporting this assumption).

A subset of five stores was selected to serve as depots for the store-based delivery service. These stores are distributed throughout Seattle and are illustrated in Fig. 3. An existing warehouse location in Kent, Washington was selected to serve as the depot location for the warehouse-based delivery service, as well as the warehouse serving the grocery stores themselves.

Household Data

Geographic data regarding households and parcels were gathered from the Washington State Geospatial Data Archive (WAGDA) and the Urban Ecology Lab at the University of Washington. The pre-processing of this data is described in Wygonik and Goodchild (2012).

As personal communication with local delivery providers indicate each truck can hold approximately 35 households worth of orders, 35-household samples are used here. Five 35-household samples were taken for each of the five grocery store depots, for a total of 25 samples for each local depot service area. These samples were used for all three travel types: household travel to their proximate store, delivery service from their assigned store-based depot, and delivery service from the regional warehouse.

Vehicle Travel

To estimate the distances traveled and the associated emissions, routing tools within ArcGIS Network Analyst were used.

To complete the routing estimates, the Network Analyst Closest Facility tool was used to calculate the distance traveled to each grocery store for each household in the sample for the Passenger Vehicle scenario. The StreetMap network was loaded for use with Network Analyst. Output from Network Analyst includes the one-way distance traveled for each residential unit and the one-way emissions associated with each residential unit's grocery store trip when the trip is optimized for shortest time. These outputs were doubled, to reflect round trip distances and emissions. Using round trips for the Last-Mile Passenger Vehicle scenario represents a

Fig. 3. Warehouse, Depot, and Stores Locations.

simplification, as some grocery shopping does occur within chained trips. However, the available data do indicate most grocery shopping occurs via passenger vehicle making exclusive trips (Wygonik & Goodchild, 2012). Not all trips would be replaced by this type of service, but it is a reasonable estimation of the impact of replacing main household stocking trips.

To complete the routing estimates, the Network Analyst Routing tool was used to calculate the distance traveled by a delivery vehicle starting and ending at the depots and serving a sample of 35 households. The StreetMap network was loaded for use with Network Analyst. Network Analyst was run to identify the fastest path to serve the given households. The analysis reordered the stops to identify the fastest route, but kept the first and last stops (the depot) constant. Output from Network Analyst includes the distance traveled for each delivery vehicle and emissions associated with each tour, with the route optimized for shortest time.

Vehicle travel to stock the grocery stores from the regional warehouse was also included to maintain a constant system boundary for all scenarios. For the personal travel, 10 tractor trailers were required to stock the 49 grocery store locations. The Network Analyst Routing tool was used to calculate the distance traveled and emissions for 10 tractor trailers leaving the regional warehouse and each serving 5 stores. The results were then divided by 10 to represent the average values for one truck. For the scenario involving the local, store-based depots, the Network Analyst Routing tool was used to calculate the distance traveled and emissions for one tractor trailer serving the five store-based depots. Fig. 2 illustrates the three scenarios.

Assumptions

A number of assumptions were required within the modeling system. First, all optimizations used hard time windows, guaranteeing that promised delivery times would be met.

Customer orders are delivered in nestable, stackable plastic bins. These bins are picked up on subsequent orders. Because they nest, they take up little space and are not considered in the capacity limits of the trucks. In addition, because the bins are returned by customers during their next order, no additional stops occur to pickup bins. This problem is therefore simplified to an urban delivery system, disregarding pickup.

The model does not consider real-time routing changes. It is a planning tool and is not intended to provide dynamic routing information. In addition, this model currently assumes uncongested conditions.

RESULTS

Figs. 4 and 5 illustrate the service areas for the grocery stores and local depots. The 35-household samples were drawn from the households within each depot service area. One of the household samples and the associated routes for last-mile passenger travel and for local-depot-based truck delivery are shown in Fig. 5.

Looking at the results from the three different delivery structures (Fig. 6), the relative contributions of the different legs of the supply chain become apparent. Personal travel requires the largest number of VMT but generates low levels of pollutants. Any use of a combination short-haul truck within a supply chain involves significant emissions production, while the passenger cars contribute very small amounts of the studied emissions and practically no PM_{10}. Combination short-haul trucks have particularly high rates of NO_x emissions, relatively. By leveraging the efficiency of a delivery structure, delivery from a local depot has the lowest VMT and travel time of any of the cases, but high levels of pollutants.

Table 2 displays the data that supports Fig. 6. Local Depot Truck Delivery service – where a single-unit short-haul truck delivers to homes from a local depot – requires the lowest amount of VMT. The single-unit

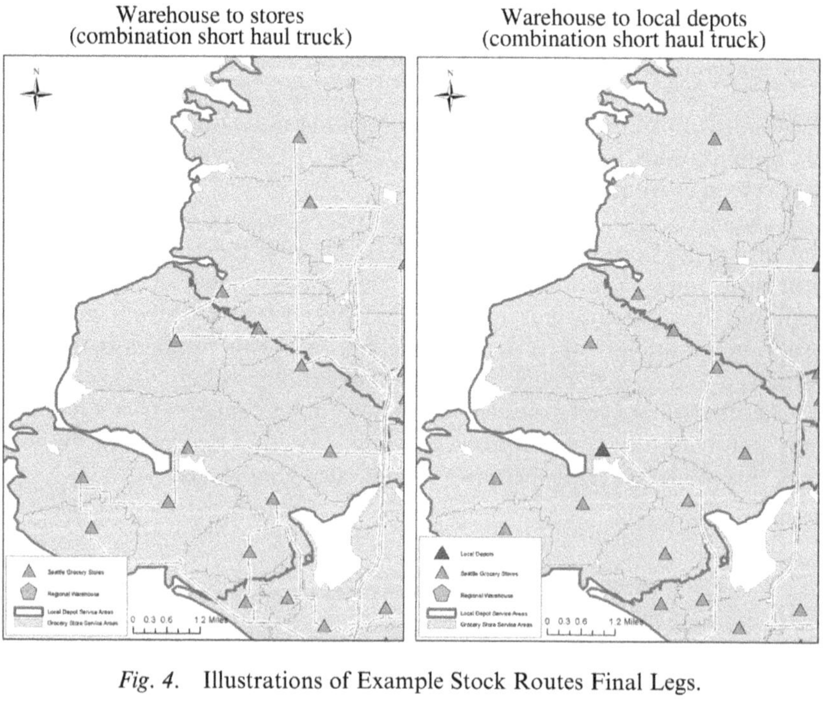

Fig. 4. Illustrations of Example Stock Routes Final Legs.

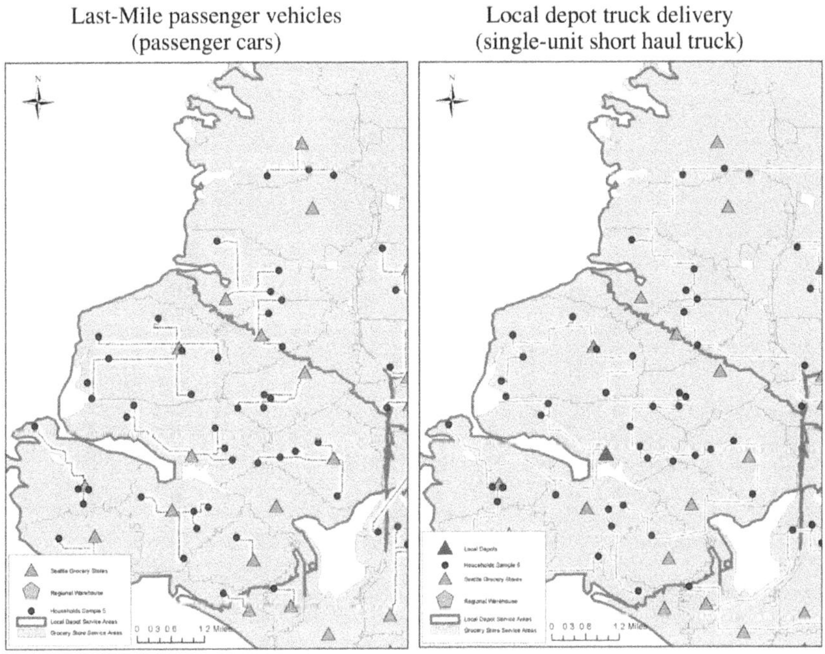

Fig. 5. Illustrations of Example Final Travel to Homes.

short-haul trucks generate higher pollutant levels than the passenger cars, so while the routing efficiency reduces VMT it is not enough to compensate for the higher rate of pollutant generation. The efficiency of delivery is highlighted by comparing the amount of VMT generated by passenger cars compared to the corresponding final-leg delivery vehicle. Even when the delivery vehicle is serving homes from a regional warehouse, it still requires fewer VMT than if individual homes travel directly to their closest grocery store.

The results in Table 2 also highlight the benefit of delivering to stops that are clustered together. While the combination trucks all serve 5 stores or depots, the stores are clustered together in the routes. The depots are spread throughout the city and require 30% more travel to serve from the warehouse. In addition, the personal travel requires twice as much travel to get from the homes to the stores as the delivery vehicle requires to serve those homes from a local depot even though the personal travel goes to the closest store and the local depot is serving an entire quadrant of the city.

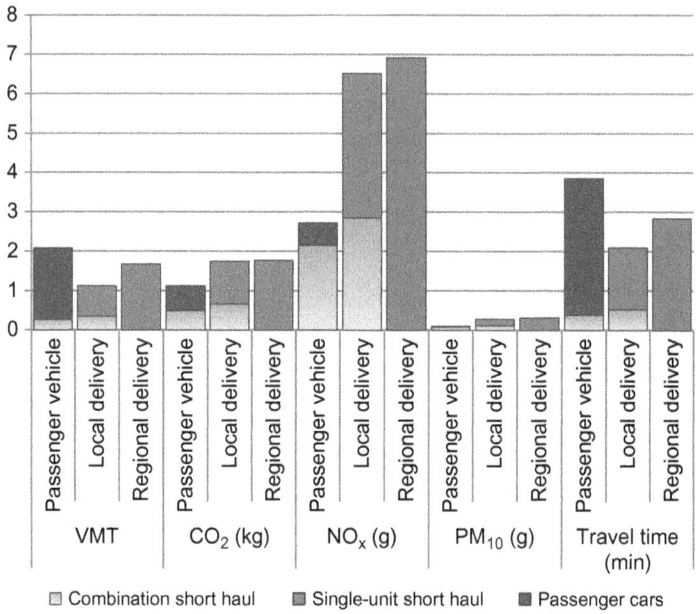

Fig. 6. Results for Each Delivery Structure, by Vehicle Type.

Table 2. Vehicle Miles Traveled, Emissions, and Travel Time by Supply Chain Leg and Design.

		VMT	CO_2 (kg)	NO_x (g)	PM_{10} (g)	Travel Time (min)
Last-mile personal travel	To store	0.3	0.5	2.2	0.08	0.4
	Personal trip	1.8	0.6	0.6	0.02	3.5
	Total	**2.1**	**1.1**	**2.7**	**0.10**	**3.8**
Local depot truck delivery	To depots	0.3	0.7	2.8	0.12	0.5
	To home	0.8	1.1	3.7	0.16	1.6
	Total	**1.1**	**1.7**	**6.5**	**0.28**	**2.1**
Regional truck delivery	**To home**	**1.7**	**1.8**	**6.9**	**0.31**	**2.8**

The shading indicates the goods movement strategy with the most/least impact.

The emission rates are higher for the trucks than for the passenger cars and are noticeably higher for combination trucks than single-unit trucks. As such, while the Local Depot Truck Delivery requires the lowest amount of VMT, it still generates more of all three pollutants than passenger

Table 3. *t*-Test Results.

		VMT	CO_2 (kg)	NO_x (kg)	PM_{10} (kg)	Travel Time (min)
Passenger vehicle vs. Local depot	*t* statistic	27.18	10.85	54.23	57.34	27.32
	d.f.	461	26	27	26	406
	p-value	0.000	0.000	0.000	0.000	0.000
Passenger vehicle vs. Regional warehouse	*t* statistic	5.42	11.53	19.27	19.84	10.07
	d.f.	36	26	24	24	52
	p-value	0.000	0.000	0.000	0.000	0.000
Local depot vs. Regional warehouse	*t* statistic	8.09	0.39	1.73	2.96	8.43
	d.f.	26	48	29	28	30
	p-value	0.000	0.701	0.094	0.006	0.000

vehicles. When comparing the Local Depot Truck Delivery to Regional Warehouse Truck Delivery, Regional Warehouse Truck Delivery generates significantly more criteria pollutants. When comparing Local Depot Truck Delivery to Last-Mile Passenger Vehicles, the Local Depot Truck Delivery has less total VMT, but higher amounts of combination truck VMT.

The results were also evaluated for significance using the two-tailed *t*-test. All comparisons were significantly different with *p*-values ($p \leq 0.001$), except for the differences in CO_2 (not significant) and NO_x (significant at $p \leq 0.1$) pollution generation between Local Depot and Regional Warehouse Truck Delivery. The VMT and travel time associated with personal travel is significantly longer than either of the delivery models. The Local Depot Truck Delivery has the lowest VMT and shortest travel time, while the Regional Warehouse Truck Delivery has the highest emissions levels. Detailed results of the *t*-tests are included in Table 3 and illustrate that variations across samples are small compared to the variation between scenarios.

CONCLUSIONS

The results show there is trade-off between VMT and pollutants. While the Local Depot Truck Delivery requires the lowest amount of VMT, it still generates more of all three pollutants than passenger vehicles. When comparing the Local Depot Truck Delivery to Regional Warehouse Truck Delivery, Regional Warehouse Truck Delivery generates significantly more criteria pollutants (NO_x, PM_{10}). Frequently, transportation policies and operating systems are designed to address VMT and congestion. If a region is also concerned with pollution, it will have to decide how to value the different impacts to decide how to shape policy. In the case studied here, delivery service – regardless of the structure – generates less VMT, less congestion, and more pollution.

Combination trucks produce exceptionally high levels of NO_x and PM_{10}. These criteria pollutants have localized impacts. Policies that limit big trucks near population centers may increase VMT, but they may be worth it to ameliorate local health impacts from NO_x and PM_{10}.

These results show notable sensitivity to the structure of the depot, the depot location, routes traveled, and business model. Earlier work by Wygonik and Goodchild (2012) found delivery services generally reduced VMT and CO_2 emissions when used in lieu of passenger vehicle travel. These results contradict those findings. Understanding operational details and including them in modeling efforts is necessary to evaluate the efficacy of these services. On-going work should pursue customer density thresholds and regional warehouse location sensitivity. Further, this analysis relied on data provided by a local supplier, in which 35-households are served from a regional warehouse using one truck within the necessary time constraints. Different regional land use patterns with higher levels of sprawl might require significantly more travel from the regional warehouse to the urban center to restrict the number of households that can be served by each truck.

REFERENCES

Allen, J., & Browne, M. (2010). Considering the relationship between freight transportation and urban form. Green Logistics Project: Work Module 9 (Urban Freight Transport). University of Westminster.

Allen, J., Tanner, G., Browne, M., Anderson, S., Chrisodoulou, G., & Jones, P. (2003). *Modelling policy measures and company initiatives for sustainable urban distribution*. Final Technical Report. Transport Studies Group, University of Westminster, London.

Andreoli, D., Goodchild, A., & Vitasek, K. (2010). The rise of mega distribution centers and the impact on logistical uncertainty. *Transportation Letters*, *2*(2), 75–88.

Cairns, S. (1997). *Potential traffic reductions from home delivery services: Some initial calculations*. TSU Working Paper No. 97/45. UCL, London.

Cairns, S. (1998). Promises and problems: Using GIS to analyse shopping travel. *Journal of Transport Geography*, *6*(4), 273–284.

Cairns, S. (2005). Delivering supermarket shopping: More or less traffic? *Transport Reviews*, *25*(1), 51–84.

Crainic, T. G., Ricciardi, N., & Storchi, G. (2004). Advanced freight transportation systems for congested urban areas. *Transportation Research Part C*, *12*, 119–137.

Dablanc, L., & Rakotonarivo, D. (2010). The impacts of logistics sprawl: How does the location of parcel transport terminals affect the energy efficiency of goods' movements in Paris and what can we do about it? *Procedia – Social and Behavioral Sciences*, *2*(3), 6087–6096.

Davis, S. C., Diegle, S. W., & Boundy, R. G. (2013). Transportation Energy Data Book: Edition 32. Retrieved from http://cta.ornl.gov/data/tedb32/Edition32_Full_Doc.pdf

Environmental Systems Research Institute. (2006). *ESRI data and maps*. CD-ROM. Redlands, CA: Environmental Systems Research Institute.

Filippi, F., Nuzzolo, A., Comi, A., & Delle Site, P. (2010). Ex-ante assessment of urban freight transport policies. *Procedia – Social and Behavioral Sciences*, *2*(3), 6332–6342.

Hesse, M. (2002). Shipping news: The implication of electronic commerce for logistics and freight transport. *Resources, Conservation and Recycling*, *36*, 211–240.

Holguín-Veras, J., Cruz, C. A. T., & Ban, X. (2013). On the comparative performance of urban delivery vehicle classes. *Transportmetrica A: Transport Science*, *9*(1), 50–73.

Punakivi, M., & Saranen, J. (2001). Identifying the success factors in e-grocery home delivery. *International Journal of Retail & Distribution Management*, *29*(4), 156–163.

Punakivi, M., & Tanskanen, K. (2002). Increasing the cost efficiency of e-fulfilment using shared reception boxes. *International Journal of Retail & Distribution Management*, *30*(10), 498–507.

Punakivi, M., Yrjola, H., & Holmstrom, J. (2001). Solving the last mile issue: Reception box or delivery box? *International Journal of Physical Distribution & Logistics Management*, *31*(6), 427–439.

Quak, H. J., & de Koster, M. B. M. (2007). Exploring retailers' sensitivity to local sustainability policies. *Journal of Operations Management*, *25*(6), 1103.

Quak, H. J., & de Koster, M. B. M. (2009). Delivering goods in urban areas: How to deal with urban policy restrictions and the environment. *Transportation Science*, *43*(2), 211.

Siikavirta, H., Punakivi, M., Karkkainen, M., & Linnanen, L. (2002). Effects of E-commerce on greenhouse gas emissions: A case study of grocery home delivery in Finland. *Journal of Industrial Ecology*, *6*, 83–98.

United States Environmental Protection Agency. (2008). National Emissions Inventory 2008 v1.5 GPR. Retrieved from http://www.epa.gov/ttn/chief/net/2008inventory.html

United States Environmental Protection Agency. Office of Transportation and Air Quality. (2013a). MOVES (Motor Vehicle Emission Simulator) [2010b model and user guide]. Retrieved from http://www.epa.gov/otaq/models/moves/

US Environmental Protection Agency. (2013b). Nitrogen dioxides: Health. Retrieved from http://www.epa.gov/air/nitrogenoxides/health.html

US Environmental Protection Agency. (2013c). Particulate Matter (PM): Health. Retrieved from http://www.epa.gov/airquality/particlepollution/health.html

Van Rooijen, T., Groothedde, B., & Gerdessen, J. C. (2008). Quantifying the effects of community level regulation on city logistics. *Innovations in City Logistics*, 387–399.

Wenger, Y. (2013). Proposed Baltimore warehouse fits Amazon's growth: Shipping industry experts say online retailer expanding in urban markets. *Baltimore Sun*. Retrieved from http://articles.baltimoresun.com/2013-08-23/news/bs-md-ci-amazon-site-20130823_1_online-retailer-amazon-same-day-delivery-distribution-center

Wygonik, E., & Goodchild, A. (2011). Evaluating CO_2 emissions, cost, and service quality trade-offs in an urban delivery system case study. *IATSS Research*. *35*(1), 7–15. doi:10.1016/j.iatssr.2011.05.001

Wygonik, E., & Goodchild, A. (2012). Evaluating the efficacy of shared-use vehicles for reducing greenhouse gas emissions: a case study of grocery delivery in Seattle. *Journal of the Transportation Research Forum*, *51*(2), 111–126.

CHAPTER 4

THE SHADES OF GREEN IN RETAIL CHAINS' LOGISTICS

Maria Björklund and Helena Forslund

ABSTRACT

Purpose – *This study aims to illustrate how retail chains with a green image align sustainable logistics actions, logistics measurements and contracts with logistics service providers (LSPs), and to develop a classification model that allows for a description of the various shades of green within companies.*

Design/methodology/approach – *We carried out a multiple case study of four retail chains with a green image operating in the Swedish market, collecting empirical data from the retail chains' sustainability reports and home pages and conducting interviews with logistics, transportation and supply chain managers.*

Findings – *Based on the literature, we developed a classification model for judging green image, green logistics actions, green measurements and green contracts. The model is used to illustrate the different shades of green found within the respective retail chains. A green image seems well-aligned with green logistics actions. However, there are more levels to judge, and the measurement systems are not sufficiently developed to track green logistics actions. Contract handling is more developed*

Sustainable Logistics
Transport and Sustainability, Volume 6, 83–112

ISSN: 2044-9941/doi:10.1108/S2044-994120140000006005

among retail chains than measurements, which is positive, as this is a way of ensuring that LSPs are involved. In our classification model, greenwashing can be judged in a more nuanced way, delving deeper under the surface.

Research limitations/implications – *The provided classification model adds to our knowledge and illustrates the alignment within companies' sustainable logistics. The robustness of the model can be strengthened by applying it to a larger number of cases and by continually validating its content and evaluation criteria.*

Practical implications – *The study's main practical contribution is the classification model, which may potentially serve as a method for managers to easily judge the green alignment of a retail chain's logistics.*

Originality/value – *Few empirical studies capture how retail chains measure environmental logistics performance, and even fewer concern contracts stipulating the environmental demands placed on LSPs.*

Keywords: Sustainable logistics; environment; performance measurement; green retail chain; contract

INTRODUCTION – GREEN RETAIL CHAINS

Even though retail chains might not have a large environmental impact by themselves, they play an important role in securing sustainable behaviour in their supply chains (Kolk, Hong, & van Dolen, 2010). Due to the large size and consolidation of bargaining power, retailers are described as having the power to change practices along the supply chain (Jones, Comfort, Hillier, & Eastwood, 2005). Furthermore, retailers can build environmental awareness along the supply chain and are often held responsible for the actions of other supply chain actors, as they have contact with both consumers and suppliers (Kolk et al., 2010; Wiese, Kellner, Lietke, Toporowski, & Zielke, 2012). In retail research, however, there seems to be a time lag of more than 10 years in using the term 'sustainability' as compared to other fields in research and industry (Wiese et al., 2012). Based on a survey study, Elg and Hultman (2011, p. 454) found that 'a significant share of the Swedish retail sector does not follow best practices' with regard to corporate social responsibility (CSR) and that 'few provide systematic reporting of CSR performance and practices'. Furthermore, most sustainability-related retail articles have been published recently (Wiese et al., 2012), and European retailers' sustainability initiatives are described as both fragmented and unsystematic (European Commission, 2010).

Logistics plays an important role in moving towards sustainable development (Chunguang, Xiaojuan, Kexi, & Pan, 2008; Lieb & Lieb, 2010). The transportation function, for instance, is often described as the most environmentally damaging operation within logistics (Rogers & Weber, 2011; Wu & Dunn, 1995). Min and Kim (2012) concluded that research on sustainable transportation is scant relative to research focusing on other factors such as manufacturing. They call for further research within this area, inviting empirical research methodologies in particular. Transportation is most commonly outsourced to logistics service providers (LSPs). Therefore, retailers have to work together with their LSPs to improve environmental performance. Including green logistics measurement in the contracts with LSPs can be a way of formalizing actions and ensuring that performance improvements are achieved. A study by Rogerson (2013) suggests a link between logistics measurements regarding fuel efficiency and the contract. Some studies, however, show that even if shippers put effort into investigating the environmental performance of LSPs, these investigations show little influence on the content in the contract (Björklund, 2005; Wolf & Seuring, 2010). Björklund and Forslund (2013a) have argued that a complete contract should include not only environmental aspects and measurement, but also handling of non-compliance. *Do retail chains with a green image also have green logistics actions? Are green logistics actions and considerations regarding non-compliance included in contracts with LSPs?*

As companies increasingly advertise their concern for the environment, they have been accused of 'greenwashing', that is, using green words rather than taking green actions (Carbone, Moatti, & Vinzi, 2012). Much scepticism has been directed towards the content of companies' sustainability reports (Tate, Ellram, & Kirchoff, 2010). These reports might only focus on the positives and are used as a marketing tool for the company to enhance its image among stakeholders. As Kotzab, Munch, Faultrier, and Teller (2011, p. 677) point out, 'It seems that political environmental correctness does not allow for reporting on failures or inadequate initiatives'. As a consequence, there may be discrepancies between the actual corporate commitment to CSR issues and the actions described in the reports (Ingenhoff & Sommer, 2011). It would be relevant to study whether retail chains do in fact greenwash and put on a green image (i.e. use green language to make external actors deem the company considerate of the environment, for example implementing green actions and showing a large

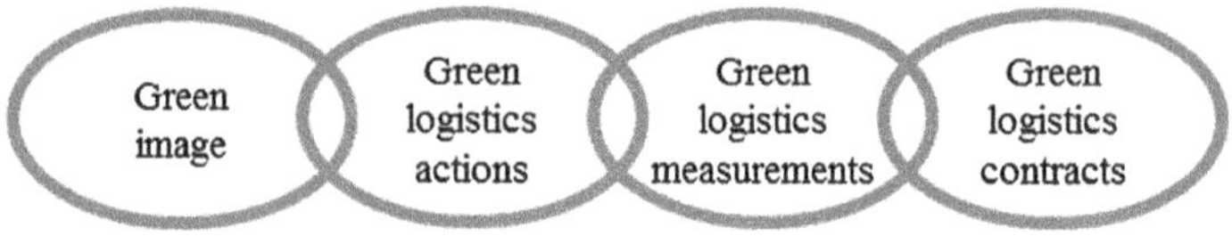

Fig. 1. Alignment between Image, Actions, Measurements and Contract Agreements as a Sign of No Greenwashing.

responsibility for the environmental impact). Greenwashing studies have been limited to studying image and the corresponding actions. Neely, Gregory, and Platts (2005) describe performance measurement as one way of quantifying the effect of actions. Björklund and Forslund (2013a) found that companies that focus on environmental performance do not necessarily consider how to measure it. *Are the green logistics actions taken by retail chains followed up in green performance measurements?*

To secure that no greenwashing is taking place, we argue that there needs to be an alignment between the image, the actions taken, the measurements applied and the contract agreements, as illustrated in Fig. 1. Not only is it important to fully align image, actions, measurements and contracts when it comes to research. It would also be valuable for managers to be able to easily describe their own awareness and also to judge the sustainable logistics awareness of supply chain partners. This study intends to determine what can be found under the green surface. The chapter aims to illustrate how certain retail chains with a green image align logistics actions, logistics measurements and contracts with LSPs. Furthermore, the chapter is intended to develop a classification model that illustrates the 'shades of green', that is, the different degrees of environmental ambitions in image, logistics actions, logistics measurements and contracts. The classification model that we propose can help to increase the awareness of other stakeholders linked to the retail chain – including consumers, investors, employees and society at large – about the retail chains' sustainability performance in logistics.

LITERATURE REVIEW

The literature review is divided into four sections: green image, green logistics actions, green logistics measurements and green logistics contracts. Each section ends with a suggested classification scale to determine the four different shades of green in each area.

Green Image

The literature describes multiple benefits from having a green image. A green image can be an important component in the value creation framework of a green business model (Hao & Jiang, 2013). A green image can also generate success, create financial savings, be a safeguard against future environmental hazards, improve sales and brand goodwill and the value of the company (Björklund, 2010; Reuter, Foerstl, Hartmann, & Blome, 2010).

Few attempts have been made to formalize what a green image actually consists of. Some of these attempts are production-oriented, such as that proposed by Azevedo, Carvalho, and Cruz Machado (2011), who describe green image as the 'number of fairs/symposiums related to environmentally conscious manufacturing in which the organization participates'. Other authors relate a green image to environmental manufacturing mode and methods. Such studies are less directly relevant to retailers.

A green image is linked to marketing (Georgiadis & Besiou, 2010). Noci (1997) identifies 'how the public sees the company' as an important part of green image. From the perspective of retailers, the image from 'providing the market with green products or brands' becomes central (Chkanikova & Mont, 2012. Environmental labelling (eco-labelling) is one important example of products' environmental marketing, as it represents an attempt to convey a green image to the consumers. A more active communication of image is mirrored in the 'type of environmental information found on the companies' webpages or in their CSR reports'. To communicate social and environmental awareness, companies are increasingly issuing annual, easily accessible CSR reports, often found on company websites (Tate et al., 2010).

From a purchasing perspective, Noci (1997) describes the importance for companies with a green image to 'use suppliers with a green image'. Suppliers' images are measured in terms of the type of relationships between the supplier and its stakeholders and whether the supplier takes environmental considerations into account. Considering the environmental impact of the retailers' suppliers is important when valuing the green image of a retailer. From a reverse logistics perspective, Georgiadis and Vlachos (2004) identify aspects such as 'demands for reused material' and 'level of reused items' as influencing a company's green image.

As the review above indicates, retail chains' green image can be described as: (1) how the public sees the company; (2) the offering of green products and brands; (3) how the company actively communicates their environmental awareness (by publishing CSR reports, providing information about their environmental practices on their websites, etc.) and (4) the green image of suppliers (their use of green manufacturing or reused material and their relationship with suppliers).

The following shades of green image are suggested for retailers:

- image shown on all four levels (i.e. the public sees the retail chain as green; the company offers green products or brands; the company actively communicates its environmental awareness; and it uses suppliers with a green image) – dark green
- image shown on three levels – medium green
- image shown on two levels – pale green
- image just on one level – no green/white.

Green Logistics Actions

Despite the frequent appearance of green logistics in the literature, there is still no undisputed definition (Chunguang et al., 2008). Green logistics has been described as an environmentally friendly and effective logistics system (Chunguang et al., 2008) and as a logistics system responsible for the environment (Wu & Dunn, 1995). In this chapter the term 'green logistics' describes the inclusion of environmental considerations in the design and management of logistics systems. As with green image, consensus has also not been reached for the categorization of green logistics actions.

McKinnon (1998) presents a four-level model of logistics decisions which influence the environmental impact of transport. The levels, ranging from strategic to tactical/operational, are

1. The physical structure of the logistics system (number, capacity and location of production units, warehouses and terminals).
2. The pattern of sourcing and distribution (selection of suppliers, customers and markets).
3. Scheduling of freight flows (time scheduling and movements of the freight).
4. Management of transport resources (selection of modes, fleet management and consolidation).

Aronsson and Huge-Brodin (2006) further develop the McKinnon's model by elaborating on how decisions made at different levels influence each other, limiting the scope of action to the lower levels. Aronsson and

Huge-Brodin also describe a level above these, taking into the overall conditions such as the company's overall strategy and product and packaging design, resulting in the following four levels:

1. Choices on an overall level, concerning product design
2. Choices concerning logistics structures/organizational forms (McKinnon's level 1 and 2)
3. Choices concerning planning/management (comparable with McKinnon's level 3)
4. Choices concerning the operative work (comparable with McKinnon's level 4)

Decisions made at different levels are related to green logistics actions, which can be applied to the following grouping (within brackets are the levels from the Aronsson and Huge-Brodin framework presented above). Wu and Dunn (1995) put forward logistics actions influencing the green impact such as mode and carrier selection (level 4), consolidation (level 4), network design (level 2) and selection of service levels (level 2). Another way of grouping the green logistics initiatives, suggested by Martinsen and Huge-Brodin (2014), is through transport-related actions: fuels, vehicle technology and mode choice (in line with level 4) and transport management (in live with level 3), and beyond-transport actions such as logistics system design and choice of partners (level 2) and environmental management systems (level 1).

In order to classify the green shades of logistics actions for retailers, we have used the following levels, adapted from McKinnon's (1998) and Aronsson and Huge-Brodin's (2006) models:

- actions taken on all four levels (regarding overall conditions, physical structure, scheduling of transport flows and management of transport resources) – dark green
- actions on three levels – medium green
- actions on two levels – pale green
- actions on one level – no green/white.

Green Logistics Measurements

Manning, Baines, and Chadd (2006) argue that performance measurements need to be developed for CSR to deliver quantifiable benefits. However, there is no agreement regarding which measurements to apply. Pioneering companies within the field have developed their own measurements for environmental performance on a stand-alone basis (Carbone et al., 2012). Following are examples of environmental performance measurements found in the logistics literature:

- *Energy use* was pointed out by Veleva, Hart, Greiner, and Crumbley (2003) and Hervani, Helms, and Sarkis (2005). Energy efficiency is also a key parameter in the analytical framework for green logistics used by McKinnon (2010).
- *Technology* such as information systems and computer models for fleet management (Wu & Dunn, 1995), as well as the age of vehicles and the fuels, engines, refrigerants and tires being used (Björklund, 2005; Martinssen & Björklund, 2011). The selection of engine determines energy intensity (kWh/vehicle-km), while the selection of fuel can heavily influence emission intensity (emissions/kWh) (Kveiborg & Fosgerau, 2004; McKinnon, 2010). The size, weight and type of vehicles have also been described as important determinants when it comes to energy consumption and other external impacts such as noise (Martinsen & Huge-Brodin, 2014; McKinnon, 2010).
- *Fill rate/loading factors* (Wu & Dunn, 1995). A survey study by Björklund (2005) indicated that fill rate and consolidation are two environmental aspects that receive significant attention when Swedish shippers purchase transport services. The average percentage of empty running and average load on laden trips are presented as key parameters in the framework for green logistics presented by McKinnon (2010). Fill rate/loading factors and the size of vehicle are examples of measurements connected to traffic intensity (vehicle-km/tonne-km) (Kveiborg & Fosgerau, 2004; McKinnon, 2010).
- *Air emissions and outlets of* CO_2 (Björklund, 2005; Hervani et al., 2005; Rogers & Weber, 2011; Veleva et al., 2003). A study by Björklund and Forslund (2013a) among Swedish LSPs and shippers (both manufacturers and retailers) found CO_2 emissions to be the most commonly applied environmental performance measurement. Emission per unit of energy is one key parameter in McKinnon's (2010) analytical framework. Carbon footprint has been analysed only rarely in retail research, thus constituting a promising field for future research (Wiese et al., 2012).

Most green logistics measurements presented in the literature target logistics activities on the lower logistics decision levels (level 4, as described in the 'green logistics actions' section above). However, there are examples

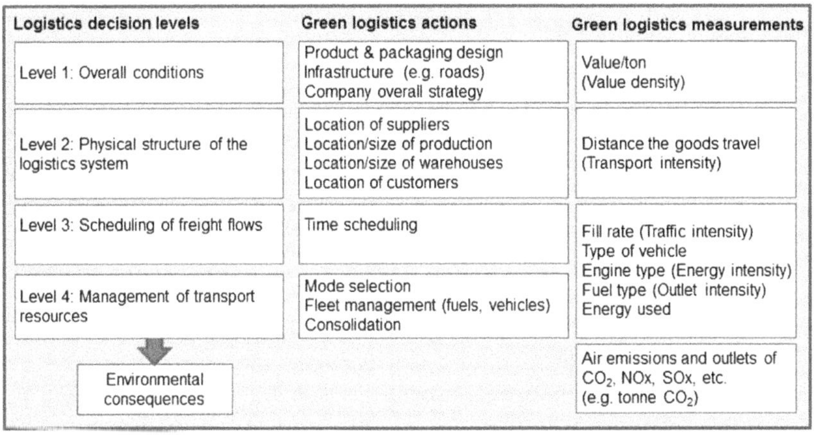

Fig. 2. A Framework Linking Green Logistics Actions at Different Decision Levels to Green Logistics Measurements.

of measurements that do target the higher levels. Value density (value/ton), for example, is one measurement that can capture how environmentally efficient the design of products and packages are (level 1), and transport intensity (tonne-km/tonne) can be decreased by changes in the physical structure level (level 2), for instance a decision to use local production (Kveiborg & Fosgerau, 2004; McKinnon, 1998). Green logistics measurements addressing activities regarding scheduling of freight systems (level 3) and management of transport resources (level 4) are difficult to separate. Therefore, their corresponding measurements are shown as one category of measurements. Air emissions can be seen as the consequence of all types of green logistics actions and is therefore put outside the decision levels in Fig. 2.

In order to classify the green shades of logistics measurements for retailers, we suggest the following four shades:

- measurement in all four levels (regarding overall conditions, physical structure, scheduling of freight flows and management of transport resources and environmental consequences) – dark green
- measurement in three levels – medium green
- measurement in two levels – pale green
- measurement in one or no levels – no green/white.

Green Logistics Contracts

Contracts are used to provide incentives to coordinate the supply chain, in the absence of an ability to exert direct control over supply chain partners (Brun & Moretto, 2012). Contracts between Swedish LSPs and shippers commonly consist of two parts: one written, formal 'frame contract' with the agreements expressed in general terms and one complementary, more important but less formal agreement regarding responsibilities. This indicates that a 'gentleman's agreement' is often used in the Swedish transport industry (Lammgård, 2007). Environmental performance measurement can be a critical aspect in LSPs' environmental offerings (Martinssen & Björklund, 2011). LSPs have reported increased interest in green concerns from their customers; however, buying decisions are still based upon performance variables other than the environment (Wolf & Seuring, 2010). Contracts in supply chains are positively related to performance improvements (Jammernegg & Kischka, 2005). Brun and Moretto (2012) made an overview of contracts in supply chain management literature and found no environmental content. Rogerson, Andersson, and Johansson (2012) identify a number of contextual dimensions that can influence the contract, such as the purchasing situation, number of services, supplier strategy, supplier relationship approach, variation in location and size of shipments.

Forslund (2009) showed the importance of clearly defining which performance measurements should be included in the contract. The findings by Rogerson et al. (2012) suggest a connection between one of the measurable logistics variables put forward by McKinnon (2010): fuel efficiency and the contract stage. This was based on identified verbal agreements regarding vehicle type and driving style in one of the cases studied. Findings from a survey study of Swedish shippers in the food and forest sector (Björklund, 2005) showed that even if most of the studied shippers considered environmental aspects during the purchasing process, only 27 of the 50 respondents actually included environmental aspects in the contract. Another more recent survey study among shippers and LSPs in Sweden (Björklund & Forslund, 2013a) showed that out of the 163 responses, 84 (52%) included environmental performance in the contract. Björklund (2005) found that 7 of the 50 respondents (13%) stated that they had measurements included in the contracts, while Björklund and Forslund (2013a) found that number to be 47%.

Non-compliance can be viewed both in a positive and in a negative way, resulting in incentives or penalties to reinforce performance improvements (Forslund, 2009). Findings from the abovementioned survey study by

Björklund (2005) showed that only 2% of the companies had written agreements regarding how to handle non-compliance. Björklund and Forslund (2013a) found that 42% of the companies had non-compliance included in contracts. To include green aspects in the contract does not necessarily mean that measurement and non-compliance are included. As the contracts are not complete (Björklund & Forslund, 2013a), this may in turn imply that the expected performance improvement effects are not reached. Altogether, a complete contract should include environmental aspects, how to measure those aspects and how to regulate non-compliance.

To classify the green logistics contract we suggest the following four shades:

- contracts which account for non-compliance (contracts taking non-compliance into consideration directly or indirectly also include green aspects and ways to measure the performance in order to identify non-compliance) – dark green
- contracts which include environmental measurement – medium green
- contracts which include green aspects – pale green
- no green considerations in the contracts – no green/white.

METHODOLOGY

The literature was searched using Google Scholar and Business Source Premier, with the keywords 'green image' (and synonyms such as 'environment'), 'green logistics actions', 'green logistics measurements' and 'green contracts'. In early stages of research, a case study is often appropriate (Yin, 2014). Barratt, Choi, and Li (2011) suggest clearly defining the study object in case studies. The study object in this study is retail chains, the intermediaries between producers and consumers (Wiese et al., 2012). The studied companies consisted of a centralised wholesaling management and a number of stores targeting consumers with the same name and image.

The retail chains we selected to illustrate the use of the framework had to fulfil certain selection criteria. First, they had to have a green image. In order to identify the green image, we first asked the question, 'How does the public sees the company?' We used two independent groups of logistics researchers at two large Swedish universities in order to single out potential Swedish retail chains. Even if it can be expected that information on web pages and CSR reports is biased and that retail chains should overemphasize their green image towards the public, this search resulted in the identification of only a few retail chains. In order to provide a variety of shades of green, we selected companies from different industries and with different contexts. The analysis aims to identify the scope of the retailers' practices within image, actions, measurements and contracts, not to compare and identify best practice in terms of environmental friendliness. In practice, the relevant analysis dimension is the internal alignment between each retailer's image, actions, measurements and contracts. For access reasons, they had to each be operating within the Swedish market. Furthermore, they had to be willing to participate in the study.

We then searched for respondents with an awareness of logistics and environmental issues. Effort was made to identify the most suitable respondents at each of the four retail chains, taking into consideration their title and areas of responsibility. In order to decrease single response bias, we also asked for respondents that could provide additional information. However, only in one case were we able to identify additional respondents. The retail chains, their characteristics and the respondents are shown in Table 1.

The data collection followed the procedures recommended by Yin (2014), such as basing interview questions on the literature review, providing definitions, following a semi-structured interview guide and letting each respondent validate his or her case description in order to increase validity. To increase reliability, method triangulation (Yin, 2014) was applied by letting other data sources complement the interviews, such as homepage information and the latest available annual environmental/CSR/sustainability reports. Telephone interviews lasting between 30 and 60 minutes were conducted with each respondent. The interviews were followed up with email conversations.

Table 1. Retail Chains, Characteristics and Respondents.

Retail Chain	Industry/ Products	Interview Respondent	Other Data Sources
Retail Chain A	Daily groceries	Logistics manager (strategic change)	Homepage, sustainability report 2011, master thesis
Retail Chain B	Outdoor apparel	Managing director	Homepage, sustainability report of mother company 2011
Retail Chain C	Cosmetics	Logistics Reporting Manager, International Transportation Manager	Homepage, sustainability report 2011
Retail Chain D	Health products	Supply chain director	Homepage

The interviews and emails resulted in descriptions of green logistics actions, performance measurement and contracts with LSPs. This structure made it possible to carry out a cross-case analysis with a pattern-matching approach (Yin, 2014). In order to capture the different shades of green in the four retail chains' logistics, the authors classified each area (i.e. green logistics actions applied, the use of green measurements and green aspects in the LSP contract) into four different shades of green. The classification and its results were then discussed in a workshop among logistics and supply chain management researchers from three large Swedish universities. Altogether, some ten researchers scrutinized and discussed the authors' classification. A presentation of the study in a preliminary form (Björklund & Forslund, 2013b) at a research conference garnered feedback, and five blind reviewers scrutinized the classification in different phases of its development. These discussions showed that in a few cases it was possible to interpret the empirical data somewhat differently using the classification model. However, the discussion always concerned the positioning between two close shades. The authors' choice to emphasize scope over depth was accepted and confirmed, as this is the first study with the aim to develop this type of classification model. Within our classification framework, a company that puts forward rather unambitious actions but covers a wide scope reaches a greener shade than a company that works very ambitiously within a more narrow scope. As a result, we have not investigated the finer details regarding the measurement method, the assumptions, the simplifications and the system limitations inherent in the measurement of the environmental impact of the logistics system. Altogether, established procedures (Barratt et al., 2011; Yin, 2014) were followed throughout the study.

EMPIRICAL STUDY – FOUR GREEN RETAIL CHAINS

Retail Chain A

Retail Chain A is a large grocery chain, with over 700 stores of different size and character located around the country (Homepage of Retail Chain A, 2013). According to its environmental policy, the company is a leading company within the sector when it comes to sustainable development. Retail Chain A's logistics is an affiliated company responsible for the distribution of goods to all the stores.

The logistics manager at Retail Chain A states, 'In our role as a retailer we have possibilities to influence other supply chain actors'. The consumers are seldom aware of the environmental impact from the logistics system, even if the company's largest contribution to CO_2 emissions comes from transportation. Instead, they focus on the production phase, for instance the type of cattle food used. The environmental aspect is discussed with all LSPs. However, the logistics manager states that the LSPs are expecting Retail Chain A to push environmental performance forward by placing environmental demands.

The company's *green image* is clearly shown in their green products and brands. Both in 2011 and 2012, Retail Chain A received the title 'Most sustainable brand in Sweden' from the organization Sustainable Brands (based on a Global Reporting Initiative [GRI]-inspired internet survey with 8,000 respondents). In the third green image aspect (how the company communicates its environmental awareness), we found both extensive homepage information and an easily accessible CSR report. Retail Chain A uses green suppliers and mentions that for product suppliers with low environmental ambition, the company pushes environmental performance forward.

The following *green logistics actions* are found. *Overall conditions:* Sustainability performance is an integrated part of the company's strategy as well as of its operational work within all business areas such as purchasing, logistics, marketing and sales (according to environmental policy). The sustainability report presents eight overall goals, two of which concern logistics. The first is a minimum 40% decrease of climate gases by 2020 compared to 2008 figures (adjusted to economic turn over); freight transport stands for more than half of these outlets. The second logistics-related goal is to place high demands on environmental and social responsibility in the selection of suppliers and business partners, with the provision that the demands must be followed up. Suppliers develop both product and packaging design. The logistics manager reported, 'Large suppliers have good knowledge and develop products and packages taking into consideration the size of load carrier to provide good fill rates'. According to the respondent, three centralised distribution centres (DCs) – one for fresh food, one for dry food and one for cold food – are used to ensure as much consolidation as possible. *The physical structure of the logistics system:* Environmental demands are placed on all suppliers, including the LSPs. Outsourcing of the transport function, completed in June 2011, was described as a way to increase delivery service and to facilitate an increased level of consolidation. *Scheduling of freight flows:* The stores place demands on lead-time and delivery windows, reducing the potential for Retail Chain

A to fully optimize the transports. Retail Chain A has considered shifting from placing the same delivery service demands for all products towards using a differentiated service level depending on the type of product (for example, differentiating between meat and a grill). Due to the large environmental ambitions among the stores, acceptance could be gained by increasing awareness of the environmental benefits from differentiated service levels. *Management of transport resources:* Considerable effort was made to change the transport mode from road to rail. Retail Chain A has seen both economic and environmental benefits from this change and from purchasing 'green electricity'. For its home delivery operations, the company has increased its use of bio fuel in the trucks. Furthermore, Retail Chain A tries to get a large degree of backhauling to decrease the transport work (tonne-km) and uses green transports when building new buildings.

When it comes to *green logistics measurements*, since all transports are purchased, Retail Chain A has abandoned the old measurement fuel consumption in favour of measurements aimed at capturing the use of the logistics system, expressed in terms of, for example, route optimization and consolidation. The measurement covers all LSPs. The LSPs report the environmental data to the transport department, which compiles the information into emissions records. Historically, the calculation was based on traffic work (vehicle-km), but due to the increased amount of products transported, the goods weight is now included when measuring the transport work (tonne-km). Retail Chain A followed up on the amount of preliminary booked pallet places in the vehicle and the actual amount of pallets transported. They followed up these two measurements in order to decrease the gap between them. In some geographical areas this figure is quite critical, since it can be difficult for the transport company to fill the vehicle with short notice, resulting in decreased fill rate. However, increasing the fill rate is mostly seen as the suppliers' task, and since this level is high, it is not described as an area with large potential. On the other hand, Retail Chain A has identified significant potential in increasing the 'fill rate of the parcel', according to the logistics manager. Increasing this fill rate requires the stores to change their ordering routines. To do this, the stores need better IT support and improvement in their ordering competency.

The environmental aspect is included in the written *green logistics contracts*. The same demands are placed on all LSPs; trucks can be no older than 8 years and can drive a maximum of 1,000,000 km, can only use diesel fuel of environmental class 1, can only use oils recommended by the vehicle manufacturer and are advised not to use tires with HA-oils in their tread area. Additionally, all LSPs must report on environmental data. There are no demands on eco-driving, but since it is often economically beneficial for the suppliers to educate their drivers, Retail Chain A knows that several LSPs employ eco-driving strategies. Retail Chain A has not considered including other demands, but instead ensures that the current demands are updated. When a new contract arises, the LSP is given a predetermined time frame to invest in the new required vehicles. Controls at the terminals are used as one part of Retail Chain A's measurement and follow-up routines. It is hard to cheat since the LSPs know the demands apply to all of them equally, and thereby the LSPs keep track of each other. Furthermore, the demands are not difficult to meet. The respondents are not aware of any LSPs failing to follow the environmental demands placed in the contract, but such a violation would be considered a very serious breach. It would be a violation of contract, independent of the area. Thus, there are no penalties formally in place, although all LSPs are incentivized through the possible loss of contracts due to non-compliance.

Retail Chain B

Retail Chain B is a retail chain for outdoor and travel apparel, with 31 owned and franchised stores around Sweden (Homepage of Retail Chain B, 2013). With the aim to become the sustainability leader in the industry, the company employs a sustainability manager has established a common approach. The company has six sustainability-oriented focus areas, with targets formulated and followed up on an annual basis. Of these six areas, 'climate action' and 'engaging suppliers' are logistics-related. The managing director believes that greenwashing exists in retail. It is difficult to point to 'best-practice' retailers; perhaps they are found among retail chains that have their own production.

The fact that Retail Chain B was awarded 'Retail chain of the year' in 2012 by the organization Svensk Handel (Swedish Retail) – an award based on their selected high-quality product assortment, high service level and environmental work – gives it a *green image* both in terms of how the public sees them and in the products they sell. When studying how the company communicates their environmental awareness, we found webpage information and the first sustainability report for Retail Chain B's mother-company, as well as a first GRI report for 2012 (Homepage of the mother company of Retail Chain B, 2013). Even if the large suppliers work with codes of conduct, there is significant potential for improvement. The managing director reported, 'Green logistics is not on top of the agenda; we are more of business men'. Retail Chain B is a relatively small customer for their brand suppliers, and it is therefore difficult to make demands, but 'this is a constant intensive process ... based upon long-term relations', according to the managing director. In that sense, the large brand suppliers have more power and responsibility in the supply chain. Retail Chain B is currently conducting an environmental development project on how to place demands on suppliers.

The following *green logistics actions* are found. *Overall conditions:* Product design is important, as a high-quality product is sustainable in itself. Retail Chain B uses the label 'a greener choice' to point out especially environment-friendly products to consumers. Packaging design is a supplier decision. However, Retail Chain B openly expresses their opinion, as they believe that this will pay off in long-term development. If a consumer orders from the online shop, they can select recycled packaging. *The physical structure of the logistics system:* Most of this is not affected by Retail Chain B. They have a Swedish DC to ensure as much consolidation as possible. *Scheduling of freight flows*: The suppliers are responsible for the freight to the Swedish DC. Retail Chain B can only affect the inbound freight by ordering far in advance to enable a large share of boat freight. The stores have weekly orders and deliveries to ensure few transports. *Management of transport resources:* The transports that Retail Chain B controls are between the DC and the stores. They strive for using ISO14001-certified LSPs, which use more environmentally friendly diesel.

No *green logistics measurement* is currently taking place. The LSP provides Retail Chain B with some measurement information, albeit with low frequency. Emissions are perceived to be a complicated, mathematical calculation, which, according to the managing director, makes little sense to use 'unless you are a chemist'. The only quantitative logistics-related measurement takes into account how the employees travel to their job.

When it comes to *green logistics contracts* between Retail Chain B and LSPs, every contract deals with environmental issues as a soft, qualitative element, which is taken into account if price and lead-time are equal. The environmental aspects are then included in a simple but understandable way, like an ISO14001 certificate. Measurement and non-compliance are not included.

Retail Chain C

Retail Chain C is an international retail chain with some 80 franchised stores in Sweden, providing original, ethical and natural cosmetics (Homepage of Retail Chain C, 2013). Retail Chain C shows a lot of green information on its homepage, as well as sustainability reports available for download. 'Protect the planet' is the green part of Retail Chain C's five CSR values. The company is well-known for its green products and brands (the second aspect of *green image*). It communicates its environmental awareness with numerous environmental aspects on its homepage and in its extensive CSR report. The company takes into account the environmental impact of suppliers by using as much local sourcing as possible.

The following *green logistics actions* are found in Retail Chain C. *Overall conditions:* The company claims to invent and recycle packaging materials. *The physical structure of the logistics system:* Retail Chain C sources and produces locally in order to minimize lead times, cost and CO_2 impact. Distribution centres have also been moved closer to consumers. *Scheduling of freight flows:* Transport scheduling is based on the use of lower carbon alternatives wherever possible. *Management of transports:* All transport is outsourced, with all transport providers using Euro 4/5 standard vehicles in tandem with vehicle driver technology to ensure optimal fuel consumption and efficiency.

The *green logistics measurements* are the following. All 'energy use' is measured and has targets for reductions. Retail Chain C ensures that all LSPs are utilizing the specified types of driver-supporting technology and complying with the latest possible versions of Euro standard vehicles. 'Fill rate' receives special focus, with the goal of optimizing all inbound loads by road and the target of 80% of all movements being full truckloads. On outbound shipments, containers are optimized to achieve at least 75% fill rate, and store deliveries are undertaken via hub and spoke shared networks using double-decker trailers where possible, which reduces the number of vehicles running. The logistics manager reported, 'As a business we record our CO_2 emissions and report it each month, broken down on markets, inbound and outbound'.

When it comes to *green logistics contracts*, all LSPs are under fixed-term contracts. Selection and agreements are only reached once their sustainability credentials are confirmed. Environmental aspects included in the contracts are responsibility to the environment and impact on communities and the wider world, absolute commitment to achieve bona fide reduction targets, etc. The performance measurements include reduction in CO_2, together with its measurement, continuous improvement in vehicle standards and transport management and actions to support genuine environmental initiatives. The international transportation manager stated that 'no formal penalties are in place, although all LSPs are incentivized through the possible loss of contracts'.

Retail Chain D

Retail Chain D is a market-leading chain in healthy foods and products. Operating in Sweden, Finland and Norway, the retail chain consists of some 100 owned stores and more than 300 franchised stores (Homepage of Retail Chain D, 2013). The company's *green image* can be seen in its green products and brands. For the third green image aspect (how the company communicates its environmental awareness), Retail Chain D does not explicitly mention environmental aspects on its homepage or in its annual report. The supply chain director stated, 'Greenwashing is a common practice but we are very careful'. Retail Chain D will include suppliers in its environmental work but have not yet started to do so.

Retail Chain D is in a developmental stage, shifting from being an independent wholesaler to becoming part of a modern integrated retail chain. Centralised logistics management (i.e. logistics management managed by Retail Chain D's headquarters) is an important part of this development. The suppliers should not manage the logistics, and neither should the customers or the owners. According to the supply chain director, 'Creating an efficient supply chain will lead to improved environmental performance'. Centralised logistics management provides Retail Chain D with the opportunity to both develop its logistics carefully and to reduce its environmental impact. International retail chains such as Tesco, Wal-Mart and HM are benchmarked as models to follow.

The following *green logistics actions* are found.

Overall conditions: Both product design and packaging development are traditionally handled by the suppliers. Retail Chain D is developing products under private labels to be able to influence packaging design.

The physical structure of the logistics system: A single DC serves all stores in Sweden and Finland. A future step is to start development projects with suppliers, where the terms of delivery will be changed so that Retail Chain D can also control inbound deliveries from suppliers. Many suppliers are Swedish. Centralised and IT-supported purchasing will also lead to more consolidated purchasing patterns and consequently fewer large inbound deliveries.

Scheduling of transport flows: Projects have been conducted to find ways to reduce the number of deliveries to each store in Finland. In Finland, the stores purchase directly from the suppliers and from several small wholesalers, operating with 10–30 deliveries per week. The target is to lower delivery frequency to once per week per store by using consolidated deliveries. In Sweden, the stores are close to the target of one delivery per week per store.

Management of transport resources: Operational decisions are made by the LSPs.

Green logistics measurements

A performance measurement system will be developed when the new logistics system is up and running. The previous experience of the supply chain director suggests that performance measurement is an important aspect to support the logistics system. A metric that will be prioritized is 'consolidation/fill rate'.

When it comes to *green logistics contracts*, Retail Chain D has contracts with some LSPs and has chosen not to use environmental demands on transports yet. This may be relevant when there is a stable logistics system. The benefit is not perceived to cover the additional cost in the current situation.

ANALYSIS

The characteristics of the cases are summarized in Table 2.

Green Retail Chains' Image

Identifying what constitutes a 'green image' is not an easy task, and the literature provides a rather scattered picture of how to operationalize green image. The companies under review display their green image in different ways. Being viewed by the public as green (Noci, 1997) was a qualifying condition for the study, and all four retail chains furthermore offer green products and brands, as suggested by Li and Zhang (2008) and Chkanikova and Mont (2012). As for the third aspect of green image – how the company communicates its environmental awareness (Tate et al., 2010) – it was found that Retail Chain D does not explicitly mention environmental aspects on its homepage or in its annual report, while the other three retail chains did and furthermore made available their CSR reports. An interesting observation is that the sustainability report in Retail Chain B was not found on its homepage but on the lesser known homepage of the mother company. This may imply that such reports are judged to be relevant only for stakeholders such as owners. Skepticism has been directed towards the contents of sustainability reports. However, in the case of these companies, they seem to have described their green logistics actions adequately in their sustainability reports, which to some extent would

Table 2. Case Characteristics Concluded.

	Retail Chain A	Retail Chain B	Retail Chain C	Retail Chain D
Green image	Green public image Offering green products/brands Homepage, CSR report Green suppliers	Green public image Offering green products/brands Homepage, CSR report Green suppliers	Green public image Offering green products/brands Homepage, CSR report Green suppliers	Green public image Offering green products/brands
Logistics actions				
Overall conditions	Environment in company strategy	Quality products purchased	Environment in company strategy, packaging material	
Physical structure of logistics system	Centralised DCs	Centralised DC	Local sourcing, DC close to customers	Centralised DC
Scheduling of freight flows	Consolidation	Consolidated purchasing and distribution	Lower carbon alternatives	Consolidated purchasing and distribution
Management of transport resources	Backhauling, mode of transport, bio gas		Mode of transport, truck and driver technology	
Measurement				
Overall conditions				
Physical structure of logistics system	Transport work			
Scheduling of freight flows/ Management of transport resources	Fill rate		Inbound and outbound fill rate	
Environmental consequences	CO_2 emissions		Energy use, CO_2 emissions	
Contracts	Aspects and measurements included	Aspects included	Aspects and measurements included	

contradict the findings of Tate et al. (2010) and Ingenhoff and Sommer (2011). Judging the retail chains' work with green suppliers, as suggested by Noci (1997), was more challenging. Retail Chains A, B and C mention in one way or another how they include suppliers in their environmental work. Retail Chain B, for instance, tries to engage suppliers in their environmental work, and Retail Chain C focuses on local sourcing.

Based on the findings in the literature review presented in the 'Green image' section, we found that green image is classified as dark green in Retail Chains A, B and C, shown as dark grey in Fig. 3. Retail Chain D is classified as pale green, shown as pale grey in Fig. 3.

Green Logistics Actions

All companies were found to perceive retailers as responsible for green logistics actions, a finding which is in line with Kolk et al. (2010) and Wiese et al. (2012). This factor refers to the second and third decision levels of Fig. 1. The retailers were also found to consider the highest level of Fig. 1, *overall conditions* such as products and packages, largely to be the responsibility of product suppliers. In line with the findings of Aronsson and Huge-Brodin (2006), this implies that retailers' scope of action is limited by the decisions made at higher levels. In order to master and improve all decision levels, retailers' collaboration with product suppliers seems necessary and could be carried out, possibly based on their power to change, as suggested by Jones et al. (2005).

When it comes to *the physical structure of the logistics system*, the locations that can be affected by retailers are those that can centralize distribution centres for a larger market in order to make storing and distribution efficient. *Scheduling of transport flows* seems to be the level where the case study companies have started, with concerted efforts to consolidate the distribution to stores as well as inbound flows from suppliers. Receiving fewer, larger deliveries may, on the one hand, be interpreted as decreased customer service by the stores. On the other hand, it can also be perceived as increased customer service, as the stores do not have to be frequently interrupted in store activities/selling by receiving a large number of deliveries. These two levels seem to be the most green and the most commonly perceived as the retail chains' responsibility.

The management of transport resources was consistently mentioned as the LSPs' responsibility, as all cases have outsourced their transportation. Retail Chain A strives for changing fuel to bio gas. According to Fig. 2, a number of decisions are then left for the LSP to make. If the retail chain wants to have some influence on those decisions, performance measurement and contract design seem critical. It is interesting to note that McKinnon (1998) mentions consolidation as an activity on the lowest level. This can then refer to the LSP operationally using their possibilities to coordinate goods in their network. However, when the companies under review decide on purchasing or distribution patterns, they refer to the actions they undertake as consolidation efforts. However, this type of consolidation operates on a higher level, definitely affecting lower-level actions, as mentioned by Aronsson and Huge-Brodin (2006).

We suggest that the 'Green logistics actions' of our retail chains is classified as dark green in Retail Chains A and C, shown as dark grey in Fig. 3. Retail Chain B is medium green, shown as medium grey in Fig. 3 and Retail Chain D is pale green, shown as pale grey in Fig. 3.

Green Logistics Measurement

There were large differences between the four companies when it came to green logistics measurement. Retail Chains A and C stand out as experienced users of measurements, indicating that their measurements have changed and developed over time. Both have a number of measurements in use. Energy use and technology have previously been used by Retail Chain A and are currently used by Retail Chain C. To some extent this contradicts the findings of Hervani et al. (2005) and McKinnon (2010). Fill rate/loading factors are used by Retail Chains A and C and is a planned metric for Retail Chain D, similar to Björklund (2005) and McKinnon (2010). Air emissions and outlets of CO_2, found to be common measurements by Björklund and Forslund (2013a, 2013b), are used by Retail Chain A and Retail Chain C.

Related to Fig. 2, all but one measurement was found on the two lowest levels. Our retail chains do not address a number of higher-level measurements. This implies that changes at the higher level of decision-making are not fully captured by the measurement systems. Hence, this state of green logistics measurement does not encourage higher-level changes, which, according to our model, should have high impact on the environmental performance of the logistics systems. Failing to quantify the effects of actions as suggested by Neely et al. (2005) may imply that performance improvements are not revealed.

Based on the findings in the literature review presented in the 'Green Logistics Measurements' section, we suggest that none of the studied retail chains qualify as dark green in measurement. Retail Chain A is medium green, shown as medium grey in Fig. 3. Retail Chains B and D are white and Retail Chain C is pale green, shown as pale grey in Fig. 3.

Green Contracts

In all cases, the companies have outsourced transportation to LSPs, which implies that the way to coordinate the supply chain and follow-up the LSPs' logistics actions and environmental performance measurement is by the use of contracts (Brun & Moretto, 2012). Retail Chains A, B and C include environmental aspects in their contracts with all of their LSPs, which, according to Jammernegg and Kischka (2005), should be related to performance improvements. Retail Chain B includes green aspects such as ISO14001-certificates, but it does not include measurement, which is similar to the practice identified by Björklund (2005) and agrees with the findings of Wolf and Seuring (2010). Retail Chains A and C regulate some measurements. Björklund and Forslund (2013a, 2013b) found that the fact that environmental aspects are included in the contract does not necessarily mean that measurement is included. Including incentives or penalties should further reinforce performance improvements (Forslund, 2009). None of the companies under review were found to regulate non-compliance, which parallels the findings of Björklund and Forslund (2013a, 2013b). However, LSPs are incentivized through the possible loss of contracts through non-compliance.

None of the retail chains is classified as dark green in the 'Green logistics contracts' category. Retail Chains A and C are medium green, shown as medium grey in Fig. 3. Retail Chain B is pale green, shown as pale grey in Fig. 3 and Retail Chain D is white.

The Shades of Green in Retail Chains' Logistics

The classification model revealed differences among the companies in our study. One aspect that hinders comparison of this kind is that the context of the studied retailers is different and that their operations are influenced by many factors beyond their control, such as their geographical spread, product type and sales volumes. In this study we have described the scope of the retail chains' actions, not their depth. This makes our suggested classification model simple to use, not demanding an extensive amount of data collection. As managers need to understand and describe green logistics,

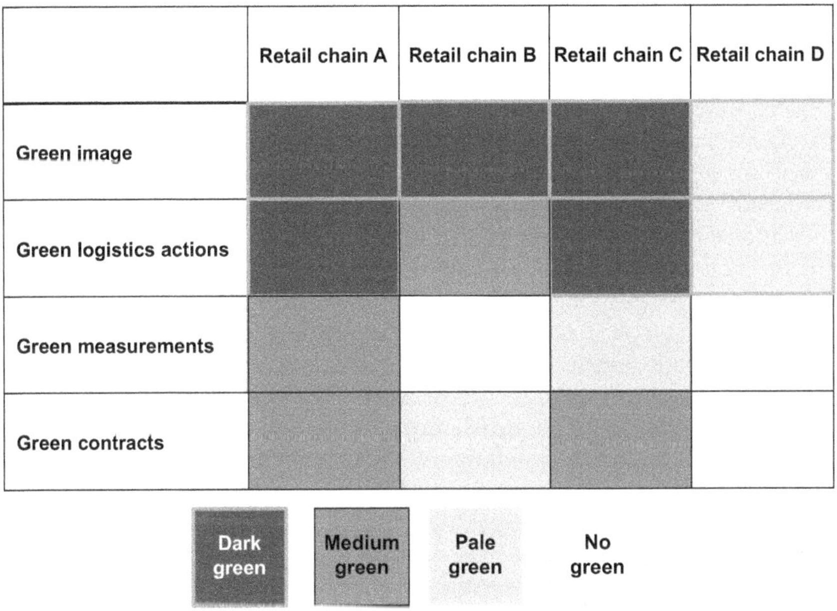

Fig. 3. The Shades of Green in Retail Chains' Logistics.

this can be seen as a first step in developing a classification model. Different shades of green – which imply different degrees of alignment – were also found within the respective retail chains, as shown in Fig. 3.

CONCLUSIONS, CONTRIBUTIONS AND FURTHER RESEARCH

This chapter aims to illustrate how some retail chains with a green image align logistics actions, logistics measurements and contracts with LSPs, as well as to develop a classification model that can describe the shades of green within a company. We conclude that greenwashing – the failure to align image with actions – was not found among the evaluated companies. The green images of the evaluated companies are mostly well aligned with their green logistics actions. However, there are more levels to judge, as the measurement systems, for instance, are not yet developed to follow up these actions. Contract handling is more developed than measurements among the studied retail chains, which is positive, as this is a way of ensuring that LSPs are involved. Altogether, full alignment is not found. Greenwashing is in our study more nuanced, delving deeper under the surface.

This study makes a theoretical contribution by adding to our knowledge of environmental ambition among retail chains, with a special focus on green image, logistics actions, performance measurement and contracts. A framework that links logistics actions at different decision levels to green measurements has been presented. The study reveals interesting insights on the division of roles within a supply chain, for example the fact that retailers seem to focus more on the middle levels in the provided framework while LSPs contribute to the lower levels. We also provide a classification model for green logistics, which has been tested on four companies and validated by several logistics researchers. Few empirical studies capture how retail chains measure environmental logistics performance, and even fewer concern contracts stipulating the demands on LSPs. The study' primary practical contribution is the classification model, which may be one way for managers to easily judge the 'greenness' of their own and/or other retail chains' logistics.

One major limitation of the study is that only four cases are addressed. The number of cases does not allow for statistical analysis. Due to the role of the empirical data in this study, the size and scope of the companies are still valid and relevant, as they can illustrate the usability of the framework and underpin the generalizability of the findings. However, few retail chains in Sweden were found to be seen as green by the public, and the purpose of the chapter is illustrative, not descriptive. Future research could aim to compare environmental behaviour between companies that are and are not perceived as green. Another limitation in this study is the fact that the focus was on retail chains operating in the Swedish market. However, retail chains are operating in a global competitive situation, and Retail Chain C is an international chain, which would imply that the findings could also be valid in other geographical settings. However, companies in other countries can be expected to be less green than Swedish companies, and environmental awareness may be less appreciated in other countries in the first place. These cultural conditions may limit the possibilities of broadening the study geographically.

The robustness of the findings can be strengthened in further research by expanding the study with more cases, possibly from other national contexts, or by conducting a larger retail chain homepage review. Other aspects in the classification model may be added, for example how the green image may be emphasized in advertising. It would be relevant to further study whether product suppliers, regarding their responsibility for higher-level decisions, also use higher-level measurements as indicated in Fig. 2. Furthermore, it would be interesting to analyse suppliers' perceptions of a dialogue with retail chains on the division of roles when it comes to environmental issues. This study did not explicitly deal with the contemporary topics of city distribution or urban logistics, which should be important for retail chains. Such green logistics actions deserve more attention in further research. Among the green logistics actions, it would be relevant to also study the perceptions of the retail stores. Do they think consolidated distribution implies higher or lower customer service? Do they bring in market demands, as suggested by Chkanikova and Mont (2012)? Including the stores can lead to additional insights from a supply chain perspective. Another supply chain partner to include could be LSPs, as they are responsible for many logistics actions that affect environmental performance. How do they handle environmental performance measurements, and what are their perceptions of them? A dyadic research approach, similar to that of Wolf and Seuring (2010), or even a triadic or network-based approach, could be fruitful and account for different actors' insights. This line of research would be in line with the retailers' role in the supply chain, which has been described by Kolk et al. (2010), Wiese et al. (2012), and Jones et al. (2005) as being capable of changing practices and building awareness along the supply chain. Finally, we see further studies in the area of environmental performance measurement as being especially relevant to carry out, as this level was the least green in our classification model. It is obvious that companies lack knowledge in what measurements to use in order to capture different actions and changes in their logistics systems.

REFERENCES

Aronsson, H., & Huge-Brodin, M. (2006). The environmental impact of changing logistics structures. *The International Journal of Logistics Management*, *17*(3), 394–415.

Azevedo, S. G., Carvalho, H., & Cruz Machado, V. (2011). The influence of green practices on supply chain performance: A case study approach. *Transportation Research Part E*, *47*(6), 850–871.

Barratt, M., Choi, T. Y., & Li, M. (2011). Qualitative case studies in operations management: Trends, research outcomes, and future research implications. *Journal of Operations Management*, *29*(4), 329–342.

Björklund, M. (2005). *Purchasing practices of environmentally preferable transport services, guidance to increased shipper considerations*. PhD thesis, Department of Industrial Management and Logistics, Lund University.

Björklund, M. (2010). Benchmarking tool for improved corporate social responsibility in purchasing. *International Journal of Benchmarking*, *17*(3), 340–362.

Björklund, M., & Forslund, H. (2013a). The inclusion of environmental performance in transport contracts. *Management of Environmental Quality: An International Journal*, *24*(2), 214–227.

Björklund, M., & Forslund, H. (2013b). *The shades of green in retail chains' logistics. The 25th International NOFOMA conference*, Gothenburg, Sweden.

Brun, A., & Moretto, A. (2012). Contract design and supply chain management in the luxury jewellery industry. *International Journal of Retail & Distribution Management*, *40*(8), 607–628.

Carbone, V., Moatti, V., & Vinzi, V. E. (2012). Mapping corporate responsibility and sustainable supply chains: An exploratory perspective. *Business Strategy and the Environment*, *21*(7), 475–494.

Chkanikova, O., & Mont, O. (2012). Corporate supply chain responsibility: Drivers and barriers for sustainable food retailing. *Corporate Social Responsibility and Environmental Management*. Retrieved from http://onlinelibrary.wiley.com/journal/10.1002/(ISSN)1535-3966/earlyview. doi: 10.1002/csr.1316

Chunguang, Q., Xiaojuan, C., Kexi, W., & Pan, P. (2008). Research on Green Logistics and Sustainable Development, *Proceedings of ICIII '08 international conference on information management and industrial engineering*, pp. 162–165.

Elg, U., & Hultman, J. (2011). Retailers' management of corporate social responsibility (CSR) in their supplier relationships – Does practice follow best practice? *The International Review of Retail, Distribution and Consumer Research*, *21*(5), 445–460.

European Commission. (2010). Services on Monitoring Retailers REAP commitments. Retrieved from http://ec.europa.eu/environment/industry/retail/pdf/report_monitoring 2009.pdf. Accessed on January 10, 2013.

Forslund, H. (2009). Logistics service performance contracts; design, contents and effects. *International Journal of Physical Distribution & Logistics Management*, *39*(2), 131–144.

Georgiadis, P., & Besiou, M. (2010). Environmental and economical sustainability of WEEE closed-loop supply chains with recycling: A system dynamics analysis. *International Journal of Advanced Manufacturing Technology*, *47*(5–8), 475–493.

Georgiadis, P., & Vlachos, D. (2004). The effect of environmental parameters on product recovery. *European Journal of Operational Research*, *157*(2), 449–464.

Hao, J., & Jiang, B. S. (2013). The framework of green business model for eco-innovation. *Journal of Supply Chain and Operations Management*, *11*(1), 33–46.

Hervani, A. A., Helms, M. M., & Sarkis, J. (2005). Performance measurement for green supply chain management. *Benchmarking: An International Journal*, *12*(4), 330–353.

Homepage of Retail Chain A. Accessed on February 28, 2013 and December 6, 2013.

Homepage of Retail Chain B. Accessed on February 14, 2013 and December 5, 2013.

Homepage of Retail Chain C. Accessed on February 5, 2013 and December 5, 2013.

Homepage of Retail Chain D. Accessed on February 5, 2013 and December 5, 2013.

Homepage of the mother company of Retail Chain B. Accessed on February 27, 2013.

Ingenhoff, D., & Sommer, K. (2011). Corporate social responsibility communication. *Journal of Corporate Citizenship*, *42*(Summer), 73–91.

Jammernegg, W., & Kischka, P. (2005). Dynamic, customer-oriented improvement of supply networks. *European Journal of Operational Research*, *167*(2), 413–426.

Jones, P., Comfort, D., Hillier, D., & Eastwood, I. (2005). Corporate social responsibility: A case study of the UK's leading food retailers. *British Food Journal*, *107*(6), 423–435.

Kolk, A., Hong, P., & van Dolen, W. (2010). Corporate social responsibility in China: An analysis of domestic and foreign retailers' sustainability dimensions. *Business Strategy and the Environment*, *19*(5), 289–303.

Kotzab, H., Munch, H., Faultrier, B., & Teller, C. (2011). Environmental retail supply chains: When Goliaths become environmental Davids. *International Journal of Retail & Distribution Management*, *39*(9), 658–681.

Kveiborg, O., & Fosgerau, M. (2004). Explaining the decoupling of freight growth and economic growth. Traffic day on Aalborg University 2004.

Lammgård, C. (2007). *Environmental perspectives on marketing of freight transports.* PhD thesis, University of Gothenburg.

Lieb, K. J., & Lieb, R. C. (2010). Environmental sustainability in the third-party logistics (3PL) industry. *International Journal of Physical Distribution & Logistics Management*, *40*(7), 524–533.

Manning, L., Baines, R. N., & Chadd, S. A. (2006). Ethical modelling of the food supply chain. *British Food Journal*, *108*(5), 358 370.

Martinssen, U., & Björklund, M. (2011). Matches and gaps in the green logistics market. *International Journal of Physical Distribution & Logistics Management*, *42*(6), 562–583.

Martinsen, U., & Huge-Brodin, M. (2014). Environmental practices as offerings and requirements on the logistics market. *Logistics Research*, *7*(115). doi:10.1007/s12159-014-0115-y

McKinnon, A. (1998). Logistical restructuring, freight traffic growth and the environment. In D. Bannister (Ed.), *Transport policy and the environment* (pp. 97–109). London: Publ. E & FN Spon.

McKinnon, A. (2010). Environmental sustainability: A new priority for logistics managers. In A. C. McKinnon, S. Cullinane, M. Browne, & A. Whiteing (Eds.), *Green logistics: improving the environmental sustainability of logistics*. London: Kogan Page Publishers.

Min, H., & Kim, I. (2012). Green supply chain research: Past, present, and future. *Logistics Research*, *4*(1–2), 39–47.

Neely, A., Gregory, M., & Platts, K. (2005). Performance measurement system design. *International Journal of Operations & Production Management*, *25*(12), 1228–1263.

Noci, G. (1997). Designing 'green' vendor rating systems for the assessment of a supplier's environmental performance. *European Journal of Purchasing & Supply Management*, *3*(2), 103–114.

Reuter, C. C., Foerstl, K., Hartmann, E., & Blome, C. (2010). Sustainable global supplier management: The role of dynamic capabilities in achieving competitive advantage. *Journal of Supply Chain Management*, *46*(2), 45–63.

Rogers, M. M., & Weber, W. L. (2011). Evaluating CO_2 emissions and fatalities tradeoffs in truck transports. *International Journal of Physical Distribution & Logistics Management*, *41*(8), 750–767.

Rogerson, S. (2013). *Purchasing process for the freight transport services and influence on CO_2 emissions*. Licentiate thesis, Chalmers University of Technology.

Rogerson, S., Andersson, D., & Johansson, M. I. (2012). *Sustainable freight transport purchasing*. Proceedings of the 21st Annual IPSERA Conference, Naples.

Tate, W. L., Ellram, L. M., & Kirchoff, J. L. (2010). Corporate social responsibility: A thematic analysis related to supply chain management. *Journal of Supply Chain Management*, *46*(1), 19–44.

Veleva, V., Hart, M., Greiner, T., & Crumbley, C. (2003). Indicators for measuring environmental sustainability. *Benchmarking: An International Journal*, *10*(2), 107–119.

Wiese, A., Kellner, J., Lietke, B., Toporowski, W., & Zielke, S. (2012). Sustainability in retailing – A summative content analysis. *International Journal of Retailing & Distribution Management*, *40*(4), 318–335.

Wolf, C., & Seuring, S. (2010). Environmental impacts as buying criteria for third party logistical services. *International Journal of Physical Distribution & Logistics Management*, *40*(1–2), 84–102.

Wu, H. J., & Dunn, S. C. (1995). Environmentally responsible logistics systems. *International Journal of Physical Distribution & Logistics Management*, *25*(2), 20–38.

Yin, R. K. (2014). *Case study research – Design and methods*. Thousand Oaks, CA: Sage.

AVOIDANCE

CHAPTER 5

ASSESSING URBAN LOGISTICS POOLING SUSTAINABILITY VIA A HIERARCHIC DASHBOARD FROM A GROUP DECISION PERSPECTIVE

Jesus Gonzalez-Feliu and Joëlle Morana

ABSTRACT

Purpose – *Urban logistics pooling is seen as a serious alternative to imposed urban consolidation centers. However, such strategies are quite new in urban distribution and merit to be evaluated using adapted methods that take into account the group decision nature of resource pooling. This chapter aims to propose, via an experimental collaborative decision support method, to define a grid of indicators and a reference situation database to measure the sustainable performance of urban logistics pooling systems.*

Methodology – *The proposed methodology combines a systematic literature analysis of Key Performance Indicators and a group decision support method to choose a suitable set to define a dashboard. First, we*

Sustainable Logistics
Transport and Sustainability, Volume 6, 113–135

ISSN: 2044-9941/doi:10.1108/S2044-994120140000006004

identify the main sustainability indicators from an overview of the literature, and class them into the categories of the 4As Sustainable Transport vision (i.e., Awareness, Act and shift, Avoidance, and Anticipation). Then, a group of 20 experts is solicited for an iterative experimental group decision-making method to converge to the concordance of a set of indicators.

Findings – *The method allowed us to define a hierarchic dashboard agreed by all experts with seven main indicators and nine secondary indicators. Moreover, the experts signaled the need of defining a unified basis of comparison to estimate initial situations. To do this, we proposed a database of urban routes from the French Surveys on Urban Goods Transport.*

Research limitations – *The proposed dashboard is an example, and to provide a more unified one, the experience has to be iterated using different groups of decision-makers.*

Practical implications – *This method has the advantage of proposing a dashboard agreed by all involved stakeholders. Therefore, this chapter shows the patterns to reproduce it since the method is able to be replicated in any context of group decision in urban logistics.*

Originality/value – *The originality of the chapter arises on the use of an experimental group decision method using a group with a majority of practitioners, and to validate it by consensus.*

Keywords: Logistics pooling; sustainable development; group decision support; urban distribution; consensus

INTRODUCTION

Sustainable supply chain management is of growing significance not only to organizations to gain a competitive advantage but also to industries sensitive to environmental problems or social issues (Seuring & Müller, 2008). Thus, the adoption of sustainable supply chain management requires that particular attention be focused on performance management, accounting, auditing, and management control. In our view, designing a sustainable dashboard provides a tool for encouraging the application of sustainable supply chain management and allowing stakeholders to discriminate positively in favor of sustainable products and services.

The organizational aspects of urban logistics schemes must be considered in the global sustainable supply chain (Allen & Browne, 2010). Indeed, with increasing urban traffic, some organizations are faced with the problem of ensuring efficient urban freight distribution (Morana, 2014). Additional constraints include relations with public authorities whose motivations for managing product flows are different (i.e., no deliveries to the city center by modes of transport considered highly pollutant). Urban logistics has been studied as a specific research topic for more than 20 years (Gonzalez-Feliu, Semet, & Routhier, 2014; Macharis & Melo, 2011; Taniguchi, Thompson, Yamada, & Van Duin, 2001) but has been traditionally associated with decision support for public authorities and single decision-makers. However, as pointed out by Boudouin, Morel, and Gardrat (2014), urban logistics involves several stakeholders of different natures and with different aims and goals, which need to communicate and collaborate.

The work proposed focuses on the performance of urban logistics, and more specifically the performance of a pooled distribution system in urban areas. Indeed, in downstream or distribution logistics, urban logistics is increasingly being organized according to a rationale of pooling, a form of logistics collaboration characterized by, a mutual usage of material or immaterial resources by two or more stakeholders of different supply chains (Gonzalez-Feliu & Morana, 2011). Collaboration is one of the most promising areas of study in supply chain management (Stefansson, 2004). In this context, we observe that collaborative logistics requires sharing both common goals and resources throughout the lifecycle of such collaboration. This lifecycle can be divided into four stages (Simatupang & Sridharam, 2002): (1) the engagement process, (2) interdependence management, (3) the implementation of operations, and (4) the evaluation of collaboration. Moreover, collaboration in logistics can take place at both longitudinal and transversal levels. Longitudinal collaboration can be defined as the treatment of management issues in a supply chain common to stakeholders of different echelons, mainly those in direct relation with each other. This common process management is based on complementary knowledge and resource-sharing aimed at the efficient use of synergies to accomplish the different tasks of supply chain planning, deployment, follow-up, and control. It is characterized by cooperation and synchronization to ensure better joint planning of a common supply chain. Horizontal collaboration can be defined as "*the collaboration between a group of stakeholders of different supply chains acting at the same levels and having analogous needs*" (Gonzalez-Feliu, Salanova Grau, Morana, & Ma, 2013). The aim of horizontal collaboration is not a common goal of optimizing the same supply chain but finding synergies between different supply chains at certain stages to obtain gains and savings individually (e.g., cost, lead time, service quality, or their combinations). In freight transport, the main collaborative issues are related to this second type since transport ensures the links between the different echelons of supply chains.

Although many works have focused on evaluating the performance of urban logistics (Melo & Costa, 2011; Patier & Browne, 2010), with some dealing more specifically with multiactor multicriteria decision support (Macharis, Turcksin, & Lebeau, 2012), the selection of criteria and indicators generally relies on the researchers' experience, without taking the angle of group decision support into account. Indeed, in most cases the choices concerning evaluation indicators are made on the basis of a single decision-maker, mainly a public authority, that sometimes give the possibility to different stakeholders to define different weights to different criteria. But in urban logistics pooling, the stakeholders most interested in knowing the impacts of such systems are the different transport carriers that have to collaborate with each other. In our context, which derives from that of group decision, decisions are not taken by an individual decision-maker but by a group that must reach agreement or consensus (Raifa, Richardson, & Metcalfe, 2002). Furthermore, there is an increasing demand from the freight transport sector[1] for standard tools to evaluate such impacts in a unified way.

The aim of this chapter is, on the basis of a field experiment, to draw up a sustainable dashboard applicable to logistics pooling. This chapter is structured as follows. The following section presents an overview of sustainable development after which the chapter proposes an overview of the notion of performance and a synthesis of performance indicators in both supply chain management and urban logistics. The position taken with respect to scientific literature is also defined. Then, the group decision methodology for building a unified set of indicators suitable for urban logistics pooling is presented. Finally, a sustainable dashboard for evaluating urban logistics is proposed in the light of the feedback gained from the group decision-making approach. To complete our proposal, a database of reference situations for this sustainability evaluation is also presented and discussed.

VISIONS OF SUSTAINABLE DEVELOPMENT IN LOGISTICS POOLING

Sustainable development is at base a principle for organizing human life in a context of noninfinite resources (Stivers, 1976). This notion has been well integrated in society since the 1997 Kyoto Protocol of the United Nations Framework Convention on Climate Change[2] (Depledge, 2000). Although the basic idea of sustainable development is shared by all, its interpretation and applications tend to diverge in the various contexts that can be found. In logistics and goods transport, sustainability is related to taking the main elements into account:

- The economic viability of logistics systems, which can mainly take place to the notion of Supply Chain Management (SCM) (Christopher, 1992). In this context, several indicators have been defined to evaluate the economic suitability of supply chains (Griffis, Goldsby, Cooper, & Closs, 2007; Gunasekaran & Kobu, 2007).
- Respect for the environment, seen as an opportunity rather than a constraint, in the perspectives of Green SCM (Srivastava, 2007). In this context, we find several concepts like ecodesign (Michelini & Razzoli, 2004) and reverse logistics (Rogers & Tibben-Lembke, 1999).
- Last but not least, the social and societal implications of logistics in a Social SCM approach (Morana, 2013). In this context, it seems important to consider both the intraorganizational stakeholders (the employees of the company) and the interorganizational ones (account taken of Stakeholder Theory).

An alternative vision of sustainable development has emerged recently (Macharis, 2014). Instead of taking a classic approach, the author proposes to consider sustainability on the basis of four characteristics, noted as "the four As." This concept, more related to the freight transport component of logistics than industrial or warehousing operations, is organized around the following four elements:

- *Awareness* is defined as the state or ability to perceive, feel, or be conscious of events. The first step to sustainable development is therefore to be aware of the need to act.
- *Act and shift* can be intended as the reactivity and capacity to ensure a modal split in order to reduce the nuisance of freight transport. The second step of sustainability can be seen as the will to transform current modes of transport modes[3] and organization into cleaner and more socially equitable ones.
- Such changes will not however be efficient without the notion of *Avoidance*, seen as the capacity to avoid increasing the nuisances of logistics and freight transport. Indeed, the third step of sustainability is that of acting to avoid instead of needing to repair later.

- Finally, such avoidance must be combined with *Anticipation*. Indeed, the fourth step of sustainability is that of forecasting and identifying the possible nuisances of logistics and freight transport in advance in order to avoid them.

MEASURING SUSTAINABLE PERFORMANCE IN LOGISTICS AND FREIGHT TRANSPORT

According to Morana (2013), performance in a supply chain is not limited to cost efficiency (in other words, profitability for the company or its shareholders). It is important to consider several aspects of performance consistent with both a strategic dimension that federates actions taken to ensure sustainability and with the aspect of competitive performance that consists in seeking solutions that go beyond a one-dimensional perception of the structure. Moreover, we also need to take into account not only the perspective of socioeconomic performance, reflecting on internal reconfiguration of organizational and social approaches, but also the inter-organizational and environmental aspects. In urban logistics pooling, performance can be measured according to two viewpoints that do not always converge, despite being complementary: that of supply chain performance and that of urban logistics sustainability.

Generally, the aim of measuring logistics performance is to ensure permanent improvement that leads to the conceptualization and implementation of measurement systems combining diagnostics and decision aids. If we focus on evaluating SCM with key indicators (KPI: Key Performance Indicators), reference can be made to two key works on the subject, both topical and considered as equally important by both researchers and practitioners. The first is the work by Gunasekaran and Kobu (2007), who proposed a list of 26 indicators to measure logistics performance, completing a previous work (Gunasekaran, Patel, & Tirtiroglu, 2001). The second is that of Griffis et al. (2007) who proposed a list of 14 indicators. Although other works have been proposed in the literature, we take into account only the 40 indicators proposed in these two works since they are representative of SCM and most works on the subject refer systematically to at least one of them, so they can be taken as a reference. However, these indicators are defined from an SCM perspective and do not necessarily take into account all the transport component,[4] so they do not fully represent the vision of sustainability given by the 4As. Table 1 gives a classification according to

Table 1. Main Key Indicators in Supply Chain Management, Consolidated into "Awareness" and "Anticipation" Indicators.

Awareness indicators ($N = 33$)	
Precision of scheduling	Percentage of deliveries within lead times
Average time for fulfilling pending orders	Variability of order cycle time
Average rate of fulfilling order by item	Process cycle time
Rate of fulfilling whole order	Product development time
Delivery reliability	Variety of products/services
Precision of forecasts	Return on investment
Inventory costs	Sales price
Procurement lead time	Cost of inventory shortage
Production lead time	Supply chain response time
Ratio of logistics costs over sales	Transport costs
Logistics costs per unit	Added value
Obsolescence costs	Cost of guarantee
Conformity with specifications	Operating expenses
Conformity with regulations	Perceived quality
Items picked up per person and per hour	Perceived value of product
Labor efficiency	Percentage of pick-up errors
Anticipation indicators ($N = 8$)	
Average order cycle time	Rate of inventory turnover
Order management cycle time	Weeks procurement
Utilization of capacities	Sales lost due to inventory shortage
Days delay in fulfilling order	Flexibility of production

the 4As approach, showing that in SCM only the Awareness and the Anticipation components are well represented. Avoidance and Act and Shift indicators are absent in such works because they do not focus on transport. Indeed, transport modes are not considered, making it impossible to define indicators of modal shift without previously defining transport mode usage. Moreover, environmental aspects are not usually addressed in SCM (Gunasekaran & Kobu, 2007); in this context, the avoidance component is also difficult to fill-in.

This nonexhaustive panorama of the use of logistics performance indicators in companies is, according to us, in line with the evaluation constraints in private organizations but it omits the role of freight transport in logistics. Moreover, sustainability is only partially taken into account, since Act and Shift and Avoidances are not reflected by these indicators.

Whereas global logistics obeys business management principles, urban logistics is generally linked to the actions of several actors, so that "business" perceptions are confronted by those of "local authorities," that is, the actions and objectives of public authorities. Furthermore, urban logistics projects involve very different sectors, raising questions of feasibility, acceptability, and impact of very different natures. Consequently, it is important to take these sectors into account in the quest for performance indicators and choose those that respond to the needs and objectives of each of the parties involved. These indicators are sometimes difficult to measure and access.

Many of the indicators of urban logistics concern goods transport. Indeed, the question of urban logistics inevitably includes that of last kilometer deliveries to recipients. However, the traditional indicators of long distance goods transport (tons transported, tons*kilometer, quantity of energy consumed per ton*kilometer, kilometer traveled empty, etc.) appear relatively irrelevant in the urban environment. The number of shipments, the number of packages or the variety of actors concerned, among others, change the way in which this measure is seen.

Table 2. Selected Key Indicators for Freight Transport, Classified Under The 4As typology.

Awareness indicators ($N = 16$)	
Distance traveled	Service rate
Travel time	Turnover generated
Loading/unloading time	Number of packages/pallets delivered/picked up
Vehicle fill rate	Number of positions/stops
Warehouse fill rate	Trading area
Ratio of loaded distances over traveled distances	Trip regularity
Return on investment	Distance between stops
Stopping time	Freight vehicle kilometers
Act and shift indicators ($N = 3$)	
Number of vehicles entering the city per mode	Vehicle size
	Vehicle capacity
Avoidance ($N = 10$)	
Energy consumption	Road occupancy by running vehicles
Greenhouse gas emissions	Road occupancy by stopped vehicles
Pollutant emissions	Number of accidents
Noise level (driving)	Number of fatalities
Noise level (loading/unloading)	Involvement of freight vehicles in accidents
Anticipation ($N = 5$)	
Customer satisfaction	Ergonomic design/user acceptability
Retailers' satisfaction	Number of jobs created, destroyed or converted
Rate of absenteeism from employees.	

Different authors have proposed sets of transport sustainability indicators for urban logistics (Behrends, Lindholm, & Woxenius, 2008; Melo & Costa, 2011; Morana, Gonzalez-Feliu, & Semet, 2014; Patier & Browne, 2010; Taniguchi et al., 2001; Vaghi & Percoco, 2011). By compiling the main indicators of these works, we can form the basis of a core set of indicators to evaluate urban logistics. Based on these works, we propose Table 2.

The 34 indicators proposed here are classified as a function of the 4As of sustainable development. We observe that most indicators are related to public authority decision support, while the vision of private companies is underrepresented. In conclusion, supply chain management indicators do not take into account all the elements of sustainable development (being essentially related to the economic performance of the enterprise) and urban logistics indicators mainly reflect public authorities' vision of sustainability. However, it seems important for us to integrate urban logistics in sustainable supply chain management, then to define a grid of indicators capable of measuring sustainability with all its components. Consequently, and to take into account the collaborative nature of logistics pooling, we propose to keep the two sets of indicators (the first focused on supply chain management and the second on transport management) presented here as the basis for establishing a sustainability dashboard applied to urban logistics pooling.

THE METHOD PROPOSED: REASONING THROUGH A COLLECTIVE AND COLLABORATIVE APPROACH

As shown above, KPIs are widely used in literature and practice, but are mainly related to individual decisions. However, and since in logistics pooling decisions are not made unilaterally, the search of consensus is an important stage of the decision-making process in our context. Since logistics pooling is a specific type of horizontal collaboration, we want to focus on group decision-making. In horizontal collaboration, stakeholders have similar functions (in this case, they are transport carriers or transport organizers[5]). They are often competitors, but when they converge to find a collaboration strategy, they need to share at least one common interest (e.g., the collaboration between Bridgestone and Continental to have a common warehouse in France follows a logic of economic performance, i.e., reducing both inventorying and transport costs) and will react like a group when taking decisions (Raifa et al., 2002). Thus, it appears that they require a tool related to group decision and reasoning communities (Evangelou & Karacapilidis, 2007; Yearwood & Stranieri, 2010) and not to individual decisions since the decision processes of reasoning communities are more complex and require particular attention since they combine two types of decision: individual decisions made by each member of the community and a group decision taken by the group as a whole, with or without negotiation. According to Yearwood and Stranieri (2010), the group decision process of a reasoning community involves three main components:

- In the *individual reasoning phase* each individual decision-maker seeks evidence, organizes it, and finally forms claims that represent his or her preferred position or beliefs in order to take an individual decision.
- The *communication of the reasoning* phase describes the transmission of all the aspects of individual reasoning, from the decision itself to the ways each individual arrived at it, and starts a discussion that will support the coalesced reasoning phase.
- Finally, in the *coalescence of the reasoning* phase a form that represents the reasoning processes acceptable to the entire community is obtained. A coalescing of reasoning does not mean that an agreement on a solution is reached. Rather, coalescing of reasoning reflects the state where each individual's reasoning is understood and accepted as valid by the community even if views diverge to the extent that agreement is impossible.

Therefore, when defining a tool for evaluating and assessing urban logistics pooling, it is important to consider the nature of the reasoning community. Therefore, here, we propose a group decision support methodology based on consensus research (Raifa et al., 2002). To do this, it is important to include in the collective reasoning the need to represent sustainability in a suitable way. We thus propose to reason along the lines of the 4As approach developed for logistics and transport, then to follow an interactive method that shows explicitly the three phases of group reasoning decision processes.

We base the proposed methodology on the consensus reaching notion (Yearwood & Stranieri, 2006). However, instead of simulating consensus via quantitative multicriteria methods, like collaborative AHP (Ammarapala & Luxhøj, 2007) or classification methods (Gonzalez-Feliu et al., 2013), we propose an interactive method, such as MAMCA (Macharis, De Witte, & Turcksin, 2010). However, contrary to MAMCA, which uses an analytic-based procedure to support decisions, we propose a consensual method (Raifa et al., 2002). To do this, it is important to constitute a commission of experts that will interact and collaborate to choose the most suitable indicators. Thus, we aim to use a basic group-decision process structure (Yearwood & Stranieri, 2006). First, the stakeholders will make individual decisions without interactions with the others, if possible. Second, a decision communication phase will be organized to allow the stakeholders to exchange on their choices and the importance of using the different indicators. Third and lastly, a consensus research phase will take place to allow all the stakeholders to take a common decision on which they all agree. In this phase, not finding a common solution but agreeing on the fact it has not been possible to find it can also be an alternative to consider.

To perform the above, we proceeded as follows. A set of experts was constituted that included four scientists belonging to different universities, six technical and research representatives from nonuniversity public research institutions, three operational managers from private companies, five project managers from logistics consulting and transport planning software development companies, and two representatives from freight transport standardization organizations, thereby making a total of 20 experts. In order to guide and support the decision process, we developed the following action plan:

1. Creating a common definition of the scope, goals, and targets to reach. The scope here was to measure the sustainability of a logistics pooling system, in a way that the viewpoints and goals of the different stakeholders can be reflected in the resulting evaluation grid. Thus, a group decision-making method to establish a sustainability dashboard was envisaged. This dashboard had to contain four sets of indicators (one per element of the 4As sustainability approach), and the main targets here were to limit the number of indicators to ensure good readability, as well as to propose at least one per category (to cover all of them). This principle had to been presented to a panel of experts that would validate and possibly modify it, then participate in the decision-making process to define an operational sustainability dashboard.
2. *Individual decision* phase. After validating the scope and targets, the experts were asked to make individual choices to fill a grid of indicators using a list of indicators from which they chose those they considered most pertinent to reach the expected goal, initially without limiting the number of indicators so as to allow adding indicators not on the list. This was done so as not to limit the decision to a choice of a specific number of indicators, and also to allow the decision-makers to identify other suitable indicators to comment on them in the following phases.

3. A meeting had to be organized to discuss the results and choose the most suitable set of indicators, in the sense of the *communication of reasoning* (Yearwood & Stranieri, 2006). This meeting would allow identifying the common indicators, the important indicators for each stakeholder and feed the communication and collaboration to find a set of indicators and agree on it. In this second part of the meeting, the *coalescing of reasoning* phase starts.
4. After choosing the indicators, the goal was to present the set of indicators to all the experts in order to reach consensus and validate or modify the group decision. This was to be done to allow the stakeholders to revise and eventually modify the set of indicators and validate the final choice of dashboard. This phase corresponds to the *consensus reaching*, which can be seen as the logical conclusion of the *coalescing of reasoning*.

To do this, a first meeting was scheduled to agree on the common scope, goal and targets. Then, an initial list of 74 indicators (both quantitative and qualitative) was proposed to the expert panel 21 days after the first meeting. A second meeting was scheduled one week after presenting the indicators to state on the suitability of the proposed list and to launch the decision communication phase. However, most of the experts agreed during the meeting that the proposed list was too exhaustive; they preferred to decide on the basis of a reduced list of 30 indicators and to have more time to decide on their choices. Thus, a reduced list was sent to the experts who had 45 days to examine them after the second meeting. The choices of each expert were discussed and a principle of agreement was sought. The following conclusions were drawn from the meeting:

After that, a third meeting was then scheduled 30 days later to update the indicators and present a final version of the document. A consensus process was defined and after reaching consensual agreement, a final dashboard was presented. This dashboard was accompanied by a technical note reporting the main indicators and its method of estimation. Then, both the final dashboard and the technical note was validated by the same group of experts. This validation, performed by experts external to the indicator definition process, led to a discussion on the interest and suitability of each indicator and the global dashboard. Marginal modifications (small changes in secondary indicators and details of form or calculation methods) were introduced but the essence of the proposed dashboard was maintained. We proposed a hierarchical dashboard composed of seven main indicators and nine secondary indicators. These indicators were classified according to the 4As typology of sustainable development. This hierarchy works as follows:

for each A (Awareness, Act and shift, Avoidance, and Anticipation), one or more primary indicators are defined. The combination of indicators had to be considered as sufficient to state on the sustainability of a logistics pooling scheme for all the stakeholders (either in the internal or the external validation process). However, some stakeholders (mainly logistics managers) showed an interest in having complementary indicators associated with certain main indicators. Consequently, we defined the sets of secondary indicators. They were not essential but could help certain stakeholders to detail several aspects of sustainability, mainly service rates, which were selected as a priority for logistics managers.

RESULTS

We propose in Table 3 the final hierarchical dashboard obtained after the double validation. To offer an easy-to-read tool usable by the different stakeholders and for different projects, we propose the following classification that presents seven main indicators (3 for awareness, 1 for Act and shift, 1 for Avoidance, and 2 for Anticipation) and nine secondary indicators (4 for awareness, 3 for Avoidance, and 2 for Anticipation). Although no secondary indicators for Act and shift are defined, the main dashboard (composed of seven main indicators) covers the 4As of sustainability. This is due to the fact that logistics pooling mainly concerns road transport, but we can link the change from pollutant vehicles to clean ones, which is the case of urban logistics pooling systems.

We observe that the indicators are in general more specific than those proposed in SCM and transport management literature. Logistics indicators are related to transport loading rates, with and without linking them to traveled distances. However, such indicators must be associated with warehousing performance (in terms of loading rates) and to the general financial balance. No inventorying performance indicators were assigned since the system was aimed at generating collaboration between transport carriers or their directly associated parties (i.e., mainly 2PL and 3PL), so inventorying was not included in the decisions here. All these indicators as well as those related to service performance can be considered as awareness indicators. Gains in congestion are linked to a reduction in number of trucks, which speaks more to private stakeholders and can be directly associated with Act and shift. Moreover, this indicator was considered useful for both public and private stakeholders. Avoidance indicators underscore

Table 3. The Key Indicators of Urban Logistics in the 4As Sustainability Approach.

Category	Nature	Main indicator	Stakeholder	Secondary Indicator
Awareness	Economic	Ratio of loaded km over traveled kilometers (weight)	Public or private	Ratio of loaded km over traveled km (volume) Loading rates (weight and volume)
	Economic	Warehouse fill rate	Private	–
	Service quality	Service rate	Private	Percentages of deliveries at goods destination Percentage of deliveries in time
Act and shift	Environmental	Saving in number of trucks used	Public or private	–
Avoidance	Environmental	Greenhouse gas emissions	Public or private	Emissions of CO_2 Emissions of CH_4, CO Emissions of NOx
Anticipation	Economic	Financial indicators (10-years IRR)	Private	–
	Social	Rate of jobs to be converted	Public	Rate of jobs that could be destroyed Rate of potential jobs that could be created

the importance of greenhouse gas and pollutant emissions (noise was not selected since transport and logistics practitioners are less sensitive to societal issues than are public authorities). Finally, Anticipation indicators are related to two main questions: the economic viability of the system proposed (via financial indicators like internal rates of return over a 10-year time frame – 10 years IRR) or the importance of converting the potential number of employees to be reduced into new and added-value jobs.

These indicators must then be applied to real logistics pooling systems and deal with the Sustainable Supply Chain Management principles of urban logistics (Morana & Gonzalez-Feliu, 2010) and the 4As vision of sustainability (Macharis, 2014). The aims of this work were to define the list of indicators, but it is also important to give an indication of how they can be used. The data needed to define the indicators is in general obtained using company data, mainly general information on routes (kilometers and times of each route), and carrier financial and commercial data, so they can be calculated at carrier level to produce output indicators that are anonymous and do not violate rules relating to company competitiveness and secrecy. Greenhouse gas and pollution emissions can then be estimated from route details. However, if each company estimates these indicators using a common basis of comparison, part of the possible advantage for the community will be lost since one of the aims of evaluation is also to establish a comparison between logistics schemes. To ensure this, it is important to have a common reference to compare the different projects and initiatives. Consequently, a reference database was built. This reference can be used to compare projects to a current situation (without taking new actions) and to find which sectors are the most suitable for entry into logistics pooling systems. The aim of the database proposed is to define their main characteristics in terms of size in number of delivery points, type of vehicle in weight capacity, type of freight delivered, mode of management, traveled distances, and loading factors, making it possible to define and propose indicators for different types of routes in an initial situation. This database was obtained from the National Survey Database on Urban Goods Transport in France, which contains data on 2,111 routes collected in three different cities between 1995 and 1999. Taking into account the quantity and quality of the data collected in the different surveys, 778 of the 2,111 routes were selected. A typology of routes taking into account the criteria presented above was established, and the main results are summarized in Table 4.

We observe that own transport deliveries mobilize less commodities than third-party transport. TL transport and small LTL routes have similar

Table 4. Main Characteristics of Urban Routes According to the Database Proposed.

Route Size Category (in Number of Deliveries)	Classical Third-Party Transport	Small Parcel LTL Transport	Consigner Own Account	Consignee Own Account	Classical Third-Party Transport	Small Parcel LTL Transport	Consigner Own Account	Consignee Own Account
	Average Load of a stop (in kg)				*Average load of a route (in kg)*			
TL routes (1 delivery)	3,937	–	3,469	3,940	3,937	–	3,469	3,940
2–10 deliveries	1,336	25	352	1,023	5,804	113	1,794	61
11–20 deliveries	489	16	103	56	3,898	247	1,440	116
21–30 deliveries	62	11	30	–	1,140	249	1,753	–
31 or more deliveries	52	8	4	–	1,181	406	1,143	–
	Average number of stops				*Average capacity of vehicles (in kg)*			
TL routes (1 delivery)	1	–	1	1	9,708	–	7,646	9,000
2–10 deliveries	6	8	6	3	8,479	7,669	6,469	5,643
11–20 deliveries	15	16	16	11	5,824	5,145	6,658	5,872
21–30 deliveries	25	25	25	–	5,721	6,373	4,976	–
31 or more deliveries	37	42	36	–	5,815	4,832	3,200	–

deliveries in weight but the vehicles are sometimes different. Concerning third-party transport, the average capacity of vehicles traveling FTL and LTL routes comprising up to10 stops is about 9t, that is, a total weight of about 19t, whereas LTL routes with more than 11 stops are traveled with single trucks of smaller capacity, about 6–7t, that is, a single truck of 13t. Concerning the average quantity of freight unloaded at each stop (in weight), FLT and small routes (up to 10 stops) involve about 4t and 1t per stop, respectively, an average weight that decreases considerably for longer routes (for 11–20 stops, the average weight is about 500 kg and for longer routes, 50–60kg). Regarding small parcel deliveries, no FTL routes were found, and the average number of stops is higher than that of pallet and parcel deliveries (the first category). Weights are low (15 kg on average) with a decreasing trend when the number of stops increases (express deliveries, in general characterized by routes with more than 30 stops, involving average weights of 8 kg/stop, whereas small parcel non express routes involve weights from 10 to 25 kg).

Own account transport follows different trends, and depends strongly on the activity of the sender. Sender's own account transport follows similar trends to those of third-party pallet and parcel deliveries, but with lower weights, except for FTL transport, where weights are similar. However, since the number of routes for this category is higher than that of classical third-party transport, the numbers of sender's own account deliveries are higher. Receiver's own account presents two trends: FTL transport (half of the total number of routes in this category) presents higher weights (about 4t/stop) than small pickup routes (up to 20 deliveries) with collected weights of about 61 kg for routes up to 10 pickup points and higher ones (about 116 kg in average) for routes from 11 to 20 pickup points.

From this database, and taking into account the statistical distribution of routes in a considered spatial zone, a reference situation can be defined. We report as an example the dashboard obtained in the case of a logistics pooling system resulted of merging 100 routes delivering Paris into 300 other routes going to the same destination. In other words, having 400 routes at the beginning, 100 of them have been canceled (those routes correspond to the second worst quartile in terms of loading factor, i.e., not the 100 emptiest routes but the 100 following). The deliveries of such routes have then been affected to the closest remaining routes in order to simulate a logistics pooling system. The simulations have been made in the context of the LUMD project (Morana et al., 2014). From those simulations, we can assess the proposed dashboard, which contains only the main indicators (Table 5):[6]

Table 5. An Example of Use of the Proposed Dashboard.

Category	Nature	Main Indicator	Values
Awareness	Economic	Ratio of loaded km over traveled kilometers (weight), per route	0.51 (+21%)
	Economic	Warehouse fill rate, per warehouse (average)	N.a.
	Service quality	Service rate	95% (=)
Act and shift	Environmental	Savings on the number of trucks used	15%
Avoidance	Environmental	Greenhouse gas emissions	−22%
Anticipation	Economic	Financial indicators (10-years IRR)	N.a.
	Social	Rate of jobs to be converted	18%

Concerning awareness, we observe that this type of logistics pooling allows for increasing the loading rates of each route (an average of 21% of improvement), but being based on only transport pooling there is no impact on warehouse performance. The simulations have been made in a perspective of maintaining a target service rate of 95%. With 25% of the routes eliminated (which is traduced to 15% of trucks not entering the city according to current practices of urban freight transport), it is possible to reduce CO_2 by 22%, which is a good indicator that shows the capacity of logistics pooling to avoid environmental nuisances. Moreover, it is important to note that the reduction of the number of trucks (and then the capacity to act) is traduced by an increase of each route's length, so in consequence an increase of traveled distances, and then polluting emissions. However, using in a good way empty capacities allow to reallocate deliveries to the routes in which distance increase is minimal with respect to others, and then contains this distance increase. In consequence, the overall traveled distance of remaining routes is lower than the total length of the routes canceled). Finally, the two anticipation indicators have to be analyzed carefully. Since the pooled system does not imply high investment costs, we can consider that finances are not impacted highly because of the cost reduction due to the usage of residual capacities instead of specific trucks. However, those costs have not still been estimated in detail, mainly for confidentiality reasons. The number of jobs having to be converted, taking into account the current structure of freight transport professionals, is estimated to be 2%, taking into account the current scheduling practices in urban goods transport, the current vehicle uses, and the potential of using the canceled vehicles for other types of transport. In this context, the number of jobs to convert, which is about eight, can be destined to assistance and added-value tasks and operations related to pooling management, land assistance (mainly for cross-docking and eventual consolidation), or transferred to inventorying and warehousing, when applicable. In any case, the proposed results are a realistic example of logistics pooling that illustrate the usage of the proposed dashboard and show the potential of transferability to the various stakeholders of logistics in urban contexts.

CONCLUSION

The evaluation of urban logistics pooling should be seen from the perspective of sustainable development. In this context, we think that the 4As vision of sustainability (Awareness, Act and shift, Avoidance, and Anticipation) is a good approach for evaluating the sustainability of urban logistics pooling systems. Likewise, it is advisable to enumerate a limited though sufficient number of indicators for decision-making, according to the principle of fast but efficient reading. In addition, the specific characteristics of urban logistics and the two predominant visions (those of private enterprise and public authority, respectively) regarding the problem of goods mobility in urban zones confer strong potential to the evaluation and communication of urban logistics projects.

As seen in the literature analysis, supply chain management indicators do not cover all four components of sustainability in transport but show several interesting indicators for the private sector. Urban logistics evaluation indicators are more focused on public decision-makers and on the case of one decision-maker needing support, and although they propose indicators included in the 4As of sustainability, not all of them seem adapted to measure the sustainability of a system in which a group of private stakeholders must agree on a collaborative logistics system in the context of group decision dynamics. Consequently, using a group decision support methodology, we proposed a grid of indicators to evaluate the sustainability of urban logistics pooling systems. Since several stakeholders are involved in such systems and effectively take decisions, it was important to define a dashboard that shows indicators with which the different stakeholders could understand and analyze whether such systems satisfied their aims or not. A group of experts deliberated sequentially on the different indicators to include in the dashboard (first, individual decisions were made; second, a set of meetings took place to ensure communication between the experts, and their agreement on and validation of a final dashboard). In this way, the indicators of the proposed dashboard are organized on two levels. First, a set of seven indicators that cover all 4As of sustainable transport was defined to provide the general state of the sustainable performance of the system with respect to a reference state (this reference state was also estimated using a database of logistics practices in urban areas). Second, a set of nine indicators was proposed to specify some aspects not included in the first set of indicators but which could be added at the request of certain stakeholders to meet their needs.

This work showed that group decision-making can be applied to evaluate sustainability in urban logistics pooling and provide a set of suitable indicators from an extensive list. However, our aim in this work was not to establish a "standard" grid of indicators as we think that more work must be done to define a unified methodology and a core set of indicators to evaluate the sustainability of urban logistics and ensure the scientific comparison of experiments. To do this, it is important to get both researchers and practitioners (private and public) to collaborate in an open-minded way to find the most suitable set of indicators, taking into account the ways data can be produced and the interpretations that the different stakeholders can make of such indicators. Since the method is directly applicable to practice, it is important to convince users of the power of group decision-making and consensus search and encourage the dialog instead of imposition of "solutions" by public bodies, which is still privileged in some urban logistics contexts.

NOTES

1. Statement pronounced in a set of interviews given in the context of the ANR-MODUM project in France in 2013.
2. The Kyoto Protocol was adopted in 1997 at the third Conference of the Parties to the UNFCCC (COP 3) in order to stabilize greenhouse gas concentrations in the atmosphere at a level that would prevent dangerous anthropogenic interference with the climatic system. This protocol established that the countries which had signed it were to reduce CO_2 emissions by 5% by 2010, a target not yet met at global level.
3. This is a vision "transport" and not "supply chain management." For this reason, the author speaks about transport modes and does not necessarily considers other aspects of SCM like production or inventorying.
4. We can see that only one indicator uses the word "transport" [transport costs]. Moreover, this list does not take environmental indicators and few social/societal indicators.

5. Mainly 3PL and 4PL, sometimes combining own account transport with subcontracting, or Freight forwarders and 5PL not having vehicles and contracting all transport operations to specific transport carriers.

6. The proposed results are an illustration obtained from simulated data and can be seen as an example of utilization of the dashboard. In this sense, only main indicators are presented, but secondary indicators can be estimated as well.

ACKNOWLEDGMENTS

Part of this work was funded in the framework of the LUMD (Logistique Urbaine Mutualisée Durable, Sustainable Urban Pooled Logistics) project by the Inter-Ministerial Unified Funds for Research and Innovation, in its program FUI 2008. The authors would like to thank the book's editors and two anonymous reviewers for their comments and suggestions to improve this book chapter.

REFERENCES

Allen, J., & Browne, M. (2010). Sustainability strategies for city logistics. In A. McKinnon, S. Cullinane, M. Browne, & A. Whiteing (Eds.), *Green logistics: Improving the environmental sustainability of logistics* (pp. 282–305). London: Kogan Page.

Ammarapala, V., & Luxhøj, J. T. (2007). A collaborative multi-criteria decision making technique for risk factor prioritization. *Journal of Risk Research*, *10*(4), 465–485.

Behrends, S., Lindholm, M., & Woxenius, J. (2008). The impact of urban freight transport: A definition of sustainability from an actor's perspective. *Transportation Planning and Technology*, *31*(6), 693–713.

Boudouin, D., Morel, C., & Gardrat, M. (2014). Supply chains and urban logistics platforms. In J. Gonzalez-Feliu, F. Semet, & J. L. Routhier (Eds.), *Sustainable urban logistics: Concepts, methods and information systems* (pp. 1–20). Heidelberg: Springer.

Christopher, M. (1992). *Logistics and supply chain management*. London: Pitman Publishing.

Depledge, J. (2000). *Tracing the origins of the Kyoto Protocol: an article-by-article textual history*. Technical Paper. United Nations, UNFCCC Framework Convention on Climate Change.

Evangelou, C. E., & Karacapilidis, N. (2007). A multidisciplinary approach for supporting knowledge-based decision making in collaborative settings. *International Journal on Artificial Intelligence Tools*, *16*(06), 1069–1092.

Gonzalez-Feliu, J., & Morana, J. (2011). Collaborative transportation sharing: From theory to practice via a case study from France. In J. L. Yearwood & A. Stranieri (Eds.), *Technologies for supporting reasoning communities and collaborative decision making: Cooperative approaches* (pp. 252–271). Hershey: Information Science Reference.

Gonzalez-Feliu, J., Salanova Grau, J. M., Morana, J., & Ma, T. Y. (2013). Design and scenario assessment for collaborative logistics and freight transport systems. *International Journal of Transport Economics*, *40*(2), 207–240.

Gonzalez-Feliu, J., Semet, F., & Routhier, J. L. (Eds.). (2014). *Sustainable urban logistics: Concepts, methods and information systems*. Heidelberg: Springer.

Griffis, S. E., Goldsby, T. J., Cooper, M., & Closs, D. J. (2007). Aligning logistics performance measures to the information needs of the firm. *Journal of Business Logistics*, *28*(2), 35–56.

Gunasekaran, A., & Kobu, B. (2007). Performance measures and metrics in logistics and supply chain management: A review of recent literature (1995–2004) for research and applications. *International Journal of Production Research*, *45*(12), 2819–2840.

Gunasekaran, A., Patel, C., & Tirtiroglu, E. (2001). Performance measures and metrics in a supply chain environment. *International Journal of Operations & Production Management*, *21*(1–2), 71–87.

Macharis, C. (2014). Innovative solutions for sustainable logistics. Presentation at Logistics Day 2014 – Cluster for Logistics, Luxembourg, April 28, 2014, invited speech.

Macharis, C., De Witte, A., & Turcksin, L. (2010). The Multi-Actor Multi-Criteria Analysis (MAMCA) application in the Flemish long-term decision making process on mobility and logistics. *Transport Policy*, *17*(5), 303–311.

Macharis, C., & Melo, S. (Eds.). (2011). *City distribution and urban freight transport: Multiple perspectives* (pp. 120–149). Northampton, MA: Edward Elgar.

Macharis, C., Turcksin, L., & Lebeau, K. (2012). Multi Actor Multi Criteria Analysis (MAMCA) as a tool to support sustainable decisions: State of use. *Decision Support Systems*, *54*(1), 610–620.

Melo, S., & Costa, A. (2011). Definition of a set of indicators to evaluate the performance of urban goods distribution initiatives. In C. Macharis & S. Melo (Eds.), *City distribution and urban freight transport: Multiple perspectives* (pp. 120–149). Northampton, MA: Edward Elgar.

Michelini, R. C., & Razzoli, R. P. (2004). Product-service eco-design: Knowledge-based infrastructures. *Journal of Cleaner Production*, *12*, 415–428.

Morana, J. (2013). *Sustainable supply chain management*. New York, NY: ISTE-Wiley.

Morana, J. (2014). Sustainable supply chain management in urban logistics. In J. Gonzalez-Feliu, F. Semet, & J. L. Routhier (Eds.), *Sustainable urban logistics: Concepts, methods and information systems* (pp. 21–35). Heidelberg: Springer.

Morana, J., & Gonzalez-Feliu, J. (2010). Sustainable supply chain management in city logistics solutions: lessons learned from the case of Cityporto Padua (Italy). *Proceedings of the 3rd International Conference on Information Systems, Logistics and Supply Chain*, April 14–16, 2010, Casablanca, Morocco.

Morana, J., Gonzalez-Feliu, J., & Semet, F. (2014). Urban consolidation and logistics pooling. Planning, management and scenario assessment issues. In J. Gonzalez-Feliu, F. Semet, & J. L. Routhier (Eds.), *Sustainable urban logistics: Concepts, methods and information systems* (pp. 187–210). Heidelberg: Springer.

Patier, D., & Browne, M. (2010). A methodology for the evaluation of urban logistics innovations. *Procedia – Social and Behavioral Sciences*, *2*(3), 6229–6241.

Raifa, H., Richardson, J., & Metcalfe, D. (2002). *Negotiation analysis – The science and art of collaborative decision making*. Harvard, MA: Harvard University Press.

Rogers, D., & Tibben-Lembke, R. (1999). *Going backwards: Reverse logistics trends and practices* (Vol. 2). Pittsburgh, PA: Reverse Logistics Executive Council.

Seuring, S., & Müller, M. (2008). From a literature review to a conceptual framework for sustainable supply chain management. *Journal of Cleaner Production*, *16*(15), 1699–1710.

Simatupang, T. M., & Sridharan, R. (2002). The collaborative supply chain. *International Journal of Logistics Management*, *13*(1), 15–30.

Srivastava, S. (2007). Green supply-chain management: A state-of the-art literature review. *International Journal of Management Reviews*, *9*(1), 53–80.

Stefansson, G. (2004). *Collaborative logistics management – The role of third-party service providers and the enabling information systems architecture*. Ph.D. dissertation, Dept. of Logistics and Transportation, School of Technology Management and Economics, Chalmers University of Technology, Sweden.

Stivers, R. (1976). *The sustainable society: Ethics and economic growth*. Philadelphia, PA: Westminster Press.

Taniguchi, E., Thompson, R. G., Yamada, T., & Van Duin, R. (2001). *City logistics. Network modelling and intelligent transport systems*. Amsterdam: Elsevier.

Vaghi, C., & Percoco, M. (2011). City logistics in Italy: Success factors and environmental performance. In C. Macharis, & S. Melo (Eds.), *City distribution and urban freight transport: Multiple perspectives* (pp. 151–175). Northampton, MA: Edward Elgar.

Yearwood, J., & Stranieri, A. (2006). The generic/actual argument model of practical reasoning. *Decision Support Systems*, *41*, 358–379.

Yearwood, J., & Stranieri, A. (2010). Group structured reasoning for coalescing group decisions. *Group Decision and Negotiation*, *19*(1), 77–105.

CHAPTER 6

PACKAGING FOR ECO-EFFICIENT SUPPLY CHAINS: WHY LOGISTICS SHOULD GET INVOLVED IN THE PACKAGING DEVELOPMENT PROCESS

Katrin Molina-Besch and Henrik Pålsson

ABSTRACT

Purpose – *For packed products, packaging affects every logistical activity and thus the overall economic and ecological efficiency (eco-efficiency) of supply chains. The purpose of this research is to explore how integrated approaches are used in packaging development processes to increase eco-efficiency along supply chains and how a set of pre-selected factors influences the adoption of practically integrated approaches within companies.*

Methodology/approach – *The research approach is explorative and based on nine cases in the food and manufacturing industries in Sweden. In total, 26 semi-structured interviews were conducted.*

Findings – *The chapter describes the way in which companies work with 'integrative' packaging development process elements. It explores how*

Sustainable Logistics
Transport and Sustainability, Volume 6, 137–163

ISSN: 2044-9941/doi:10.1108/S2044-994120140000006006

four factors – product characteristics, packaging requirements, logistical conditions and environmental efforts – influence their approach.

Research limitations/implications – *The study analyses the packaging development processes at a limited number of companies in Sweden.*

Practical implications – *The findings can help logistics managers to better understand how integrated approaches can be applied in packaging development processes to increase eco-efficiency of logistical processes along the supply chain. The study provides logistic managers also with information about which influencing factors can serve as facilitators or barriers to these approaches in their organisations.*

Originality/value – *Previous research has demonstrated the potential economic and environmental benefits of integrating a logistics perspective into the packaging development process. This study complements existing knowledge by presenting extensive empirical data on the practical application of integrated approaches in packaging development processes in industry.*

Keywords: Green logistics; packaging; packaging development process; research paper; supply chain integration; supply chain eco-efficiency

INTRODUCTION

Packaging is a central component in logistics as it follows the product from the point of filling to the point of consumption. For packed products, packaging affects every logistics activity (Bowersox, Closs, & Cooper, 2002) and thus the overall economic and environmental efficiency of the supply chain (Klevås & Saghir, 2004). The environmental impact of logistics is influenced by packaging in several ways. It influences the logistical efficiency as it adds weight and space to products during handling and transport, but it also facilitates volume utilisation (due to e.g. stackability) and handling efficiency (Grant, Lambert, Stock, & Ellram, 2006). In this way, packaging has a direct impact on the energy use of transportation, handling and storage processes along supply chains. Packaging also influences the amount of product waste along supply chains. To keep product waste and the related environmental impacts at a minimum during logistical processes, both the protection function and information function of packaging (e.g. printings/labels) are important (Grönman et al., 2013). In addition, packaging influences the product's sensitivity to temperature changes along supply chains because it acts as an isolation barrier to the surroundings. Packaging formats that remove the need for refrigeration (Garnett, 2007) or heating can reduce energy consumption during transportation and storage. Finally, packaging choices such as material selection and the choice between one-way or returnable packaging have a clear impact on waste handling (Pålsson, Finnsgård, & Wänström, 2013) and reverse logistics.

By integrating logistical and supply chain considerations into the packaging development process, the economic and environmental performance of supply chains should be improved. The economic benefits of integrating logistical aspects into packaging and product development have been demonstrated in a number of studies (Bramklev & Hansen, 2007; Klevås, 2005). Several authors indicate that an integrated packaging, product and logistics development process entails not only economic benefits for companies but also provides opportunities to reduce the environmental impact along supply chains (e.g. Bramklev, Bjärnemo, & Jönson, 2001; Hellström & Nilsson, 2011). The potential for reduced environmental impact from packaging design is illustrated in a report by the World Economic Forum which identified 'packaging design initiatives' as one of the top four areas to decarbonise supply chains (Doherty & Hoyle, 2009). Based on a literature review and an industrial packaging supply chain research programme, Verghese and Lewis (2007) concluded that a cooperative supply chain approach during industrial and transport packaging development reduces environmental and commercial costs and increases efficiency over the whole chain.

This research clearly shows great potential for increased efficiency of supply chains by integrating packaging design with product design and logistical considerations. Some companies apply such an approach, but to varying extents. Even though the processes for product development and packaging development are similar and interdependent (Paine, 1991; Ten Klooster, 2002), they are usually implemented separately (e.g. Hine, 1995). The existing studies have focused on the benefits and success factors of an integration of logistical aspects into packaging development processes but little is known about why certain companies apply this approach while others do not. The purpose of this research is to fill this gap by investigating how integrated approaches are used in packaging development processes to increase eco-efficiency over the supply chain and by exploring how a set of pre-selected factors influence the adoption of practically integrated approaches within the studied companies. This is a first step in finding potential explanations for what influences the adoption of integrative approaches to packaging development in industry.

The chapter is structured as follows: firstly, we present some background on eco-efficiency and supply chain integration and introduce the analytical framework of the study. Secondly, the method is described, followed by the results section. We conclude with a discussion of findings, conclusions and implications.

AN INTEGRATED PACKAGING DEVELOPMENT PROCESS FOR ECO-EFFICIENT SUPPLY CHAINS

Eco-Efficiency, Supply Chain Integration and Logistics

Eco-efficiency is defined as 'competitively priced goods and services that satisfy human needs and bring quality of life while progressively reducing environmental impacts of goods and resource intensity throughout the entire life cycle to a level at least in line with the earth's estimated carrying capacity' (World Business Council for Sustainable Development, 2000). Eco-efficient supply chains should maximise the value creation and minimise the use of resources and emissions of pollutants in supply chains (Verfaillie & Bidwell, 2000).

Companies integrate eco-efficiency into their business strategies to reduce waste and prevent pollution at sources (Hart & Ahuja, 1996; Russo & Fouts, 1997). But to operationalise these strategies, processes and products need to be redesigned (Klassen & Whybark, 1999). One particular process that affects the economic and environmental performance of all logistical activities along the supply chain is the packaging development process (Klevås & Saghir, 2004; Tsoulfas & Pappis, 2006).

As noted in the introduction, process redesign for eco-efficient supply chains requires an integrated approach of, for instance, the packaging, product and logistics (e.g. Hellström & Nilsson, 2011). Such an approach is related to supply chain integration (SCI), a concept which declares that the different functional areas and parts of a supply chain should align their objectives and integrate resources to deliver the highest value to the customer (Ballou, Stephen, & Mukherjee, 2000; Lambert, Cooper, & Pagh, 1998). SCI has been defined in different ways (Pagell, 2004), but definitions typically include elements such as cooperation, coordination, interaction and collaboration. The broader concept of SCI includes integration of different processes such as information, physical and financial processes, and can be divided into internal and external integration. Internal integration concerns the integration of different functions within one company (e.g. Narasimhan & Kim, 2001). External integration can be divided into vertical integration (with upstream suppliers and downstream customers) and horizontal integration (between organisations placed on the same level of the supply chain) (Caputo & Mininno, 1996). In this chapter, we focus on the behavioural aspects of internal and vertical integration.

SCI is a valuable concept for the achievement of eco-efficiency along supply chains in general and for the achievement of eco-efficient logistical operations in particular. In most supply chain management (SCM) literature, it is presumed 'that integration is the best way to obtain efficiency of the supply chain' (Bagchi, Chun, Skjoett-Larsen, & Soerensen, 2005). Since SCM deals with the resources needed for the production of goods, it has not only an effect on economic resources in a supply chain but also impacts strongly on the natural environment (Mentzer et al., 2001).

Analytical Framework

Applying the concept of SCI to the area of packaging suggests that companies along the supply chain should align their packaging objectives and integrate packaging related resources so that packaging provides maximum value to all supply chain actors. A prerequisite for packaging systems[1] that support eco-efficient supply chains is that companies systematically consider relevant logistical aspects during the packaging development process. An initial literature review identified the following 'integrative' elements of packaging development processes as relevant for the eco-efficiency of supply chains:

- *Integration with product development*: Since product and packaging are transported, stored and handled together in many parts of the supply chain, an integrated product and packaging development process is a requirement for eco-efficient product and packaging systems. Nilsson, Olsson, and Wikström (2011) highlight that sustainable goods flows can only be developed if companies apply a holistic and integrated approach to packaging and product.
- *Integrated development of packaging system levels*: The different levels of a packaging system can be developed successively starting from one end or in an integrated approach. Integrated development of several

packaging levels can optimise the packaging system (Olsmats & Dominic, 2003), for instance, by decreasing the amount of air in packaging systems which contributes to transport efficiency.

- *Cross-functionality of the process*: If packaging systems are to be adapted to the different requirements along the supply chain, packaging requirements from different actors have to be considered. A cross-functional packaging development process aims to have a systems perspective on packaging. Verghese and Lewis (2007) found that companies often have a poor understanding of full packaging costs, since the different types of costs associated with packaging occur in different parts of a company.
- *Involvement of external supply chain actors*: For consideration of company external supply chain aspects, external supply chain actors have to be involved in the packaging development process. Verghese and Lewis (2007) have identified a lack of cooperation between different actors in the supply chain as one of the main barriers to environmental innovation of industrial transport packaging systems.

This research analyses why certain companies apply integrated approaches in the packaging development process to increase eco-efficiency along the supply chain while others do not. There is a multitude of internal and external factors that can influence the way companies work with packaging development. In this chapter, the influence of the following four factors on the above identified 'integrative' packaging development process elements will be explored:

- *Product characteristics*: Product characteristics that affect packaging design and the interplay between product and packaging in logistical processes include both physical characteristics, such as size, weight and perishability, and features related to value and demand. For instance, the value-to-weight ratio and the perishability of a product can affect handling requirements (Stank & Goldsby, 2000) as well as the risk of theft.
- *Packaging requirements*: Verghese and Lewis (2007) found that successful environmental innovation in industrial transport packaging requires an understanding of the full costs of packaging. Thus, it seems relevant to analyse the different requirements on packaging that companies consider in total and the requirements they prioritise.
- *Logistical conditions*: There is a relation between the logistical conditions under which companies operate and their opportunities to obtain packaging solutions that are logistically efficient and optimised for transport. For instance, fixed transport schedules can limit efficiency gains of fill-rate optimised packaging systems (Nilsson et al., 2011). Short lead times

can also limit a high packaging and loading efficiency as maximised utilisation of loading units and transport vehicles takes time (Santén, 2012).

- *Environmental efforts*: Companies often handle packaging in a simplistic way with regard to environmental issues (Nilsson et al., 2011). However, environmental assessment of the product and packaging system is necessary for successful environmental innovation of industrial transport packaging systems (Verghese & Lewis, 2007). Packaging has a direct environmental impact connected to the packaging itself and an indirect environmental impact connected to its effect on the packed product (e.g. transport efficiency and product loss along the supply chain). Thus, this chapter addresses whether companies recognise and consider the 'double' environmental impact of packaging (Svanes et al., 2010) and whether this is reflected in their environmental efforts.

METHOD

Overall Approach

The choice of a qualitative approach was derived from the exploratory nature of the research and from the aim to develop comprehensive and profound explanations of the phenomena under study (Eisenhardt, 1989). A multiple case study approach was chosen to explore how different conditions affect packaging design processes (Ellram, 1996) that are contemporary events over which the researcher has no control (Yin, 2003). Semi-structured interviews as a data collection method allowed for the collection of comprehensive and comparable information on packaging development processes in a relatively short period of time. Data were collected at nine case companies in the food and manufacturing industries in Sweden.

Case Selection

The case selection criteria were type of industry and type of products (B2B/B2C). The case sample was limited to producing companies because they have in most cases a direct influence on the packaging selection (as opposed to trading companies). The overall case sampling strategy can best be described as stratified purposeful (Miles & Huberman, 1994) as the aim was to include an interesting mixture of subgroup cases to facilitate comparisons: The food industry was selected as it consumes the greatest quantity of packaging among all industries in Sweden (Packforsk, 2000). The manufacturing industry with mainly B2B products was identified as a contrasting group to the food industry (which usually includes a high share of B2C products). The selection of individual food and manufacturing companies aimed for a wide variation of different kinds of products (maximum variation sampling) (Miles & Huberman, 1994). Products that are commonly transported in bulk have been excluded from the scope. The number of case companies was limited to 6–10 cases based on suggestions by Eisenhardt (1989) and Yin (2003), which seemed appropriate due to the explorative character of the study. Of the 11 initially contacted companies, 9 were able to participate in the study. Table 1 provides an overview of the companies and information about interview respondents. For an overview of the products produced by the case companies, see Table 2.

Data Collection

The data collection at the case companies took place from October 2012 until January 2013. In total, 26 semi-structured interviews were conducted. Data were collected from the perspectives of three different managers at each case company: the packaging manager, the logistics manager and the environmental manager. They were selected because they play an important role in influencing the outcome of the packaging development process and its effect on the eco-efficiency of supply chains. The interview questions were formulated on the basis of the analytical framework described above. The interview guide for the packaging manager focused on the packaging development process and the company's requirements on packaging. The logistics manager was mainly asked about the conditions and requirements of outbound logistical operations. The interview with the environmental manager centred on the company's environmental efforts. The first contact with the case companies was via an e-mail to logistics managers, followed up by a phone call. The logistics manager was asked to provide contact with the packaging manager and the environmental manager. Interview participants were selected based on their work function and availability. To collect data of good quality, interview participants were interviewed in personal meetings at their places of work. Interviews were recorded and lasted between 30–60 minutes. The recorded interviews were transferred into a text format. These drafts were validated by asking the

Table 1. Overview of Case Companies.

Company No.	Branch	Employees in Sweden	Employees Multi-National Group	Years of Employment in Relevant Function at Case Company		
				Packaging manager	Logistics manager	Environmental manager
1	Manufacturing	1,000–5,000	10,000–30,000	1	6	5
2	Food	500–1,000	5,000–10,000	7	12	10
3	Manufacturing	300–500	5,000–10,000	3	7	10
4	Food	1,000–5,000	10,000–30,000	13	25	27
5	Manufacturing	500–1,000	Over 50,000	25	N/A	14
6	Food	300–500	1,000–5,000	5 and 7	5	2
7	Manufacturing	Over 5,000	Over 50,000	18	20	13
8	Manufacturing	500–1,000	1,000–5,000	4	1	5
9	Food	500–1,000	10,000–30,000	2	3.5	3

N/A: not applicable (interview with this function was not performed).

informants to review them to make sure that all data were correctly understood (Ellram, 1996).

Data Analysis Procedure

The data analysis consisted of a cross-case analysis (Voss, Tsikriktsis, & Frohlich, 2002). To prepare for the analysis, all interview data were sorted into five topic areas for each company covering a description of the packaging development process and summaries of the four selected factors to explore (see Fig. 1). After the informants reviewed the drafts, the data were transferred into two tables for each case company to allow for a systematic comparison. The element table included data about the four identified 'integrative' packaging development process elements. The factor table summarised the information about the four selected factors influencing the company during packaging development. The analytical framework of the elements and factors is presented in Fig. 1. The aim of the cross-case analysis was to combine the data from different cases and to explore if common patterns could be identified (Miles & Huberman, 1994). In the first step, the element tables were compared between all companies, one element at a time. Companies with a similar approach in one element were sorted into a group. In the second step, factor tables were compared between companies of one group to see whether any of the factors could explain this group's approach to the specific element.

DESCRIPTION AND COMPARISON OF CASES

In what follows, a brief overview of the case companies is provided with regard to product characteristics, packaging requirements, logistical conditions and environmental efforts. The data used for the analysis were more detailed than the information provided here, but cannot be presented in detail due to space restrictions and confidentiality.

- *Product characteristics:* Table 2 shows that the case companies cover a wide range of different products with regard to product category, size and physical conditions. It must be added that the food companies included cover ambient, chilled and frozen distribution.
- *Packaging requirements:* Packaging managers were asked in the interviews to select the three most important packaging requirements

influencing them during packaging development (Table 2). Product protection requirements and production requirements are most important for most companies. Manufacturing companies are more influenced by logistical requirements and cost considerations while food companies are more influenced by marketing requirements. Environmental requirements on packaging were not selected as most important, but four companies reported a small number of customers/consumers with specific environmental requirements.

- *Logistical conditions:* Table 2 shows that food companies mainly have customers in Sweden who order frequently, they have short lead times (a few days), and the most common transport mode is truck. In comparison, manufacturing case companies have a bigger geographic market that requires the use of several transport modes, lead times are on average longer and orders are less frequent.
- *Environmental efforts:* When asked about the most important environmental aspects of their products in a life-cycle perspective, environmental managers mentioned production of raw materials, energy consumption during product use and impacts from production facilities. None of the companies mentioned the impact of packaging. With regard to environmental goals, most companies focus on the reduction of environmental impacts from their own operations (often focused on production facilities), but two food companies have environmental goals about packaging.

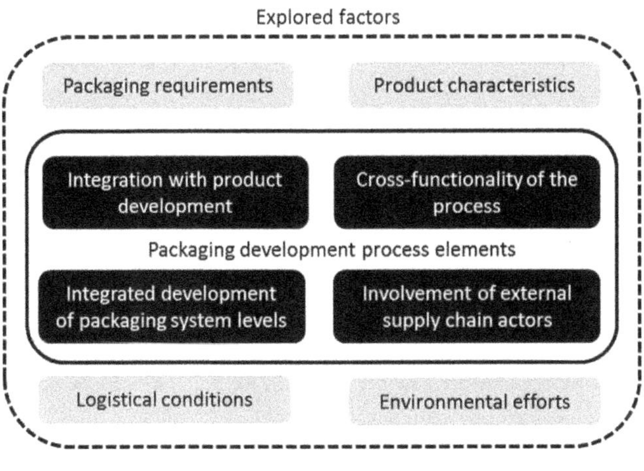

Fig. 1. The Analytical Framework.

Table 2. Product Characteristics, Packaging Requirements and Logistical Conditions at Cases.

Company No.	1	2	3	4	5	6	7	8	9
Product category	Machinery	Meat	Sanitation	Dairy	Automation	Staple foods	Telecom	Lighting	Processed food
Product characteristics									
Product type	B2B	Both	B2B	B2C	B2B	Both	B2B	B2B	B2C
Product size (PU = product unit)	pallet	>50 PU on pallet	10–30 PU on pallet	>50 PU on pallet	>pallet	From > 50 PU on pallet to > pallet	From 30 PU on pallet to > pallet	10–30 PU on pallet	>50 PU on pallet
Physical condition	Solid	Solid	Fragile	Mainly liquid	Solid	Liquid and powder	Solid	Fragile	Varies
Packaging system components	P, T	P, S, T	P, T	P, S, T	P, T	P, (S), T	P, T	P, (S), T	P, S, T
Three most important packaging requirements									
Product protection	X	X	X	X	X		X	X	X
Production requirements	X	X	X	X		X[a]	X	X	X
Logistical requirements	X		X		X		X		
Marketing requirements		X		X					X
Low costs					X			X	
Logistical conditions									
Customer location	Global	Sweden	Nordic countries	Sweden	Global	Nordic countries	Global	Mainly EU	Sweden
Lead time	Days–months	1–3 days	10 days	1 day	6–12 weeks	1 day–1 week	Varies	1 day–1 week	1 day
Order frequency	Varies	Daily	Weekly	Daily	Varies	Varies	Varies	Daily to weekly	Daily
Transport modes	Road, water, air	Road	Road	Road	Road, water, air	Road	Road, water, air	Road, water, air	Road

Packaging system components: P, primary packaging; S, secondary packaging; T, tertiary packaging; N/A, not applicable; B2B, business-to-business; B2C, business-to-consumer.
[a]Company 6 stated that production requirements were the single most important factor influencing them.

RESULTS

The results are presented firstly by describing how companies work with each packaging development process element and secondly the factors that have been identified as influencing the companies' approach.

Integration with Product Development

Based on when and how the product and packaging development processes are integrated, the level of integration were classified as:

Full: Product and packaging are developed in parallel from beginning to end.

Medium: Product and packaging development are adapted to each other to some extent.

Low: Packaging is developed towards the end of the product development process.

None: Packaging is developed once the product development process is finished.

According to this definition, all but one company have links between the two processes, but no company has a fully integrated product and packaging development process (Table 3). At company 7, for instance, packaging aspects are already discussed during the pre-phase of product development, the actual packaging development process starts however once the major product decisions are taken. At the beginning of this process packaging developers receive detailed information from the product developers such as product drawings, information about production site and production volumes, projected markets, and product protection needs.

Based on the analysis, product characteristics and packaging requirements seem to influence the level of integration between product and packaging development processes:

- *Product characteristics:* Food producers have on average a higher level of integration between packaging and product development compared to non-food companies. This is likely to stem from different product characteristics of non-food and food products. All non-food products have a defined shape while some food products were liquid or powder. Manufacturing company respondents mentioned during interviews that product dimensions were often pre-defined due to specific product

Table 3. Integration between Packaging and Product Development at Case Companies.

Company	Integration between Packaging and Product Development	Level of Integration
1	The packaging manager is involved in the product development process from the concept phase. Product developers consider but do not prioritise packaging related aspects of products.	Medium–Low
2	Packaging and product are developed in parallel for some products; for others, packaging is developed when product development is almost finished.	Medium–Low
3	Packaging development is part of the product development process. Packaging development starts when product dimensions are decided.	Low
4	For new products, packaging and product development are coordinated processes.	Medium
5	Product developers try to consider the most important logistical aspects of products.	None
6	Product developers and packaging managers work together in project groups.	Medium
7	Packaging development is part of the product development process; packaging aspects are discussed during the pre-phase. Packaging development starts when major product decisions are taken.	Medium–Low
8	Packaging development starts when major product decisions are taken.	Low
9	Product and packaging developers work in the same location and can easily discuss related aspects with each other. Decisions in both processes influence each other.	Medium

functions or by external standards, which explains why they start with packaging development when the product dimensions are defined. Company 3's products, for example, have to fulfil standard dimensions because they are often installed together with complementary products from other companies. Food producers are in many cases more flexible with regard to product and packaging dimensions, which allows for co-development.

- *Packaging requirements:* Manufacturing companies highlighted physical stress as the most important protection aspect, while food companies had a range of different protection needs (oxygen, humidity, light, odour and microorganism) to consider for different products. Manufacturing companies can consequently handle packaging development as a standard process while food companies need to organise it more flexibly to adapt to the varying packaging requirements of different products. There is another difference between manufacturing and food companies when it comes to production requirements on packaging: Food companies stressed that in most cases packaging development is limited by existing packaging machines, which means that product and packaging together have to fit with existing machinery. At manufacturing companies the most important packaging requirement from production was to allow for easy handling in manual packing processes. Since manual packing processes can be developed on a case by case basis, manufacturing companies can wait with packaging development until late in product development.

Integrated Development of Packaging System Levels

Based on when and how companies develop the different levels of a packaging system, the level of integration was classified as:

Full: Different levels of the packaging system are developed in parallel from the beginning to the end.

Partly: Different levels of the packaging system are adapted to each other to some extent.

None: Primary packaging is developed without consideration of other packaging levels.

Parallel development of different levels of packaging systems was common among non-food producers and less common among food companies with the exception of company 6 (Table 4). Company 9, for example, starts with development of primary packaging and then adapts primary packaging to secondary packaging and pallet dimensions later on if possible. Company 3, on the other hand, develops product packaging directly to fit well with pallet dimensions.

The analysis of influencing factors found that integrated packaging system development appears to be influenced by product characteristics, packaging requirements and logistical conditions:

- *Product characteristics and packaging requirements:* It is more common among producers of B2B products to integrate the development of different packaging system parts as compared to producers of B2C products. The latter have a clear focus on the development of primary packaging in the beginning of the process, probably due to the importance of

Table 4. Integrated Development of Packaging System Levels at Case Companies.

Company No.	Integrated Packaging System Development	Level of Integration
1	Products are very big and sometimes even incompatible with transport infrastructure.	N/A
2	Primary packaging developed before secondary packaging (which is often predefined by customers).	None
3	Pallet dimensions are given but they try to optimise product packaging to increase pallet utilisation.	Partly
4	Most focus on primary packaging in the beginning, but with whole packaging systems in mind.	Partly
5	They use very little packaging anyhow and volume efficiency is often limited by product dimensions.	N/A
6	They develop primary and secondary packaging in parallel.	Full
7	They use software to analyse filling rates and area utilisation of packaging systems.	Partly
8	They work with volume optimisation of packaging to increase the amount of products on full pallets.	Partly
9	Primary packaging developed before secondary packaging but adapted to secondary packaging later.	Partly

marketing requirements. Producers of B2B products have fewer trade-offs to handle and from the beginning can focus more on logistical aspects. Producers of B2C products use packaging systems with three levels of packaging while producers of B2B products mainly use primary packaging, minimal secondary packaging and load carriers. The comparison of cases indicates that it is easier to develop different packaging levels in an integrated manner if packaging systems include fewer levels.

- *Logistical conditions:* Order frequency may be another factor that influences how companies work with the mutual adaptation of packaging system levels. Case companies that have none or partial integration of packaging system levels development all receive daily orders by customers. Quick order processing and fast product delivery may be prioritised over transport efficiency by these companies.

Cross-Functionality of the Process

All of the case companies involve a group of people from different functions in the packaging development process (Table 5). As involvement of

Table 5. Functions Involved in the Packaging Development Process.

Company	Packaging dev.	Product dev.	Marketing	Purchasing	Production	Quality	Logistics	Consideration of Environmental Aspects	Consideration of Logistical Aspects	When are Logistical People Involved?	How Logistical Aspects are Considered During the Packaging Development Process
1	X	X			X		(X)			When needed	Informal contact to colleagues working with logistics.
2	X		X		(X)		(X)			When needed	Logistical optimisation of packaging under certain circumstances.
3	X	X	X	X	X	X	X		X	Late	For new products: packing tests and transport tests.
4	X		X		X		X	X	X	Late	They optimise secondary packaging from a logistics perspective and perform transport tests of packed pallets.
5	X	X			X				X	When needed	Packaging managers adapt packaging dimensions to container size.
6	X		X	X	X		X	X	X	During the whole process	They use optimisation tools to adapt new packaging solutions to pallets and container dimensions.
7	X	X		X	X			X	X	When needed	After supplier selection, they do transport tests and use software to maximise fill rates and area utilisation of packaging systems.
8	X	X			X		X		X	Early	Warehouse people can comment on suggestions from packaging suppliers. They perform also drop tests of packaging samples.
9	X	X	X	X	X		X		X	Late	For secondary packaging: packaging managers use software and perform transport tests on newly developed packaging systems.

logistics is central with regard to optimisation of packaging systems from a supply chain eco-efficiency perspective, a deeper analysis with regard to involvement of the logistics function was performed. Most companies consider logistical aspects in the packaging development process, but to a varying extent (Table 5). In general, the formalised involvement of logistical personnel in the packaging development process is limited. Most of the companies involve logistics late in the process or 'when needed'. The logistical function is used as a 'logistical check-up' of already developed packaging solutions. It is, for instance, common to involve the logistical function to perform transport tests towards the end of the packaging development process.

Involvement of the environmental function or defining specific environmental aspects to assess helps to systematically consider environmental issues in the packaging development process. Since all companies that want to sell products in Sweden have to fulfil the EU Packaging and Packaging Waste Directive (European Council, 1994), these requirements were not considered as 'specific environmental aspects' in this study. Based on this definition, three case companies have defined specific environmental aspects for new packaging. Company 4 uses a life-cycle assessment tool to systematically assess the climate impact of new packaging solutions. Company 6 has defined compulsory environmental requirements for the selection of new packaging suppliers. Company 7 ensures that new packaging solutions fulfil all packaging legislation worldwide and employs a packaging engineer who is responsible for environmental aspects of packaging.

The analysis shows that the focus of packaging requirements and environmental efforts seems to influence cross-functionality:

- *Packaging requirements:* Aspects that are mentioned under the top three packaging requirements by companies are generally represented by a related function in the packaging development process. For instance, at companies that mentioned marketing as one of their top three packaging requirements, the marketing function bears the main responsibility for packaging development together with packaging managers. An exception is the logistical function, which is not formally included in the packaging development process at three companies even though they have mentioned logistical requirements among the top three packaging requirements.
- *Environmental efforts:* Since the environmental work of case companies does not put much focus on packaging in most cases, and since customer/consumer interest in the environmental impact of packaging is

low, case companies do not involve environmental personnel in packaging development. In most case companies, packaging is considered of minor importance from an environmental perspective both since other aspects are considered to have a higher impact in a life-cycle perspective and because most of the environmental work focuses on production processes and not on the total environmental impact of the supply chain.

Involvement of External Supply Chain Actors

The involvement of packaging suppliers is very common among the case companies while the involvement of other external supply chain actors is rare. Packaging suppliers are mainly involved based on the need for technical support. Case companies influence the outcome of packaging development processes heavily themselves since packaging suppliers have to follow their requirements. Only two companies involved customers in the development process. Company 9, for instance, presents new primary packaging solutions to major customers towards the end of the process to receive feedback. Their supply chain manager is also in contact with packaging experts at some customers who sometimes test new secondary packaging solutions.

Since all case companies involve packaging suppliers in packaging development based on the need for technical support, the influence of different factors regarding this aspect was not explored. The analysis found that logistical conditions and customer requirements on packaging influence whether customers are involved in packaging development:

- *Logistical conditions:* Two food companies but no manufacturing companies reported customer involvement in packaging development. The logistical conditions of food companies (limited geographic market and regular deliveries to the same customers) support the development of close and long-term relationships to customers, which seems more difficult to establish under the manufacturing case companies' logistical conditions (global market, infrequent orders, shifting customer base).
- *Packaging requirements:* It is more beneficial for food companies to involve customers in packaging development as compared to manufacturing companies. Three (of four) food companies reported specific customer requirements on secondary and tertiary packaging to allow for easy handling in customer warehouses and shops, which they have to fulfil for all their packaging systems. Three (of five) manufacturing

companies occasionally receive specific packaging requirements from customers that they fulfil on a case by case basis charging customers for extra packaging costs.

DISCUSSION

Integration with Product Development

Integration of packaging and product development was common at the case companies, but full integration of the two processes was lacking. The risk of partially separated product and packaging development is that companies might miss out on important packaging aspects during product development that can lead to inefficient transports and handling. The analysis showed that product characteristics such as pre-defined product dimensions can force companies to wait with packaging development until the end of product development. To reduce the environmental impact of logistical operations, logistics managers should get involved in product and packaging development to make sure that product and packaging performance is critically examined from a logistics perspective.

The influence of installed packaging machinery that directs most packaging development projects at companies with automated packing processes constitutes another hinder to full integration between packaging and product development. Decisions to invest in specific packaging machines, which were in many cases taken 10–15 years ago, limit the possibilities to adapt packaging and product systems to the current supply chain conditions such as logistical infrastructure. The dependency on existing packaging machinery is a typical 'lock-in' effect that has been described by several authors (e.g. Brunsson & Jacobsson, 2002). Consequently, it is important that logistics managers are involved into decision making processes regarding new packaging machine investments as they are best placed to assess logistical aspects of new packaging types.

Integrated Development of Packaging System Levels

All case companies are aware of the benefits of parallel development of different levels of the packaging system to increase storage and transport efficiency, but producers of B2C products seem to prioritise marketing requirements. This focus seems to constitute a potential barrier for the development of eco-efficient supply chains since marketing requirements often conflict with logistical and environmental requirements on packaging, in agreement with Nilsson et al. (2011). Logistics managers should not directly accept marketing requirement on packaging but discuss with marketing and environmental managers how different requirements can be balanced against each other. Delivery requirements are another identified potential barrier to the development of packaging systems adapted for supply chain eco-efficiency. This is in line with results from Nilsson et al. (2011) who found that fixed delivery schedules of companies were a hinder towards the improvement of fill rates in packaging system.

Cross-Functionality of the Process

Case companies all have a partly integrated approach to packaging since they involve three to seven different functions during packaging development, but in most cases they focus on company-internal aspects. These results do not fully agree with Verghese and Lewis (2007) who found companies lacked a cross-functional perspective on packaging. The relation between functions involved in the packaging development process and the stated importance of packaging requirements may have been based on involving the functions that companies believe are the most important. It could also be the other way around: that the packaging requirements are first formulated by the functions currently involved. The risk of involving only a few functions in packaging development is that it could limit the perspective companies have on packaging and reduce the opportunities to adapt packaging to different supply chain needs.

Since logistics and environmental managers have similar interests in supply chain eco-efficiency, both functions should align their requirements on packaging. Formalised involvement of logistical personnel in packaging development is limited and often takes place late in the process. To support eco-efficient supply chains, early and continuous involvement of the logistics function seems the best approach to ensure that packaging developments are aligned to logistical requirements. Defined logistical requirements can grow outdated over time and in addition there is a risk that projects place a low priority on these requirements. The lack of involvement of environmental personnel in packaging development and the few specific environmental requirements on packaging at the case companies indicate that they do not fully consider that packaging affects environmental impacts along the supply chain.

Involvement of External Supply Chain Actors

Only two of nine companies involve customers in packaging development, which indicates limited insight on external packaging aspects at most case companies. These results agree with Verghese and Lewis (2007) who identified a lack of cooperation between different actors in the supply chain as one of the main barriers to environmental innovation of industrial transport packaging.

Cooperation with customers was more common among food companies than manufacturing companies which could be because Swedish food retailers typically impose requirements on secondary and tertiary packaging to be handled in their distribution centres and shops. Food companies are in this way forced to consider logistical efficiency further down the supply chain. It should be noticed that the difference between food and manufacturing companies regarding customer requirements on packaging recognised here does not seem to be generalisable. There are examples of manufacturing industries, such as the automotive industry where a high level of packaging standardisation is also common.

Verghese and Lewis (2007) highlighted the importance of early communication between different supply chain partners. Early communication with external supply chain actors was not found at the case companies. Supply chain managers have a good overview of external processes and they constitute an important link to external supply chain actors that companies should utilise during packaging development.

CONCLUSIONS

A common assumption in the SCM literature is that 'the more integration – the better the performance of the supply chain' (Bagchi et al., 2005). This study indicates that the application of integrated approaches in packaging development processes is not very widespread in practice. Most case companies have developed packaging development processes that include 'integrative' elements to allow for a systematic consideration of logistical aspects of packaging. However, these elements are in most cases not deployed in a fully integrative manner. There are a number of internal and external factors that seem to influence companies' more or less integrated approaches on packaging development and which therefore might impact on the environmental performance of supply chains. Fig. 2 shows the identified potential influence of the explored factors on the four packaging development process elements.

Internal factors influencing one or several packaging development process elements are production requirements, prioritised packaging requirements and environmental work focus. External factors are marketing requirements, customer packaging requirements, delivery requirements, composition of customer base and product characteristics. The analysis shows that some factors might constitute barriers while others are facilitators to the adaption of integrated approaches in packaging development processes at the case companies. Identified barriers include long-term dependency on packaging machinery, delivery requirements and marketing requirements. Identified facilitators include customer requirements on secondary and tertiary packaging, environmental packaging requirements by customers and a customer base with relatively few customers who are close-by.

Results from this study can help logistics managers to better understand how integrated approaches can be applied in packaging development processes to increase cost efficiency and reduce the environmental impacts of logistics. The results can also help to identify factors that can serve as facilitators or barriers to these approaches in their organisations. Logistics managers should critically assess existing packaging systems as there are most

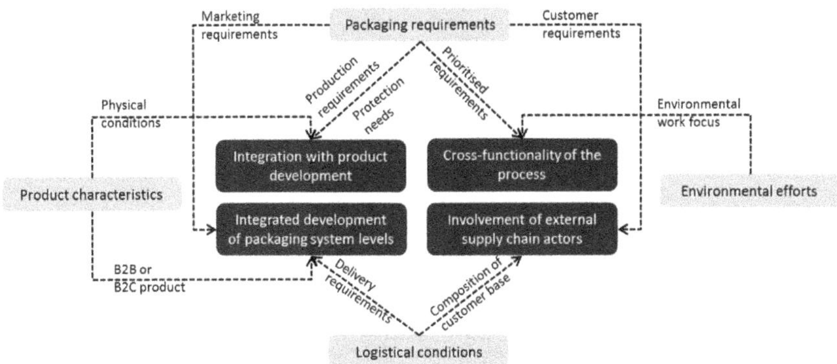

Fig. 2. A Summary of the Study Results.

probably improvement potentials in many organisations. In addition, they should get involved in packaging development processes to make sure that future packaging systems are supporting cost and environmentally efficient logistical processes.

Previous research has demonstrated the potential economic and environmental benefits of integrating a supply chain perspective into the packaging development process, but empirical data has been limited. This study fills this gap by presenting extensive empirical data about the practical application of integrated approaches in packaging development processes in industry. Thus, it contributes to existing knowledge by proposing factors that seem to influence the companies' approaches. The results indicate that the application of integrated approaches in packaging development processes is difficult to implement in practice and that there are factors that hinder the theoretical benefits of such approaches.

The findings of this chapter are derived from a limited number of cases, and the data collection and analysis are limited to the exploration of four potentially influencing factors during the packaging development process (product characteristics, packaging requirements, logistical conditions and environmental work). The relations between the factors and the packaging development process are presented on a qualitative basis. To test and verify these relations, broader empirical research should be conducted. Future research should also address possible ways of overcoming the identified barriers, such as through action research in pilot packaging development projects targeted at supply chain eco-efficiency. Since this research focused on packaging development processes at producing companies, it seems relevant to investigate the role of other supply chain actors for the development of packaging for eco-efficient supply chains (e.g. transport providers and retailers). The analysis of other influencing factors than the ones selected for this study can complement the results and allow for a deeper analysis of packaging development processes and their effects on the eco-efficiency of supply chains.

NOTE

1. A packaging system can be described in a hierarchy of three levels: primary, secondary and tertiary (Jönson, 2000). The primary packaging is in direct contact with the product, the secondary packaging contains several primary packages and the tertiary packaging (e.g. pallets or roll containers) contains a number of primary or secondary packages.

REFERENCES

Bagchi, P. K., Chun, H. B., Skjoett-Larsen, T., & Soerensen, L. B. (2005). Supply chain integration: A European survey. *The International Journal of Logistics Management*, *16*(2), 275–294.

Ballou, R., Stephen, M., & Mukherjee, A. (2000). New managerial challenges from supply chain opportunities. *Industrial Marketing Management*, *29*(1), 7–18.

Bowersox, D. J., Closs, D. J., & Cooper, M. B. (2002). *Supply chain logistics management*. New York, NY: McGraw-Hill/Irwin.

Bramklev, C., Bjärnemo, R., & Jönson, G. (2001). Concurrent design of product and package – Extending the concept of IPD. In *Proceedings of International Conference on Engineering Design (ICED 01)* (pp. 377–384).

Bramklev, C., & Hansen, T. C. (2007). On the logistics effects of integrated product and package design. In *Proceedings of the International Conference on Engineering Design 2007*.

Brunsson, N., & Jacobsson, B. (2002). *A World of standards*. Oxford: Oxford University Press.

Caputo, M., & Mininno, V. (1996). Internal, vertical and horizontal logistics integration in Italian grocery distribution. *International Journal of Physical Distribution & Logistics Management*, *26*(9), 64–90.

Doherty, S., & Hoyle, S. (2009). *Supply chain decarbonization – the role of logistics and transport in reducing supply chain carbon emissions*. Geneva: World Economic Forum.

Eisenhardt, K. (1989). Building theories from case study research, *Academy of Management Review*, *14*(4), 532–550.

Ellram, L. M. (1996). The use of case study method in logistics research. *Journal of Business Logistics*, *17*(2), 93–138.

European Council. (1994). Directive 94/62/EC on packaging and packaging waste. *Official Journal of the European Union*, *50*(365), (31/12/1994), 10–23.

Garnett, T. (2007). *Food refrigeration: What is the contribution to greenhouse gas emissions and how might emissions be reduced?* A working paper produced as part of the Food Climate Research Network, Surrey.

Grant, D. B., Lambert, D. M., Stock, J. R., & Ellram, L. M. (2006). *Fundamentals of logistics management*. Berkshire: McGraw-Hill.

Grönman, K., Soukka, R., Järvi-Kääriäinen, T., Katajajuuri, J.-M., Kuisma, M., Koivupuro, H.-K., … Linnanen, L. (2013). Framework for sustainable food packaging design. *Packaging Technology and Science*, *26*(4), 187–200.

Hart, S. L., & Ahuja, G. (1996). Does it pay to be green? An empirical examination of the relationship between emission reduction and firm performance. *Business Strategy and the Environment*, *5*, 30–37.

Hellström, D., & Nilsson, F. (2011). Logistics-driven packaging innovation: A case study at IKEA, *International Journal of Retail & Distribution Management*, *39*(9), 638–657.

Hine, T. (1995). *The total package: The secret history and hidden meanings of boxes, bottles, cans, and other persuasive containers*. Boston, MA: Little Brown & Co.

Jönson, G. (2000). *Packaging technology for the logistician*. Lund: Department of Design Sciences, Division of Packaging Logistics, Lund University.

Klassen, R. D., & Whybark, D. C. (1999). The impact of environmental technologies on manufacturing performance. *Academy of Management Journal*, *42*, 599–615.

Klevås, J. (2005). *On opportunities of integrating packaging, logistics and product development – Experiences from a case study at IKEA*. Licentiate thesis, Department of Design Sciences, Division of Packaging Logistics, Lund University, Sweden.

Klevås, J., & Saghir, M. (2004). Introducing the concept of design for packaging logistics. In *Proceedings of the International Conference on Productivity & Quality Research (ISPQR)*.

Lambert, D. M., Cooper, M. C., & Pagh, J. D. (1998). Supply chain management: Implementation issues and research opportunities. *International Journal of Logistics Management*, *9*(2), 1–19.

Mentzer, J. T., DeWitt, W., Keebler, J. S., Min, S., Nix, N. W., Smith, C. D., & Zacharia, Z. G. (2001). Defining supply chain management. *Journal of Business Logistics*, *22*(2), 1–25.

Miles, M. B., & Huberman, A. M. (1994). *Qualitative data analysis*. Thousand Oaks, CA: Sage.

Narasimhan, R., & Kim, S. W. (2001). Information system utilization strategy for supply chain integration. *Journal of Business Logistics*, *22*(2), 51–76.

Nilsson, F., Olsson, A., & Wikström, F. (2011). Toward sustainable goods flows – A framework from a packaging perspective. *Proceedings of the 23rd Annual NOFOMA Conference* (pp. 925–942).

Olsmats, C., & Dominic, C. (2003). Packaging scorecard – A packaging performance evaluation method. *Packaging Technology and Science*, *16*(1), 9–14.

Packforsk. (2000). *Packat i pocket*. Sweden: Packforsk.

Pagell, M. (2004). Understanding the factors that enable and inhibit the integration of operations, purchasing and logistics. *Journal of Operations Management*, *22*(5), 459–487.

Paine, F. (1991). *The packaging user's handbook*. London: Blackie.

Pålsson, H., Finnsgård, C., & Wänström, C. (2013). Selection of packaging systems in supply chains from a sustainability perspective: The case of volvo. *Packaging Technology and Science*, *26*(5), 289–310.

Russo, M. V., & Fouts, P. A. (1997). A resource-based perspective on corporate environmental performance and profitability. *Academy of Management Journal*, *40*, 534–559.

Santén, V. (2012). Increased load factor and sustainable logistics. *Proceedings of the 24th Annual NOFOMA Conference* (pp. 738–753).

Stank, T. P., & Goldsby, T. J. (2000). A framework for transportation decision making in an integrated supply chain. *Supply Chain Management: An International Journal*, *5*(2), 71–77.

Svanes, E., Vold, M., Möller, H., Pettersen, M. K., Larsen, H., & Hanssen, O.-J. (2010). Sustainable packaging design: A holistic methodology for packaging design. *Packaging Technology and Science*, *23*(3), 161–175.

Ten Klooster, R. (2002). *Packaging design: A methodological development and simulation of the design process*. Doctoral dissertation, Delft University of Technology, The Netherlands.

Tsoulfas, G. T., & Pappis, C. P. (2006). Environmental principles applicable to supply chains design and operation. *Journal of Cleaner Production*, *14*(18), 1593–1602.

Verfaillie, H. A., & Bidwell, R. (2000). *Measuring eco-efficiency – A guide to reporting company performance*. Geneva: World Business Council for Sustainable Development.

Verghese, K., & Lewis, H. (2007). Environmental innovation in industrial packaging: A supply chain approach. *International Journal of Production Research*, *45*(18–19), 4381–4401.

Voss, C., Tsikriktsis, N., & Frohlich, M. (2002). Case research in operations management. *International Journal of Operations & Production Management*, *22*(2), 195–219.

World Business Council for Sustainable Development. (2000). *Eco-efficiency: Creating more with less*. Retrieved from http://www.wbcsd.org/web/publications/eco_efficiency_creating_more_value.pdf. Accessed on April 23, 2013.

Yin, R. K. (2003). *Case study research: Design and methods*. Thousand Oaks, CA: Sage.

ACTING

CHAPTER 7

FREIGHT TRANSPORT MODE CHOICE AND MODE SHIFT IN NEW ZEALAND: FINDINGS OF A REVEALED PREFERENCE SURVEY

Hyun-Chan Kim, Alan Nicholson and Diana Kusumastuti

ABSTRACT

Purpose – *This study aims to identify the determinants of transport mode choice and the constraints on shifting freight in New Zealand (NZ) from road to rail and/or coastal shipping, and to quantify the trade-off between factors affecting shippers' perceptions, to assist in increasing the share of freight moved by non-road transport modes.*

Methodology – *A revealed preference survey of 183 freight shippers, including small and medium enterprises and freight agents in NZ, is used to investigate whether freight shippers' characteristics affect their ranked preference for attributes related to mode choice and modal shift.*

Sustainable Logistics
Transport and Sustainability, Volume 6, 165–192

ISSN: 2044-9941/doi:10.1108/S2044-994120140000006007

Additionally, a rank-ordered logistic (ROL) model is estimated using the ranking data.

Findings – *The results reveal several distinct types of transport mode choice behaviour within the sample and show how the preferences for timeliness, cost, accessibility, damage and loss, customer service, and suitability vary between industry groups and business types. Also, the ROL method allows us to identify heterogeneity in preferences for mode choice and mode shift factors for freight within NZ.*

The results imply that NZ shippers ranked transport time as the most significant constraint upon distributing goods by rail, while accessibility and load size were the most significant constraints upon using coastal shipping. The study also identifies how NZ shippers' modal shift constraints vary according to the firm's individual or logistical characteristics.

Research implications – *This study informs freight transport policy makers about the needs of NZ shippers by providing quantitative measures of the intensity of preference for the various mode choice factors.*

Practical implications – *Those involved in freight transport have a better basis for formulating transport policy.*

Keywords: Freight transport; mode choice; mode shift; revealed preference survey; rank-ordered logit model

INTRODUCTION

Market globalisation and developing service economies have increased the demand for reliable, flexible, cost-effective, timely, and viable door-to-door freight services, from the shippers in the world. In New Zealand (NZ) freight transport demand has grown by more than 32% during the last decade and freight transport (in tonne-kms) is expected to grow about 70% between 2005 and 2020 (Richard Paling Consulting, 2008). The Ministry of Transport expects the strong growth of freight movements to continue and to double by 2040. Concurrently, the modal share of road transport has increased substantially and is expected to increase further in the coming years. In addition, with rising fuel prices and growing awareness about the challenge of global climate change, innovative policies and technologies are being introduced for reducing the negative impacts (i.e. congestion and pollution) of the dependency on road transport.

The National Freight Demand Study (Richard Paling Consulting, 2008) was the first comprehensive freight movement study in NZ. The study involved surveys of 100 key firms across the industries, addressing the factors influencing freight mode choice qualitatively. The key factors identified by this study were cost, reliability, modal connectivity, loss and damage, mode-to-mode transfer, customer service, environmental and sustainability issues, and some logistics issues within the supply chain. Rockpoint (2009) mainly focused on NZ sea and coastal freight issues from both the shipper's and the carrier's points of view. This study provided a better understanding of how NZ shippers choose the appropriate mode of transportation through interviewing 45 firms across various industries. The study offered a choice of five service criteria, which were: product care, cost, timeliness, reliability and safety. Reliability was cited as the most important service factor, followed by product care and safety. Interestingly, this study uses 'reliability' and 'timeliness' as different service factors. However, 'timeliness' often encompasses both average shipment time (variables affecting the average include standard transit times and directness of service) and variations in shipment time (reliability of service) (Evers, Harper, & Needham, 1996). A limitation of previous NZ studies is the lack of quantitative information about how those choosing between modes make trade-offs between conflicting objectives and factors. Furthermore, there is increased emphasis on environmental issues in the transport area. However, none of the previous studies in NZ have attempted to investigate the constraints on freight mode shift from road to rail or coastal shipping.

Many nations are considering rail and coastal shipping as a sustainable economic infrastructure to transport freight. Freight mode shift offers strong benefits in terms of environmental benefits, the lower energy consumption, the economies of scale, and the lower costs needed for infrastructure expansion (Perakis & Denisis, 2008). In 2003, the European Commission launched the Marco Polo programme, which aims to ease road congestion and the associated pollution, and to promote reliable and efficient transport of goods, by switching to greener transport modes, such as railways, coastal or deep sea shipping and inland waterways (European Commission, 2009). More than 500 companies have already successfully shifted freight from road to greener modes. The Marco Polo programme target is to free Europe's roads of 20 billion tonne-kilometres of freight per annum, the equivalent of more than 700,000 trucks a year travelling between Paris and Berlin. In the United Kingdom, £19m of funding was allocated to support intermodal shift to rail in 2011 and the same amount was recommended for the following two years (European Reference Center for Intermodal Freight Transport, 2010).

While freight mode shift is identified specifically as an important element of the NZ government's goal to limit its greenhouse gas emissions to 50% of 1990 levels by 2050 (Ministry for the Environment, 2009), there is no programme like the Marco Polo programme to encourage a shift of freight from road to other modes. Transport policy in NZ has not had a clear and consistent emphasis on reducing greenhouse gas emissions, and previous freight studies in NZ (Richard Paling Consulting, 2008; Rockpoint, 2009) have not found evidence that the environmental impacts of transport are significant factors for NZ shippers when choosing transport mode. The initial phase of this study involved interviews with several practitioners across industry sectors, and these confirmed that NZ shippers generally do not consider environmental factors when choosing a freight transport mode. This study therefore did not include environmental factors as mode choice factors.

The decision-makers' perception is a major input component to the decision making process in mode selection or mode shift. The ranking approach may be seen as an attractive approach between the rating and the single-choice approaches because the respondent provides a preference ordering of alternatives but not the relative degree of preferences (Srinivasan, Bhat, & Holquin-Veras, 2006). A rank-ordered logit (ROL) model, also known as an exploded logit model, can be used to analyse data on the preferences of individuals over a set of alternatives, where the preferences are partially observed through surveys or conjoint studies. Empirical applications describing preferences using the ROL model can be found in many fields such as social science (Drewes & Micheal, 2006; Hsieh, 2005; Mark, Lusk, & Scott, 2004), business and economics (Ahn, Lee, Lee, & Kim, 2006; Dagsvik & Liu, 2006), medicine (Alava, Brazier, Rowen, & Tsuchiya, 2013; Lemanske et al., 2010), environmental economics (Kumar & Kant, 2007), and transportation studies. Calfee, Winston, and Stempski (2001) used SP data to estimate a ROL model for estimating the value of automobile travel time. More empirical applications in the field of transportation using the ROL model can be found Odeck (1996), Fridstrom and Elvik (1997), Hunt (2001), Kockelman, Podgorski, Bina, and Gadda (2012), Srinivasan et al. (2006), and Dagsvik and Liu (2009).

In this chapter, we present an ROL model to examine the freight transport mode choice determinants and mode shift constraints for NZ shippers. The data used in the empirical analysis are obtained from a revealed preference (RP) survey of 183 NZ freight shippers and freight agents conducted in 2011. The questionnaire included several questions aimed at eliciting relevant mode choice related information. The RP survey was primarily designed to capture 'current shippers' freight operations in NZ. The results of this study were used as a base to develop a subsequent Stated Preference (SP) survey, the data from which are being used to develop models to predict mode choice, calculate willingness to pay, and assess freight policies in NZ.

The remainder of this chapter is organised into sections. The section 'Rank-Ordered Logit Model' describes the structure of the ROL model used in this study. The section 'Data Analysis' provides a general description of RP survey, population and sample, and the analysis of survey data, including the demographics of survey respondents and general pattern of NZ shippers' freight operations and logistics. The section 'Rank-Ordered Logit Analyses for Mode Choice Preference' describes the estimation of a ROL model, to identify the relative importance of mode choice factors. The section 'Rank-Ordered Logit Analyses for Mode Shift' presents the estimation of the relative importance of constraints on mode shift to move freight from rail or coastal shipping instead of road transport. To capture the effects of respondents' individual or firm characteristics on the importance of mode choice and mode shift factors, the models are extended to include the firm's characteristics. The final section 'Conclusions' summarises the results and their implications.

RANK-ORDERED LOGIT MODEL

The rank-ordered logit (ROL) has been used extensively in marketing research. This model is an extended form of the conditional logit (CL) regression model introduced by McFadden (1974). The logistic model for ranking was proposed by Beggs, Cardell, and Hausman (1981) and further developed by many marketing researchers (Allison & Christakis, 1994; Chapman & Staelin, 1982; Hausman & Ruud, 1987; Punj & Staelin, 1978) under the name 'rank-ordered logit model'. An alternative specification of the logistic regression model, based on random utility models, is often used in econometrics (e.g. Maddala, 1983). In random utility models the rank of an alternative is determined by its utility. Therefore, the utility U_{ij} provided to individual i by product j is modelled as

$$U_{ij} = V_{ij} + \varepsilon_{ij} \quad (1)$$

where V_{ij} is a function of the attributes of the alternatives and the error component ε_{ij} is assumed to be independently identically distributed (IID), with an extreme value distribution, given by $\Pr(\varepsilon_{ij} \leq t) = \exp\{-\exp(-t)\}$, and the probability of ranking j higher than k is given by $\exp\{u_{ij} - u_{ik}\}$. McFadden's random utility model implies the following likelihood L_i for a single respondent, where $\delta_{ijk} = 1$ when $Y_{ik} > Y_{ij}$ and $\delta_{ijk} = 0$ otherwise;

$$L_i = \prod_{j=1}^{J} \left[\frac{e^{V_{ij}}}{\sum_{k=1}^{j} \delta_{ijk} e^{V_{ij}}} \right] \tag{2}$$

It follows that the probability of item j being the most preferred item from the set J is

$$\Pr\left(U_1 > U_2 > \cdots > U_j\right) = \frac{e^{V_1}}{\sum_{j=1}^{J} e^{V_j}} \tag{3}$$

which is the form of the classical Multinomial Logit (MNL) model. Because of the assumed independence from irrelevant alternatives (IIA) between the choices, the likelihood of a certain ranking of the alternatives in the entire choice set is thus the product of J logit probabilities (Allison & Christakis, 1994; Luce, 1959). This likelihood can be written as

$$\begin{aligned} \Pr(U_1 > U_2 > \cdots >) &= \Pr\left(U_1 > U_j, j = 2, \ldots, J\right) \bullet \Pr\left(U_2 > U_j, j = 3, \ldots, J\right) \cdots \\ &\quad \bullet \Pr(U_{J-1} > U_J) \\ &= \frac{e^{V_1}}{\sum_{j=1}^{J} e^{V_j}} \bullet \frac{e^{V_2}}{\sum_{j=2}^{J} e^{V_j}} \bullet \cdots \bullet \frac{e^{V_{J-1}}}{e^{V_{J-1}} + e^{V_J}} \\ &= \prod_{j-1}^{J-1} \frac{e^{V_J}}{\sum_{m=J}^{J} e^{V_m}} \end{aligned} \tag{4}$$

Finally, estimation of a ROL model can be accomplished with most partial likelihood procedures for estimating proportional hazard models. For a sample of n independent respondents, Eq. (4) implies a log-likelihood of

$$\log L = \sum_{i=1}^{n} V_{ij} - \sum_{i=1}^{n} \log\left[\sum_{k=1}^{j} \delta_{ijk} \exp(V_{ik})\right] \tag{5}$$

The linear model for the V_{ij}s in Eq. (1) can be substituted into Eq. (5), which can then be maximised with respect to the coefficient vectors. Beggs et al. (1981) proved that the likelihood is globally concave, which means if a maximum is found, it is a global rather than a local maximum.

DATA ANALYSIS

Ranked data on relative preference for freight transport modes for this study comes from two major groups of freight transport user; freight shippers and consigners, who actually own freight (e.g. manufacturers, wholesalers), and freight agents, such as freight forwarders, transport service providers (e.g. contracted carriers, warehousing) and 3PL (3rd party logistics) companies. Both types of information are termed 'shipper information' in this chapter. An RP survey of freight shippers and agents was conducted on-line during 2011 and 2012. The survey sample was chosen randomly from the list of firms registered with the New Zealand Stock Exchange, and firms that are members of industry associations, groups and councils. Firms were assigned to four business categories; 'primary/raw material providers', 'manufacturers', 'wholesalers/retailers' and 'logistics service providers'.

A detailed company profile (including business summary, products/services and industry/sector information) was carefully considered prior to selecting potential survey participants. We also considered the structure of supply chains for major industry sectors. A typical supply chain consists of multiple firms, both upstream (i.e. suppliers) and downstream (i.e. distribution), and the ultimate consumer (Mentzer, 2001).

Invitations to participate were sent via email to a sample population of 2000 NZ-based companies, with 207 shippers replying and completing all or almost all of the survey. Twenty four respondents did not complete the ranking questions and were excluded from analysis. Therefore, our sample for this study consisted of 183 respondents, with 146 firms from three different business divisions (primary/raw material providers, manufacturers and wholesalers/retailers) and 37 freight agents. Table 1 shows the distribution of survey respondents among the various business types.

Of the 146 firms who responded, 48% were categorised as 'durable/non-food product' shippers, with 52% being classed as 'non-durable/food product' shippers. In terms of firm size, 56% of responding firms were SMEs (i.e. Small and Medium Enterprises, with 19 or fewer employees). NZ is the third smallest national market in the OECD, with a total national market which is equivalent in scale to only a medium sized urban market in the United States. In terms of its accessibility to international markets, NZ is also one of the two most geographically isolated countries in the world (Shangquin, McCann, & Oxley, 2009).

The questionnaire for this study distinguished between four different transport modes (road, rail, air and sea) and two types of destinations (domestic and international). Not surprisingly, regardless of product types

Table 1. Sample Description.

Characteristics	Descriptions		%
Position of respondents	Top managers (e.g. CEOs, managing directors)		52.9
	Staff managers (e.g. transport, logistics)		47.1
Freight transport user	Shippers and consignors	Primary sector	24.3
		Manufacturers	37.6
		Wholesalers/retailers	17.7
	Agents (forwarders, carriers, 3PLs, etc.)		20.4
Export volume	Domestic distribution only, no exports		25.1
	Exports 1–49% of produce		37.7
	Exports 50–99% of produce		30.6
	100% exports, no domestic distribution		6.6
Transport/delivery distance	Within city/region (<100 km)		11.9
	Within south or north island (<250 km)		20.2
	All over New Zealand		67.9
Integrated supply chain	Integrated supply chain		38.7
	Not integrated		61.3
Size of company	Less than 19 employees (SMEs)		56.5
	20–99 employees		26.6
	100 or more employees		16.9
Logistics facilities	No warehouse		30.6
	One warehouse		38.8
	More than one warehouse		30.6
Use of contracted carriers	1–2 contracted carriers		47.4
	3–4 contracted carriers		35.3
	5 or more contracted carriers		17.3
Length of contract with carriers	Less than 3 years of contract		22.7
	3–9 years of contract		31.2
	10 or more years of contract		46.1
Total respondents	183 (100%)		

or business types, the most widely used mode of freight transport within NZ is road transport, followed by sea, air and rail. Firms in the primary sector are the highest road transport users with the road transport share being 11% higher than for manufacturers. Fig. 1 shows that mode shares change by business groups and with the volume of exports, respectively. When the volume of exports increases, modal shares of non-road modes (especially sea and air) gradually increase.

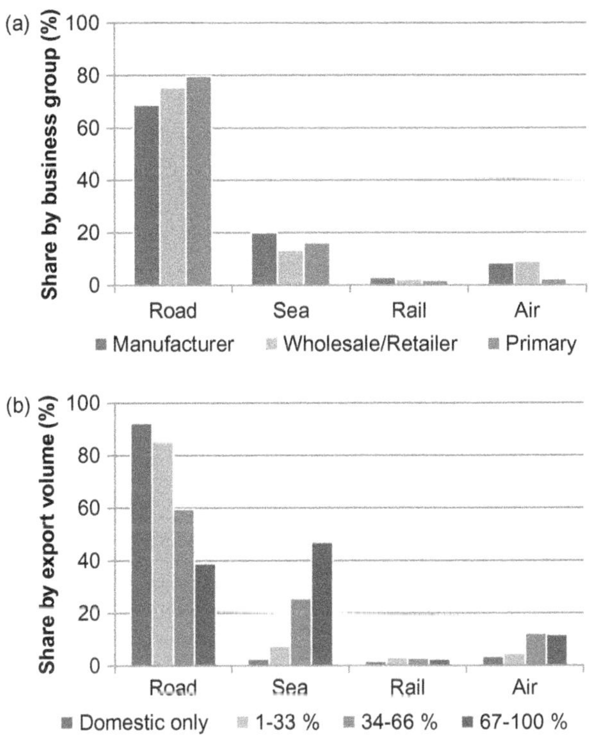

Fig. 1. (a) Mode Share by Business Division and (b) Export Volume.

NZ shippers show a strong dependence on road transport in the chemical, base-metal/glass, plastics and non-animal food product groups. However, as demonstrated in Fig. 2, nearly 40% of animal food products and 25% of wood products are moved by sea. This is understandable since NZ exports large volumes of wood, wood products and animal products.

Using rail for distributing products to domestic destinations is very limited over all product groups. However, relatively high proportions of leather/textile products and machinery/mechanical equipment products are moved by both sea and air, and food-stuffs/FMCG (Fast Moving Consumer Goods) products are moved by sea as well. The survey responses also revealed that more than half of the annual production of goods is exported.

Shippers were also asked to indicate the use of intermodal transportation to both domestic and international destinations. Table 2 indicates that the 'road with interisland ferry transport' combination is the most common

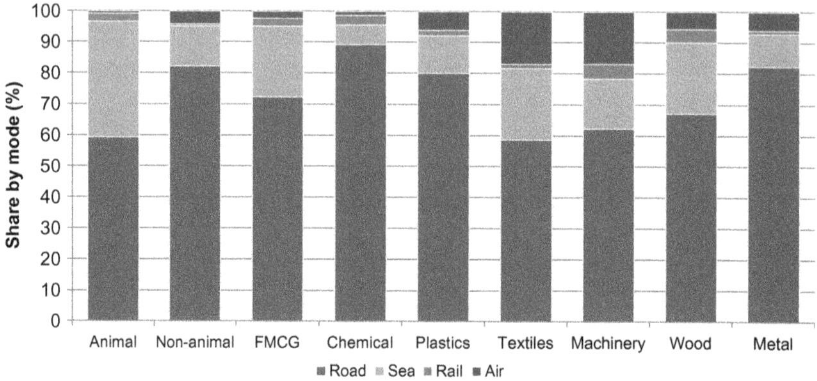

Fig. 2. Mode Shares by Product Group.

Table 2. Intermodal Use by Destination.

Intermodal Export	%	Intermodal Domestic	%
Road + Air	22	Road + Inter-island ferry	49
Road + Deep sea	55	Road + Air	11
Road + Rail + Air	1	Road + Coastal sea	13
Road + Coastal sea + Air	1	Road + Inter-island ferry + Rail	13
Road + Rail + Deep sea	12		
Road + Coastal sea + Deep sea	9	Don't know	14
	100		100

in domestic intermodal distribution. For export distribution, about 77% of shippers use two transport modes, which are either road with deep sea or air, while combinations of road and deep sea, with rail or coastal sea, account for about 21%. It is also interesting that nearly 40% of the respondents say the decisions to use intermodal transportation options are made by external professionals such as freight forwarders, freight brokers or contracted carriers, while 24% of the respondents answered that the decisions are made by them (i.e. internally).

Fig. 3a illustrates the transport distance for the domestic shipments. The survey respondents revealed that around two third of respondents are moving goods up to 250 km. The result is consistent with the findings of the National Freight Demand Study (Richard Paling Consulting, 2008), which reported that the average road trip was 144 km and two thirds of all road movements were less than 200 km.

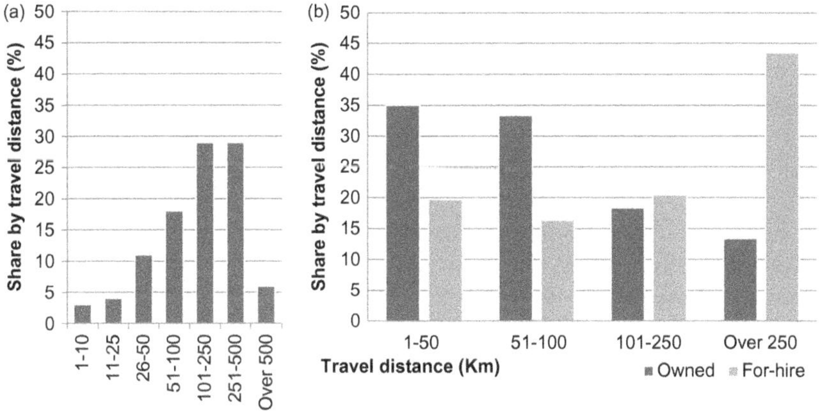

Fig. 3. (a) Distance for Domestic Shipments (km), (b) by Road Transport Options (km).

The shares of travel distance for owned-fleet and for-hire carriers are also shown in Fig. 3b. NZ shippers tend to hire carriers for longer distance deliveries.

Often the shippers perceive there to be benefits in having a long contractual agreement with contracted carriers. When medium–long-term contracts are offered by contracted carriers, shippers can get lower freight rates. Also, some contract carriers even offer dedicated equipment for a customer and tailor service to that customer. Longer term contracts also provide some security to the contract carriers to continue to provide and even increase capacity. Survey responses also revealed that 86% of NZ firms have contracts with one to four trucking companies for over 10 years.

RANK-ORDERED LOGIT ANALYSES FOR MODE CHOICE PREFERENCE

The Mode Choice Determinants[1]

In the survey, respondents were asked to rank the importance of six factors in determining freight mode choice, from 'most important' to 'least important'. These six factors were damage and loss, timeliness, cost, customer service, accessibility and suitability. A full description of the dependent variables is shown in Table 3.

In practice with any paper-based or self-administered survey, respondents often assign the same rank to two or three items, but the on-line survey tools used for this study did not allow equal or tied rankings, thereby making the data analysis more straight-forward.

The ranking data were analysed using a parametric statistical model, the rank-ordered logit (ROL) model, and the PHREG procedure in SAS 9.2. This method takes advantages of the fact that when respondents rank a series of items, they provide more information about their preferences than when they simply select the most preferred item from the list. Parameter estimates provided by these models represent the differences in the log-odds of preferring a mode choice factor compared to an omitted factor, and so provide an estimate of the size of differences within a ranked list. Table 4 provides the maximum likelihood estimates from the ROL model, where damage is the base category and is assigned a zero coefficient. The choice factors shown in Table 4 are arranged in order of increasing mean rank which are in order of decreasing importance.

The Wald Chi-square statistic for the test of the global null hypothesis (H0: there are no choice differences in shippers' preference) for the overall model is 415.79 (with 5 degree of freedom), yielding a p-value much less than 0.0001, which means that NZ shippers, in general, have statistically significantly different preferences among the six mode choice factors.

Also, all of the coefficients were significantly different from zero with p-values less than 0.0001. The overall rank order results are largely consistent with the mean ranks, except for last two factors (suitability and damage). All of the factors contrast with the reference category, damage. On average, these estimates indicate that NZ shippers considers rank timeliness, cost, customer service and accessibility as significantly more important than damage

Table 3. Description of Mode Choice Factors.

Choice Factors	Code	Descriptions
Damage and loss	damg	Amount of loss and damage, processing of loss and damage claim
Timeliness[a]	Time	Transport time, transit time and on-time reliability
Transport cost	Cost	Transport cost
Customer service	c_serv	Firm contact, after sale service and personal service
Accessibility	acces	Availability of mode or equipment at origin or destination
Suitability	Suita	Suitability for shipment size, suitability for commodity to be carried

[a]The revealed preference (RP) survey was conducted in 2011 prior to performing a stated preference (SP) survey. The surveys are part of research to develop a mode choice model for NZ freight transport. Despite its ambiguity, the term *Timeliness* was used in the RP survey to maintain consistency with the previous freight transport studies in NZ. However, two separate terms *Transport time* and *On-time reliability* were used in the later SP survey, to avoid ambiguity.

Table 4. Maximum Likelihood Estimates from ROL Model: Mode Choice.

Mode Choice Factors	Code	Estimate of Coefficient (β)	Standard Error	Exponent of Coefficient (e^{β})	Mean Rank	Model Statistics
Timeliness	time	2.5328***	0.1577	12.58	1.694	
Cost	cost	1.7987***	0.1461	6.042	2.372	
Customer Service	c_serv	0.6411***	0.1329	1.898	3.765	Wald χ^2: 415.7886
Accessibility	acces	0.6301***	0.1348	1.878	3.803	DF:5
Suitability	suita	−0.4621***	0.1412	0.630	4.295	$p < 0.0001$
Damage	damg	0.0000	0.0000	1.000	5.071	

***$p < 0.0001$.

when considering mode choice, but consider suitability lower than damage. The estimated factor coefficients can be interpreted as differences in log-odds. Thus, the exponent of the coefficient for timeliness ($e^{(2.53275)} = 12.59$) indicates that the odds of preferring timeliness are 12.59 times the odds of preferring the damage choice factor. Similarly, the odds of preferring cost, customer service, accessibility and suitability are 6.04, 1.90, 1.88, 0.63 times the odds of preferring the damage choice factor, respectively.

The Mode Choice Determinants between Groups

The preceding model assumed that every respondent included in this study had the same probability distribution of mode choice preferences and that the observed differences in their rankings were due only to random variation. We now extend the model to capture the heterogeneity in mode choice preference across the characteristics of the respondent's firm.

The freight transport task in NZ is conditioned by many factors including the geography, topography, climate, and the pattern of natural resource distribution, as well as the resultant patterns of historical settlement and varying regional economic growth (Cavana, Harrison, Heffernan, & Kissling, 1997).

NZ is a country greatly dependent on international trade, particularly in agricultural and horticultural products, fishing, forestry and mining. Geographically, NZ encompasses two main islands, the North and South Islands. The two main islands are separated by a 30 kilometre wide channel, the Cook Strait, and the road and rail networks are connected by ferry services. NZ is remote from major international markets; the trade-route between Australasia and the west coast of the United States is about 8,000 miles and is one of the longest in the world (Byrne et al., 1994). Despite this, many industries are oriented towards exports, because of the small domestic market. In 2011, NZ had an estimated population of approximately 4.4 million. About 77% of the population lives in the North Island and 32% of the country's population lives in the Auckland metropolitan area. The low and dispersed population density (16.5 people/km^2), combined with NZ's mountainous terrain and disconnected islands, makes transport systems less efficient and difficult to achieve *the economies of density* enjoyed by other countries.

In the modern supply chain environment, including the JIT (just-in-time) concept of lean production, the firms with integrated supply chains benefit from cost reductions and increased levels of reliability through reduced delivery lead times and improved inventory turnovers, supplier reliability and maintainability. Integrated supply chains also give firms more competition strategy options by gaining bargaining power, for example, negotiating better transport rates with carriers or 3PLs (Basnet Corner, Wisner, & Tan, 1999). As of 2009, 97% of firms in NZ were SMEs and the proportions have remained relatively constant over time. The small size of NZ firms makes it very difficult to include all components of the supply chain, and Boehme, Childerhouse, and Corner (2007) found that most NZ companies face high uncertainty, with weakly integrated and inefficient supply chains. Due to the unique business environment, NZ firms are under pressure to lower logistics costs. One case study (Ministry of Transport, 2010) shows that NZ firms spend 8.4% of annual turnover on total logistics cost and its major components are the direct transport cost (about 60% for both international and domestic transport).

To address these issues, we included the attributes of the respondent's firm in the model. The following eight characteristics were included:

- Number of employees
- Mode choice and modal shift decision-maker of the firm
- Supply chain management system
- Operations of logistics facilities
- Length of contract with transport service providers
- Volume of exports
- Average distance of domestic deliveries
- Lead time policy

To capture the effects of firm characteristics on their preferences for mode choice factors, the products of each of the eight choice factors and each of these characteristic variables (i.e. independent variables) are included in the ROL model. Table 5 presents the description of independent variables and the coding.

Apparently, NZ shippers' preferences for factors in determining mode choice are significantly related to the firms' 'logistics' characteristics, namely the length of contract with transport service providers, supply chain integration, logistics facilities and export volume (Table 6). ROL models related to four firms' 'logistics' characteristics (i.e. the size of firm, the status of the decision-maker, lead time policy and the average distance of domestic deliveries) do not provide any statistically significant parameters, and therefore the results are not presented in Table 6.

The results (Table 6) reveal that firms that have a shorter length of contract with transport service providers rank the mode choice factors of cost, customer service, accessibility and suitability significantly higher than the mode choice factor of damage, when compared to those firms which have longer contracts with transport service providers. Transport cost ($\beta = 1.231$, $p < 0.01$) is considered as the most important factor, followed by customer service ($\beta = 1.090$, $p < 0.01$), suitability ($\beta = 1.068$, $p < 0.01$) and modal accessibility ($\beta = 1.040$, $p < 0.01$). The exponent of the coefficient (e^{β}) of the

Table 5. Description of Firm-Related Variables.

Characteristic	Descriptions and Coding
Size of company	1 = A company has less than 19 employees (SMEs: small and medium enterprises) 0 = Over 20 employees
Decision-maker	1 = 'Top' managers (e.g. CEOs, managing director) 0 = Other staff
Integrated Supply Chain	1 = A company with integrated supply chain (vertical, horizontal) 0 = Not integrated
Logistics Facilities	1 = A company does not have logistics facilities 0 = A company has more than one logistics facility (i.e. warehouse, trans-shipments facility or distribution centre)
Length of Contract with Carriers	1 = A company has less than 3 years of contract with transport service providers or contracted carriers 0 = Contract length of 3 or more years
Export Volume	1 = A company exports less than 50% of its production 0 = Exports over 50% of its production
Transport Distance	1 = The average distance for the delivery of freight is less than 250 km or within Island 0 = All over New Zealand (over 250 km)
Lead time	1 = Order-to-shipping lead time policy of not exceeding 1 month 0 = Lead time exceeds 1 month

Table 6. Results of ROL Model for Mode Choice Factors.

Firm-related Variables	Choice Factors[¤]	Estimate of Coefficient (β)	Standard Error	Exponent of Coefficient (e^{β})	%Δ[a]	Model Statistics
Length of contract with carriers (<3years)	time	0.655	0.454	1.926		
	cost	1.231***	0.437	3.428	23.0	Wald χ^2:
	c_serv	1.090***	0.396	2.977	14.0	324.792,
	acces	1.040***	0.396	2.830	11.9	DF:10,
	suita	1.068***	0.390	2.911	13.0	$p<0.0001$
Supply chain integration (integrated)	time	−0.650*	0.349	0.522	0.24	Wald χ^2:
	cost	−0.671**	0.321	0.321	0.23	352.402,
	c_serv	−0.342	0.295	0.710		DF:10,
	acces	−0.486	0.300	0.615		$p<0.0001$
	suita	−0.511	0.319	0.600		
Logistics facilities (no facility)	time	0.359	0.394	1.432		Wald 2:
	cost	−0.088	0.358	0.915		342.784,
	c_serv	−0.273	0.327	0.761		DF:10,
	acces	0.660*	0.340	1.935	4.12	$p<0.0001$
	suita	0.787**	0.342	2.198	5.71	
Export volume (less 50%/year)	time	0.004	0.327	1.004		Wald χ^2:
	cost	−0.261	0.306	0.770		422.445,
	c_serv	0.403	0.273	1.497		DF:10,
	acces	−0.612**	0.284	0.542	0.27	$p<0.0001$
	suita	0.125	0.296	1.133		

[¤]Damage is the base factor ($\beta=0.0000$), ***$p<0.01$, **$p<0.05$, *$p<0.10$.
[a]$\%\Delta = 100(e^{\beta}-1)$ %Δ is calculated only for statistically significant β with minimum $p<0.10$.

cost factor (3.428) indicates that the odds of preferring the cost factor for the group are 3.428 times the odds of preferring the damage factor. Between the group of firms with short-term contracts and the group of firms with long-term contracts, there are significant differences in the importance of transport cost, customer service, modal accessibility and suitability factors.

For the coefficients of the mode choice factors indicated as statistically significant, the %Δ values, which are the percentage changes in the odds of preferring a choice factor for a percentage increase in coefficient, were calculated for each explanatory group. The exponent values for these differences (%Δ) are 23.0 for cost and 14.0 for accessibility, which mean that the odds of preferring the cost factor over damage among the firms with short-term contracts, are 23 times the odds for the firms with long-term contracts, and the odds of preferring the accessibility factor over the damage factor, among the firms with short-term contract, are 14 times the odds for the firms with long-term contracts.

NZ firms with an integrated logistics and supply chain are more likely to rank the damage factor ahead of timeliness ($\beta = -0.650$, $p < 0.10$) and cost ($\beta = -0.671$, $p < 0.05$) in determining transport mode when compared to the firms without such an integrated system. The estimates of the exponent values of the coefficient of cost and time factors are 0.522 and 0.321, which mean that the odds of preferring the timeliness factor are 0.522 times of the odds of preferring the damage factor. Similarly, the odds of preferring the cost factor are 0.321 times the odds of preferring the damage factor. The $\%\Delta$ values indicate that the odds of firms with an integrated logistics and supply chain preferring time to damage and cost to damage are respectively about 0.24 times and 0.23 times the odds for firms without an integrated logistics system.

Table 6 shows that the exponent of the accessibility factor coefficient ($\beta = 0.660$, $p < 0.10$) and suitability factor coefficient ($\beta = 0.787$, $p < 0.05$) are 1.935 and 2.198 respectively for firms not having logistics facilities. The odds of preferring the accessibility and suitability factors to the damage factor for the group of firms not having logistics facilities are 4.12 and 5.71 times the odds for the firms operating logistics facilities, such as warehouses and transhipment facilities. This may happen because many NZ firms not having any logistics facilities are in the retail portion of the value chain or, in particular, are small and medium size manufacturers that have adopted traditional production systems in order to reduce inventory and increase the level of customization. Thus, when making a freight mode choice decision, they value accessibility and suitability higher than damage, to compensate for not having logistics facilities.

In intermodal shipping environments, modal accessibility (especially equipment for container handling) is vital when choosing a freight transport mode, particularly for firms with high export volume. The results in Table 6, to some extent, confirm the statement above as NZ firms with low export volume are more likely to rank the damage factor ahead of the accessibility factor ($\beta = -0.612$, $p < 0.05$). This is consistent with many NZ firms being in the higher export volume rank the accessibility factor ahead of the damage factor.

RANK-ORDERED LOGIT ANALYSES FOR MODE SHIFT

The Mode Shift Drivers and Constraints

In the RP survey, shippers were asked to rank factors that constrain the shift in mode choice and discourage the shippers from using rail or coastal shipping instead of road transport to move their goods. The respondents were asked to consider two options (shifting from road to rail, and shifting from road to coastal shipping), and to rank the seven constraining factors (Table 7) from one ('most constraining') to seven ('least constraining').

Each of the mode shift records included four types of data: (1) a unique identification number for the respondents; (2) the rank assigned by the respondent to the particular modal shift constraint; (3) a set of six variables corresponding to the seven different modal shift constraints (the 'base' or 'reference' factor, transport time, is omitted); (4) eight firm-related characteristics (see Table 5).

First, a ROL model was estimated, allowing for differences between the seven modal shift constraints but no differences between respondents. That is, it was assumed that every respondent in the study had the same probability distributions for the modal shift constraints and that the observed differences in the rankings were due only to random variation.

Table 8 shows the maximum likelihood estimates of the coefficients for each constraint arranged in order of increasing mean rank, for the two mode change options. It should be noted that transport time is the 'base' (or 'reference') constraint, and is assigned a coefficient of zero, with the coefficients for the other constraints being either positive or negative. The coefficients, along with the standard errors of estimation, indicate whether the constraint has a statistically significant effect.

The overall statistical significance of the model can be assessed using the Wald chi-square statistic, and it was found that this was 109.56 for shifting from road to rail and 158.91 for shifting from road to coastal shipping, with 6 degrees of freedom in both cases. The null hypothesis is that all the explanatory variables have the same ranking or importance, but this hypothesis can be rejected at the 0.01% significance level (or 99.99% confidence level), given the very large values of the Wald chi-square statistics.

Table 7. Description of Mode Shift Factors.

Factors	Variable Code	Descriptions
Transport time	time	Total transport time and on-time reliability
Accessibility	acces	Ease of reaching transport services
Frequency	freq	Frequency of service
Transport cost	cost	Total transport cost
Load size	load	Minimum load size requirement
Modal transfer	transf	Ease of road/rail, rail/road, road/sea & sea/road transfer
Door-to-door	dtod	Door-to-door service availability

Table 8. ROL Model: Coefficients of Mode-Related Variables.

	Mode Shift Factors	Estimate of Coefficient (β)	Standard Error	Exponent of Coefficient (e^{β})	Mean Rank	Model Statistics
Road to rail	Transport time	0.000	0.000	1.000	2.901	Wald χ^2: 109.56, DF:6, $p < 0.0001$
	Accessibility	−0.250*	0.133	0.779	3.461	
	Loading size	−0.462***	0.142	0.630	3.532	
	Door-to-door	−0.901***	0.150	0.406	3.915	
	Transport cost	−0.972***	0.142	0.379	4.596	
	Modal transfer	−1.196***	0.148	0.303	4.766	
	Frequency	−1.052***	0.142	0.349	4.830	
Road to coastal shipping	Loading size	−0.056	0.153	0.946	2.944	Wald χ^2: 158.91, DF:6, $p < 0.0001$
	Accessibility	0.139	0.143	1.149	2.968	
	Transport time	0.000	0.000	1.000	3.056	
	Frequency	−0.803***	0.148	0.448	4.544	
	Transport cost	−0.940***	0.154	0.391	4.600	
	Door-to-door	−1.382***	0.167	0.251	4.824	
	Modal transfer	−1.167***	0.156	0.311	5.064	

***$p < 0.01$, **$p < 0.05$, *$p < 0.10$.

There is very strong evidence that NZ freight shippers have statistically significantly different rankings for the seven modal shift constraints.

Table 8 also shows the mean ranks across respondents for each explanatory variable. While the average mean rank is 4, as expected, it can be seen that the mean rank orderings are not the same for the two mode shift options (i.e. the relative importance of a constraint factor depends upon which mode shift is being considered). The exponent of the coefficient for each constraint factor can be interpreted as the odds of the constraint factor being ranked lower (i.e. less important) or higher (i.e. more important) than the 'base' (or reference) constraint factor, transport time.

On average, NZ shippers rank transport time as the greatest constraint upon freight modal shift from road to rail, with modal transfer and frequency being ranked much lower. These results are largely consistent with the mean ranks shown in Table 8. However, while not statistically significant, NZ shippers rank accessibility 1.149 times as important as transport time, as a constraint on modal shift from road to coastal shipping, and apparently do not feel constrained by transport cost and ease of modal transfer between road and coastal shipping.

The Mode Shift Behaviour between Groups

The next stage was to identify the effects of firm characteristics, in addition to the effects of the seven above-mentioned factors relating to the transport modes. Eight extra 'dummy' variables associated with the firms' characteristics (see Table 5) were included in the ROL model. The maximum likelihood estimates, with transport time as the 'base' category, are shown in Table 9. Note that only the results of models with statistically significant coefficients are presented in the table. Therefore, the results of interaction with two of the firm' characteristics, namely length of contract and supply chain integration, are not presented.

The overall statistical significance of the model can be assessed using the Wald chi-square statistic, and the value was found to be in a range between 57.147–132.000 ($p < 0.0001$) for shifting from road to rail and 74.831–172.481 ($p < 0.0001$) for shifting from road to coastal shipping, with 12 degrees of freedom in both cases. The null hypothesis is that all the explanatory variables have the same ranking or importance, but this hypothesis can be rejected at the 0.01% significance level (or 99.99% confidence level), given the very large values of the Wald chi-square statistics.

Table 9. ROL Model: Coefficients of Mode and Firm-Related Variables.

Firm-Related Variable	Mode Shift Factors[a]	Coastal Shipping			Rail		
		Estimate of Coefficient (β)	Standard Errors	$\exp^{(\beta)b}$	Estimate of Coefficient (β)	Standard Error	$\exp^{(\beta)b}$
Modal shift decision maker (top manager)	acces	−0.069	0.396		0.291	0.362	
	freq	−0.491	0.407		0.229	0.393	
	cost	−0.045	0.425		0.073	0.390	
	load	−0.270	0.420		0.846**	0.390	2.33
	transf	0.712*	0.429	2.04	0.820**	0.403	2.27
	dtod	0.854*	0.464	2.35	1.186***	0.416	3.27
Model statistics		*Wald* χ^2*: 74.831, DF:12,* $p < 0.0001$			*Wald* χ^2*: 57.147, DF:12,* $p < 0.0001$		
Export volume (less 50%/year)	acces	−0.125	0.294		0.398	0.275	
	freq	0.006	0.307		0.581*	0.297	1.79
	cost	−0.947***	0.312	0.39	0.256	0.296	
	load	0.284	0.314		0.928***	0.295	2.53
	transf	−0.044	0.323		0.220	0.308	
	dtod	0.294	0.350		1.292***	0.319	3.64
Model statistics		*Wald* χ^2*: 172.481, DF:12,* $p < 0.0001$			*Wald* χ^2*: 132.000, DF:12,* $p < 0.0001$		
Transport distance (<250 km)	acces	0.012	0.329		0.314	0.295	
	freq	0.110	0.337		−0.040	0.319	
	cost	0.098	0.353		0.619**	0.310	1.86
	load	−0.212	0.353		−0.031	0.312	
	transf	−0.251	0.365		0.221	0.326	
	dtod	−0.837**	0.406	0.43	−0.307	0.336	
Model statistics		*Wald* χ^2*: 146.139, DF:12,* $p < 0.0001$			*Wald* χ^2*: 111.165, DF:12,* $p < 0.0001$		
Logistics facilities (no facility)	acces	−0.607*	0.366	0.28	0.111	0.324	
	freq	−0.476	0.386		0.283	0.345	
	cost	−0.225	0.389		0.078	0.346	
	load	−0.936**	0.391	0.39	−0.405	0.353	
	transf	−0.913**	0.417	0.40	−0.278	0.373	
	dtod	−1.402***	0.462	0.25	−0.143	0.375	

Table 9. *(Continued)*

Firm-Related Variable	Mode Shift Factors[a]	Coastal Shipping			Rail		
		Estimate of Coefficient (β)	Standard Errors	$\exp^{(\beta)b}$	Estimate of Coefficient (β)	Standard Error	$\exp^{(\beta)b}$
Model statistics		*Wald* χ^2*: 135.554, DF:12,* $p<0.0001$			*Wald* χ^2*: 91.438, DF:12,* $p<0.0001$		
Lead time (<1month)	acces	0.244	0.350		0.491	0.327	
	freq	−0.094	0.363		0.447	0.359	
	cost	−0.665*	0.376	0.51	−0.009	0.347	
	load	0.225	0.369		1.216***	0.350	3.37
	transf	−0.324	0.380		0.306	0.369	
	dtod	−0.286	0.404		0.688*	0.371	1.99
Model statistics		*Wald* χ^2*: 138.113, DF:12,* $p<0.0001$			*Wald* χ^2*: 110.154, DF:12,* $p<0.0001$		
Size of firm (SMEs)	acces	−0.292	0.314		−0.112	0.290	
	freq	−0.029	0.327		0.235	0.312	
	cost	−0.047	0.344		−0.030	0.311	
	load	−0.876***	0.338	0.41	−0.535*	0.311	0.58
	transf	−0.046	0.348		−0.027	0.327	
	dtod	0.540	0.375		0.025	0.325	
Model statistics		*Wald* χ^2*: 140.510, DF:12,* $p<0.0001$			*Wald* χ^2*: 100.915, DF:12,* $p<0.0001$		

***$p<0.01$, **$p<0.05$, *$p<0.10$.
[a]Transport time is the base factor ($\beta=0.0000$).
[b]$\exp^{(\beta)}$ = Exponent of coefficient (e^{β}).

There is very strong evidence that NZ freight shippers have statistically different rankings for the seven modal shift constraints.

In general, the higher the position of the person who makes transport mode choice decisions in a firm, the greater the importance attached to the modal transferability and door-to-door capability factors in comparison to the time factor, for both rail and coastal shipping. NZ shippers' ranking of factors, when determining whether to shift from road to coastal shipping, is strongly related to the firms' logistics characteristics, such as whether or not they operate warehouses, transhipment facilities and other logistics facilities.

It can be seen from Table 9 that the following containing factors are ranked significantly higher or lower than transport time (i.e. the base factor) when considering shifting to coastal shipping from road transport:

- Modal transferability ($\beta = 0.712$, $p < 0.10$) and door-to-door capability ($\beta = 0.854$, $p < 0.10$) are significant factors if the decision-maker is a high-level manager. The odds of the top managers group preferring modal transferability and door-to-door are 2.35 and 2.04 times the odds of preferring transport time.
- Cost ($\beta = -0.947$, $p < 0.01$) is ranked significantly lower than transport time by firms which export less than 50% of their yearly production. The odds of preferring the transport cost to transport time are 0.39. In NZ, there is not much variation in the domestic transport costs because of high competition between local freight companies. However, there is a lot of variation in the transportation time between coastal shipping and road transport. Thus, companies which export a small portion of their production consider time as a more constraining factor to shifting to coastal shipping, compared to the cost factor. This fact is supported by the model estimates.
- Door-to-door service ($\beta = -0.83$, $p < 0.05$) is ranked significantly lower than transport time by firms that transport short distances (less than 250 kilometres).
- Loading size ($\beta = -0.936$, $p < 0.05$), transferability ($\beta = -0.913$, $p < 0.05$), and door-to-door service ($\beta = -1.402$, $p < 0.01$) are ranked significantly lower than transport time by firms without any logistics facilities. The odds of preferring those factors to transport time are 0.39, 0.40 and 0.25 in turn, implying that transport time is still the biggest constraint to shifting to coastal shipping. In general, road transport offers much more flexibility to firms that have no logistics facilities to distribute their goods, making it harder for coastal shipping to compete with road transport.

- For firms with a short lead time policy, transport cost ($\beta = -0.665$, $p < 0.10$) is ranked significantly lower than transport time, with odds of 0.51 preferring cost to time.
- For small-medium enterprises, the load size factor ($\beta = -0.876$, $p < 0.01$) is ranked lower than transport time, with odds of 0.41 preferring load size to time.

The results show that firms with low export volume, short distance transport, no logistics facilities, short lead time policy, and a small number of employees rank travel time as the most significant factor which constrains their decision to shift to coastal shipping from road transport. Time is only ranked as a less constraining factor when the mode choice decisions are made by the top level managers. In that case, other factors, namely transferability and door-to-door capability, are considered as more constraining.

It can also be seen from Table 9 that when firms consider shifting to rail transport from road transport, the following constraining factors are ranked significantly higher or lower than transport time:

Door-to-door capability ($\beta = 1.186$, $p < 0.01$), loading size requirement ($\beta = 0.846$, $p < 0.05$) and transferability ($\beta = 0.820$, $p < 0.01$) are considered as more significant constraining factors, in comparison to travel time, to shifting to rail transport if the decision-maker is a high-level manager.

Door-to-door service ($\beta = 1.292$, $p < 0.01$), load size ($\beta = 0.928$, $p < 0.01$) and frequency ($\beta = 0.581$, $p < 0.10$) are ranked higher than transport time if the firms export 50% or more of their production. The odds of preferring door-to-door service, loading size and frequency are 3.64, 2.53 and 1.79 times the odds of preferring transport time. These odds are considerably higher than for other factors.

Transport cost ($\beta = 0.619$, $p < 0.05$) is ranked higher than transport time for short hauling shippers. In general, rail transportation (per kilometre) is cheaper than road transportation. However, it may be more beneficial for companies that only need to transport their goods within a 250 kilometre radius to use road, as many road carriers offer lower fixed-fare rates within a city or region.

None of the mode shift factors has a statistically significant interaction with logistics facilities.

For firms with short lead time policy (less than 1 month), minimum loading size requirements ($\beta = 1.216$, $p < 0.01$) and door-to-door service ($\beta = 0.688$, $p < 0.10$) are considered as more constraining factors than travel time when considering shifting from road to rail. This may happen because many NZ businesses operate under unfavourable economic and geographical conditions, including a small domestic market and low population density.

SMEs consider the loading size requirement ($\beta = -0.535$, $p < 0.10$) as a less constraining factor than travel time to shifting to rail transport, with odds of 0.58 of preferring load size to travel time.

CONCLUSIONS

This chapter first analyses how NZ shippers choose freight transport modes. Our findings show that timeliness and transport cost were the two most important factors for shippers' mode choices, while customer service, accessibility, suitability and loss and damage were distinctly less important. There is increased emphasis on timeliness and cost as mode choice factors, compared with previous studies, which found the most important factors were reliability and product care.

Shippers' factor preferences vary with their firms' supply chain integration, use of logistics facilities and length of contract with transport service providers. For example, firms with an integrated supply chain were more likely to rank timeliness and cost ahead of damage than firms which do not have such a production management system; firms with a shorter length of contract with carriers or transport service providers are less likely to rank cost, customer service, accessibility and suitability factors higher than damage.

In the second part, this study has identified what freight shippers in NZ perceive as constraints on modal shift from road to rail or coastal shipping as transport modes for domestic shipments. Seven perceived constraints have been analysed. Our findings show that, on average, NZ shippers rank transport time as the highest modal shift constraint for shifting from road to rail, followed by accessibility, with modal transfer to/from road and service frequency ranking much lower. With shifting to coastal shipping, however, shippers rank accessibility higher than transport time as a constraint. That is, the rank ordering of constraint factors depends upon whether one is considering shifting to rail or coastal shipping.

It has also been found that NZ shippers' rank ordering of the constraint factors, when considering a shift to coastal shipping, is strongly related to the firms' logistics characteristics, such as whether they operate warehouses, transhipment facilities and other logistics facilities. When considering shifting to rail, however, the firm's lead time policy is the firm-related characteristic with the greatest influence. The outcomes further indicate that freight policies to support mode choice should not be based only on reducing the cost (e.g. subsidy for rail transport and sea transport) but also on improving transport time and accessibility.

The methodology applied to measure and analyse freight transport mode choice factors involves assumptions about the quality of service characteristics of NZ freight transport modes and infrastructures. Using a parametric modelling technique to rank-order mode choice preferences and mode shift constraints, this study lends support to some qualitative findings in previous NZ freight studies. The results of this study provide quantitative measures of the intensity of preference for the various choice factors. For instance, it has been shown that timeliness, in addition to being the most important choice factor, has a probability of 52.4% of being ranked as the most important factor, with the odds of the mode choice being based on timeliness being 12.59 times the odds of the mode choice being based on the damage factor. Such quantitative information is very useful in identifying how shippers make trade-offs between conflicting objectives and factors when choosing between freight transport modes.

It should be noted that the analysis described above is based on RP survey data, and the ROL models cannot easily be used for predicting mode choice. The RP survey and ROL models have been used to identify the importance of various factors in determining mode choice in general and to assess the barriers to mode shift, as well as providing some insight into issues that should be considered in freight policy development. The results of such an analysis provide a good basis for the design of SP surveys, to obtain data for the estimation of models for predicting freight mode choice.

As noted above, shippers in NZ generally do not consider environmental impacts when choosing the mode of transport. This might well change if transport policy in NZ had a clear and consistent emphasis on reducing greenhouse gas emissions, as is the case in the European Union with its Marco Polo programme.

Finally, the results of this study indicate that some firms are more inclined to shift from road to rail than to shipping, while other firms are more inclined to shift from road to shipping than to rail, depending upon the characteristics of the firms. This suggests that programmes to encourage the use of greener transport modes would be more effective if they were designed to encourage firms to shift to the mode to which they are already more inclined to shift. In addition, the cost of providing incentives would be less if trying to achieve a shift to a mode to which they are less inclined to shift, giving greater economic efficiency.

NOTE

1. The factors provided in the questions to consider in selecting and shifting freight transport mode were selected based on (1) international literature, (2) NZ literature and (3) comments from industry experts consulted during pilot survey, and reflect current freight operation in NZ.

REFERENCES

Ahn, J., Lee, J., Lee, J. D., & Kim, T. Y. (2006). An analysis of consumer preferences among wireless LAN and mobile internet services. *ETRI Journal*, *28*, 205–215.

Alava, M. H., Brazier, J., Rowen, D., & Tsuchiya, A. (2013). Common scale valuations across different preference-based measures estimation using rank data. *Medical Decision Making*, *33*(6), 839–852.

Allison, P. D., & Christakis, N. (1994). Logit models for sets of ranked items. *Sociological Methodology*, *24*, 199–228.

Basnet, C., Corner, J., Wisner, J., & Tan, K.-C. (1999). A survey of supply chain management practice in New Zealand. In *Proceedings of the 34th Annual Conference of the Operational Research Society of New Zealand*, Hamilton, NZ, December, pp. 309–318.

Beggs, S., Cardell, S., & Hausman, J. (1981). Assessing the potential demand for electric cars. *Journal of Econometrics*, *16*, 1–19.

Boehme, T., Childerhouse, P., & Corner, J. (2007). Supply chain integration in New Zealand: Fact or fantasy. Paper presented at the 11th Annual Waikato Management School Student Research Conference, Hamilton, New Zealand.

Byrne, P. M., & Golden, R. (1994). Transporting goods down under. *Transportation and Distribution*, *35*(12), 53–59.

Calfee, J., Winston, C., & Stempski, R. (2001). Econometric issues in estimating consumer preferences from stated preference data: A case study of the value of auto mobile travel time. *The Review of Economics and Statistics*, *83*, 699–707.

Cavana, R. Y., Harrison, I. G., Heffernan, F. E. B., & Kissling, C. C. (1997). *Freight transport industry in New Zealand, graduate school of business and government management.* Working Paper No. 2/97. Victoria University of Wellington, Wellington.

Chapman, R. G., & Staelin, R. (1982). Exploiting rank ordered choice set data within the stochastic utility model. *Journal of Marketing Research*, *14*, 288–301.

Dagsvik, J. K, & Liu, G. (2006). *A framework for analyzing rank ordered panel data with application to automobile demand.* Discussion Papers 480, Research Department of Statistics Norway.

Dagsvik, J. K., & Liu, G. (2009). A framework for analysing rank-ordered data with application to automobile demand. *Transportation Research Part A*, *43*(1), 1–12.

Drewes, T., & Michael, C. (2006). How do students choose a university? An analysis of applications to universities in Ontario, Canada. *Research in Higher Education*, *47*, 781–800.

European Commission. (2009). Lightening the load, Marco Polo Leads the Way, p. 2.

European Reference Center for Intermodal Freight Transport. (2010). *Rail freight funding will support intermodal shift.* Hamburg, Germany.

Evers, P., Harper, D., & Needham, P. (1996). The determinants of shipper perceptions of modes. *Transportation Journal, 36*, 13–25.

Fridstrom, L., & Elvik, R. (1997). The barely revealed preference behind road investment priorities. *Public Choice, 92*, 145–168.

Hausmann, J. A., & Ruud, P. A. (1987). Specifying and testing econometric models from rank-ordered data. *Journal of Econometrics, 34*, 83–104.

Hunt, J. D. (2001). Stated preference analysis of sensitivities to elements of transportation and urban form. In *Transportation research record: Journal of the transportation research board* (No. 1780, pp. 76–86). Washington, DC: TRB, National Research Council.

Hsieh, C. (2005). Age and relative importance of major life domains. *Journal of Aging Studies, 19*, 503–512.

Kockelman, K. M., Podgorski, K., Bina, M., & Gadda, S. (2012). Public perceptions of pricing existing roads and other transportation policies: The Texas perspective. *Journal of the Transportation Research Forum, 48*(3), 19–38.

Kumar, S., & Kant, S. (2007). Exploded logit modeling of stakeholders' preferences for multiple forest values. *Forest Policy and Economics, 9*(5), 516–526.

Lemanske, R. F., Jr., Mauger, D. T., Sorkness, C. A., Jackson, D. J., Boehmer, S. J., Martinez, F. D., & Taussig, L. M. (2010). Step-up therapy for children with uncontrolled asthma receiving inhaled corticosteroids. *New England Journal of Medicine, 362*(11), 975–985.

Luce, R. D. (1959). *Individual choice behavior: A theoretical analysis*. New York, NY: Wiley.

Maddala, G. S. (1983). *Limited dependent and qualitative variables in econometrics*. Cambridge, MA: Cambridge University Press.

Mark, D. R., Lusk, J. L., & Scott, D. M. (2004). Recruiting agricultural economics graduate students: Student demand for program attributes. *American Journal of Agricultural Economics, 86*, 175–184.

Mcfadden, D. (1974). Conditional logit analysis of qualitative choice behavior. In P. Zarembka (Ed.), *Frontiers in econometrics* (pp. 105–142). New York, NY: Academic Press.

Mentzer, J. T. (2001). *Supply chain management*. Thousands Oak, CA: Sage.

Ministry for the Environment. (2009). New Zealand's 2020 emissions target, Wellington, New Zealand.

Ministry of Transport. (2010). Understanding transport costs and charge, phase two – Transport costs in freight logistics, Wellington, New Zealand.

Odeck, J. (1996). Ranking of regional road investment in Norway: Does socioeconomic analysis matter? *Transportation, 23*, 123–140.

Perakis, A. N., & Denisis, A. (2008). A survey of short sea shipping and its prospects in the USA. *Maritime Policy and Management, 35*(6), 591–614.

Punj, G. N., & Staelin, R. (1978). The choice process for graduate business schools. *Journal of Marketing Research, 15*, 588–598.

Richard Paling Consulting. (2008). *National freight demand study*. Wellington, New Zealand: Ministry of Transport.

Rockpoint. (2009). *Coastal shipping and modal freight choice*. Wellington, New Zealand: New Zealand Transport Agency.

Shangquin, H., McCann, P., & Oxley, L. (2009). *Innovation in New Zealand: Issues of firm size, local market size and economic geography*. Working Paper No. 04. Department of Economics, College of Business Economics, University of Canterbury, New Zealand.

Srinivasan, S., Bhat, C. R., & Holguin-Veras, J. (2006). Empirical analysis of the impact of security perception on intercity mode choice: A panel rank-ordered mixed logit model. *Transportation Research Record: Journal of the Transportation Research Board, 1942*(1), 9–15.

CHAPTER 8

BUSINESS MODELS FOR SHIPPER-OPERATED INTERMODAL TRANSPORT SOLUTIONS

Jonas Flodén and Edith Sorkina

ABSTRACT

Purpose – *One of the main difficulties in developing new intermodal transport solutions is finding the right business model. Though business models have received limited attention in the existing intermodal research, several authors have pointed out the importance of business models in the intermodal context.*

Existing intermodal literature discusses several types of different business. The current study takes an in-depth look at The Own-Account Model, *its strengths and weaknesses through two empirical examples.*

The chapter investigates this model by analysing the business model in practice, for example actors, roles and responsibilities, risk distribution and contracts.

Methodology – *Research is conducted using a qualitative approach with two case studies. Osterwalder's (2004) framework for business models is applied to analyse the empirical cases.*

Sustainable Logistics
Transport and Sustainability, Volume 6, 193–215

ISSN: 2044-9941/doi:10.1108/S2044-994120140000006008

Findings – *The roles and responsibilities of the actors are described. For the parties to be willing to "invest into" the new intermodal solution, long-term contracts are required. The shipper controls the channel, but has to rely heavily on the transport operators for their expertise and resources.*

The analysis has found that the model can be used to avoid many of the difficulties in setting up a new intermodal solution, such as ensuring the base volume or having the overall control over the intermodal chain.

Research implications – *Better understanding of this type of business model allows authorities to better support the development of intermodal transport through policy measures. The results obtained also improve the understanding of how intermodal transport is performed in practice.*

Keywords: Intermodal transport; case study; business model; freight transport; rail

INTRODUCTION

The intermodal road-rail industry in Europe has undergone an important change in recent years. Traditionally, there has been a clear division between road transport companies and rail transport companies. Intermodal transport started as a subsidy of the rail companies where door-to-door intermodal transport was offered. Today, intermodal transport often acts as a subcontractor to the road transport industry, focusing on the rail transport (UIC, 2010). The deregulation of the European rail industry has caused this structure to change. New intermodal transport companies have emerged with new business models (BM) and the intermodal market has become more fragmented with a large number of companies offering different, specialized services. As a consequence, some transport customers have taken on a larger role in the intermodal system. Shippers with large goods volumes have taken the initiative to arrange and manage their own intermodal systems. This innovative procedure has challenged the conventional intermodal companies and has opened up new potential markets for intermodal transport. These shippers have acted in different ways: some have become rail operators and operate their own trains, while others act in a similar way to a 3PL and subcontract the physical transport operations to outside transport service providers, but maintain the control and management of the system.

This is particularly visible in the Swedish market, with Sweden being one of the first countries to deregulate the rail sector in 1996. As a result the Swedish market is one of the most developed rail markets in Europe, with an increasing number of operators and consequently increasing competition in the market (Vierth, 2012). The market has gone from two intermodal operators in 1996 to more than 10 today. However, this does not include the increasing number of own-account intermodal systems being developed where the control remains with the shipper.

From a sustainability perspective, intermodal transport foremost addresses the environmental pillar of sustainability: intermodal transport has a lower environmental impact than competing modes (Kreutzberger, Macharis, Vereecken, & Woxenius, 2003). Intermodal transport has, on average, 20–50% less CO_2 emissions than all-road transport on 19 tested European routes (IFEU & SGKV, 2002). This is valuable for the customer for marketing purposes, and can also be used for marketing by the intermodal transport company (Lammgård, 2007). To voluntarily use a more environmentally friendly transport mode also reduces the risk that external actors, for example government, will impose binding regulations. Moreover, the economic aspect is also very relevant for the context of intermodal transport. Using transport modes that generate economies of scale often results in lower costs and in the transport buying decisions, where cost is often the primary factor (Flodén, Bärthel, & Sorkina, 2010), intermodal solution should generate cost reductions or at least be able to meet the costs of a conventional unimodal solution.

Studying these new BMs is important to understand the future potential of intermodal transport. Intermodal transport by nature calls for cooperation between multiple actors, which makes the development of a successful BM challenging, but essential for commercially viable solutions. This chapter investigates how these new BMs have been developed in practice by studying two real-world cases. The BM concept is described using a 9-level typology and the studied BM's strengths and weaknesses are analysed through the typology. Getting a better understanding of these new BMs allows authorities to better support the development of intermodal transport through policy measures. The results obtained also have research implications in improving the understanding of how intermodal transport is performed in practice and practical implications for shippers considering starting their own intermodal services.

A qualitative case study has been performed during 2011 of two major Swedish shippers that have been involved in developing intermodal transport solutions and are currently managing their own intermodal transport solutions. The selection of cases has been based on their relevance and the possibility to get access to the empirical data. In both cases, at least one end-point of the intermodal solution is situated in Sweden, thus being at least partly exposed to similar institutional environments. Since examining the BM requires understanding of the internal rationale of companies and their actions, the case study is found to be a suitable methodology. Moreover, to avoid excessive focus on aspects that are contextual rather than generic, two case studies are chosen. Empirical data has been obtained through in-depth semi-structured interviews with staff involved in the development/management of the studied intermodal solutions. Six interviews have been performed.

BUSINESS MODELS

The term BM has become commonly used in recent years, although it is most often used as a 'buzz word' and some vague reference to how a company does business (Osterwalder, Pigneur, & Tucci, 2005). The term has been used as a 'loose conception of how a company does business and generates revenue' and has been criticised as 'murky at best' (Porter, 2001, p. 73). BMs are often confused with individual parts of the complete BM, for example pricing model, revenue model, channel model, commerce process model, Internet-enabled commerce relationships, organisation form, and value proposition (Linder & Cantrell, 2000).

However, BMs are the firm's framework for making money through the building of its resources and the transformation of those resources into products or services that customers want. It can be defined as the set of activities, *which* a firm performs, *how* it performs them and *when* it performs them to earn a profit (Afuah, 2004). It thus takes a holistic view on the companies' operations. The concept of BMs can be used to describe, develop and analyse how the organisation creates, delivers and captures value. By describing a company's business in a structured way it can be used to describe and classify businesses or to identify options for future development.

Though BM definition has been focusing on a single company, in today's supply chains products and services are commonly formed in cooperation with multiple firms, thus encompassing a firm's boundaries. Therefore the BM concept is applicable to product/service production by a single firm or in a chain/network of companies and activities.

Several attempts have been made to define and formalise the concept, as a need for a formal definition exists (Margretta, 2002). Shafer, Smith, and Linder (2005) identified 12 different definitions in a literature review with a total of 42 different components in the definitions. Other reviews have been made by, for example Osterwalder et al. (2005), Schweizer (2005), Osterwalder (2004), Pateli and Giaglis (2004), Voelpel, Leibold, and Tekie (2004) and Pateli (2002).

One of the most-used frameworks is Osterwalder's (2004). This divides the BM into four pillars and nine building blocks. See Table 1. Note that 'Infrastructure' in Osterwalder's model, refers to firm infrastructure and not transport infrastructure, while the term 'Distribution Channel' is defined as means of getting in touch with the customers (similar to marketing channel), rather than as a logistics term (how products are physically distributed).

Table 1. The Nine Business Model Building Blocks (Osterwalder, 2004, p. 43).

Pillar	Building Block of Business Model	Description
Product	Value proposition	A Value Proposition is an overall view of a company's bundle of products and services that are of value to the customer.
Customer interface	Target customer	The Target Customer is a segment of customers a company wants to offer value to.
	Distribution channel	A Distribution Channel is a means of getting in touch with the customer.
	Relationship	The Relationship describes the kind of link a company establishes between itself and the customer.
Infrastructure management	Value configuration	The Value Configuration describes the arrangement of activities and resources that are necessary to create value for the customer.
	Capability	A capability is the ability to execute a repeatable pattern of actions that are necessary in order to create value for the customer.
	Partnership	A Partnership is a voluntarily initiated cooperative agreement between two or more companies in order to create value for the customer.
Financial aspects	Cost structure	The Cost Structure is the representation in money of all the means employed in the business model.
	Revenue model	The Revenue Model describes the way a company makes money through a variety of revenue flows.

Business Models in Intermodal Transport

The term BM is seldom used in intermodal research. BM models cover all aspects of running an intermodal business. Although studies have been made that describe the market and its actors (Flodén, 2009b; Woxenius, 1994; Woxenius & Bärthel, 2002, 2008), they have not been put into a BM framework. Other studies exist that focus on specific parts of the intermodal system, for example drayage operations (Spasovic & Morlok, 1993), rail haul (Yan, Bernstein, & Sheffi, 1995) and door-to-door service (Tsai, Morlok, & Smith, 1994) but they focus on specific parts of the operations and do not take in the full BM perspective. The term BM is used in three studies: REORIENT (2007), DIOMIS (2007) and Flodén (2009a). The REORIENT project studied potential business BMs for a transport corridor between Scandinavia and Greece (REORIENT, 2007). Four BMs were identified with the focus on which actor controls the transport chain and who makes the transport agreement with the customer. The models developed did not meet the full definition of a BM as defined by Afuah (2004). The DIOMIS project described European and US BMs for intermodal transport. The description comes close to meeting the definition, but is limited to only describing one general model for each continent.

Flodén (2009a) uses the framework by Osterwalder to describe four general models for intermodal transport and is the only study that meets the definition by Afuah (2004). The *subcontractor model* shows the traditional intermodal BM where the intermodal actor is a rail operator who acts as subcontractor to the road transport/forwarder actors. They are the target customers of the intermodal actor. The road transport/forwarder actors are channel leaders with the end customer contact. The *complete transport company* model shows a model where the intermodal actor takes on the role as channel leader and directly approaches the end customers as the target customer. The offer includes a complete transport service including rail transport and road haulage to and from the terminals. The *local cooperation* model is a model where intermodal transport is organised by one or several local[1] actors in cooperation, often with support from local authorities. This cooperative model is based on local needs and operated by the local industry in cooperation, often on a single small destination that is too small to attract interest from the large commercial actors, towards one of the major hubs or cities. The *own-account transport* model describes a situation where an actor with large transport volumes uses intermodal transport in a closed system for own-account transport.

This model is of particular interest, as it most closely resembles the BM in the studied cases.

In the *Own-Account Transport Model*, a large company or forwarder with large volumes of freight starts operating their own intermodal transport services, where the company itself controls and coordinates the intermodal transport, for example setting the requirements and most likely subcontracting the actual transport operations. Flodén (2009a) explains that such BMs are applied because it enables better control, offers the company opportunity to be the channel leader, gives more possibilities to adapt the system to specific needs, and often helps to realise lower costs by excluding intermediaries. The system is considered closed to other shippers; however, as acknowledged by the author, the model can be mixed with other BMs by selling out the excess capacity to outside customers. An important risk connected to this model is that the company does not have transport as part of their core business and thus may lack necessary experience in intermodal transport (Table 2).

Table 2. The Own-Account Transport Model.

Pillar	Building Block of Business Model	Description
Product	Value proposition	Rail transport and terminal handling of intermodal transport units, offering lower cost and reduced environmental impact compared to all-road transport. Road haulage to and from the terminal is offered. The system is closed to outside customers.
Customer interface	Target Customer	The own company. The system is closed to outside customers.
	Distribution Channel	Internal.
	Relationship	A close internal relationship.
Infrastructure management	Value configuration	Performing intermodal road-rail transport.
	Capability	Being able to perform intermodal road-rail transport as an intermodal transport company.
	Partnership	Rail haulage and terminal handling could be subcontracted. Sometimes rental of equipment, wagons, etc.
Financial aspects	Cost structure	Main cost on rail haulage, road haulage and terminal handling.
	Revenue model	The revenue comes from fees paid for the transport. Fees are the operating cost without profit.

Business Model Challenges

Traditionally, the intermodal operator has taken on the role as a subcontractor to a forwarder/haulier. The new deregulated rail market and changes in customer requirements have opened the opportunity for intermodal actors to explore new BMs and for new actors to enter the intermodal market. A number of challenges facing these new BMs can be identified. The organisation and underlying BM in which the involved parties cooperate and define responsibilities is essential for the success of an intermodal transport system, including a fair share of costs, benefits, and risks (PROMIT[2], 2009). One can note that these challenges are rather similar to the challenges in supply chain management: why should benefits be redistributed, how power affects decisions related to transport chains, how risk exposure of different actors differs, etc.

Several authors (e.g. PROMIT, 2009) have noted the lack of attention towards economic integration and coordination of intermodal transport chains. Moreover, the existing research body has shown the importance of intra-organisational and business strategic issues for the success of new systems (Bärthel & Woxenius, 2004). Organisation or governance structure should basically explain the organisation structure of an intermodal transport chain in the sense of which actors are involved, how the chain is managed, how roles and responsibilities are distributed, how decision-making and change processes are organised, etc. However, within this topic, one can distinguish two important points: from one side, there is a definite *need for some sort of coordination in the intermodal chain*; from another side, the successful arrangements can differ from case to case. The latter means that *different BMs may exist* on how the actual management, cash flows, etc. in the chain are arranged. Choosing the right BM is important for the competitiveness and success of the intermodal transport (Flodén, 2009a).

Furthermore, as highlighted by Zomer and Krupe (2009), horizontal cooperation and advanced consolidation models increase complexity of the BM. Thus, a certain degree of simplicity in the set-up is one factor of successful BMs.

High cost for initial investments (Woxenius & Bärthel, 2008) and poor profitability (PROMIT, 2007) of intermodal solutions have been pointed out as major barriers for increasing the use of intermodal transport. Irrespective of other benefits that intermodal transport can provide (environment, quality, etc.); intermodal solutions need to be financially competitive to survive.

Finally, finding a base volume and balance in cargo flows in the intermodal set-up is another commonly discussed challenge. Irrespective of where the initiative for a new development originates, a solution should have some sort of base volume to get operationalized. Keeping the utilisation high is an on-going effort and one that is largely dependent on the ability to attract target customers. Different BMs have different target customers and thus will have to build up relationships and distribution channels in various ways.

EMPIRICAL OBSERVATIONS

The empirical study is based on case studies of two Swedish companies: Volvo and Coop. Though rather different companies in terms of size, industry, transport needs, etc., both have chosen to develop and manage their own intermodal transport solutions. The selection of cases has been based on their relevance and the possibility to get access to the empirical data. The cases represent successful and well-established implementations of shipper-operated intermodal transport and have received a lot of attention from the transport industry. While other examples can be found in the Swedish context (e.g. Stora Enso Baseport System, Stena Recycling, Ikea Rail), the two cases are rather recent and thus allow opportunity for contact with people involved in both setting up the concept and running the operations. Moreover, though both Swedish companies, one of the studied cases is a domestic solution, while the other cross-border, thus allowing capture of diverse complexities.

Volvo Group is one of the world's leading manufacturers of trucks, buses, construction equipment, drive systems for marine and industrial applications. Volvo Logistics that was in practice responsible for the whole development and management of the intermodal solution was at the time an internal business unit in the Volvo Group. Though mainly oriented towards serving internal customers, Volvo Logistics has a few external customers as well – all from the automotive industry – and other than to these customers, services are not sold to the market.

Coop is the second largest grocery retail chain in Sweden with more than 760 retail stores and three warehouses. Coop accounts for 21.5 per cent of the entire Swedish grocery retail sector. The company is owned by a Swedish consumer cooperative with more than 3 million members and has its origins in the consumer cooperative federation from 1899 (Coop, 2012).

Table 3. Basic Characteristics of the Cases.

	Volvo	Coop
Industry	Automotive	Retail
Geographical coverage	Sweden–Germany	Domestic Sweden
Transport distance	900 km	600 km
Previous solution	Road	Road, partly conventional wagon load
Cargo	Components, packaging material	Non-food; chilled; frozen foods; fruits
Part of supply chain	Inbound: from components suppliers to assembly plant	Inbound and outbound: from suppliers and importers to central warehouses and from warehouses to stores and cross-docking facilities
Load unit	Semi-trailer	Semi-trailer

The development and management of the intermodal solution has been the responsibility of the logistics unit: Coop Logistics.

Table 3 gives an overview of the basic characteristics of the two cases.

In the whole development process of the new solutions, the key issue has been finding a suitable BM. The following section will give an overview of some general observations regarding the BM in the studied cases as well as present an overview of the BMs applied, using the four main pillars of Osterwalder's BM framework (product, customer interface, infrastructure management and financial aspects).

Pillar 1: Product

The Coop intermodal service runs between Helsingborg and Stockholm, Sweden every day and consists of 36 semi-trailers of groceries from the suppliers and importers in the South of Sweden, through a terminal in Helsingborg to the three main warehouses in the Stockholm area (Bro, Västerås, Enköping). Return flows are shipments to the stores (directly or through cross-docking facilities) in the south of the country. Road haulage from warehouse to rail terminal is present in both ends of the chain. The service was initiated in 2008 by an initiative from the Coop top management and became operational in 2009. The service is intended to offer an efficient and cost-effective transport solution that is in line with Coop's strategy to appear as an environmentally aware and 'green' grocery store. It was started as a part of a major rationalisation programme that included restructuring of Coop's distribution network.

The Volvo intermodal service includes inbound transport from component suppliers in Germany to the assembly plants for Volvo Trucks and Volvo Cars in the Gothenburg area, Sweden. The train capacity is 36 mega trailers (2 trailers per wagon). Hannover, Germany is used as a rail hub where single wagon flows from Kornwestheim and Herne are consolidated into a block train destined for Gothenburg. The train has 5 departures per week. Pre- and post-haulage to the rail terminals are done by truck. Return flows include packaging material for Volvo and other goods from the road hauliers. The development work was initiated in 2006 as a pre-study and operationalised in 2008. Similarly to Coop, both Volvo Group and Volvo Logistics in particular have the environment as one of the core values and development of the new solution was a direct response to new quantitative emission reduction goals set for Volvo Logistics. Intermodal transport solutions have been developed and used previously in various domestic and cross-border transports.

Both Coop and Volvo fulfil the role as the intermodal operator and are responsible for the overall management and coordination of the intermodal transport. They also bear the commercial risk.

Pillar 2: Customer Interface

Coop and Volvo have chosen to approach the commercial openness of their intermodal systems differently. Commercial Openness (CO) relates to what extent all actors have equal access to transport goods in the system without any time or price discrimination. The openness can range from completely closed to various degrees of openness (Sjöstedt, Woxenius, & Hulten, 1994). Commercial openness can be subject to changes after the system is implemented.

In the case of Coop the solution is kept completely closed to other users, though free capacity exists – especially at the terminal. Coop has considered selling out the existing unutilised capacity, however, due to the unwillingness to engage themselves in other businesses, lack of market knowledge, and necessary network to sell out the free capacity to external actors – the capacity is kept underutilised.

Similarly, Volvo wanted to keep the set-up simple and not engage other shippers. An exception was made during the financial crisis in 2009 when cooperation was initiated with another shipper that had a similar intermodal set-up in terms of geography. During the crisis volumes dropped and to utilise the existing capacity on rail – cooperation was a necessity.

Despite the success, a solution based exclusively on Volvo cargo was desired. Consolidation of flows with another shipper was seen as a complication both in coordination and operations.

However, in the case of Volvo the material flows in their supply chain could not provide necessary volumes in the southbound direction. Though part of the southbound capacity on rail is used to transport packaging material, 2/3 of the capacity is sold to the road hauliers that are responsible for the pre- and post-haulage in the intermodal set-up. Note that the system is open only for the hauliers already involved in road haulage within the system. The idea again has been not to engage with the market, but try to keep the set-up simple. Hauliers are charged a fee per trailer transported.

However, the systems have a large technical openness (TO) (Sjöstedt et al., 1994). TO refers to the restriction in technical acceptance of different unit loads, lorries, transfer equipment, and rail wagons. The systems today only transport trailers, but use standard handling equipment and rail cars and could technically be used to transport all common types of load units. A large TO can often result in control problems for the operators (Sjöstedt et al., 1994), but this is avoided in the current cases due to the use of multipurpose standard equipment and limited CO.

While CO of the system can be changed throughout the operations, TO should be defined as early as possible during system design (Sjöstedt et al., 1994). Both systems have selected a large TO, giving them a freedom to start using other unit loads if necessary. For example, Coop considered opening their system for ISO containers, but realised that most containerised cargo arrived through the Port of Gothenburg and not Helsingborg or other ports in southwest Sweden. Although they have decided to not utilise this option for the time being, this shows that designing the system with a large TO increases flexibility.

Both Volvo and Coop are their own 'customers' in the intermodal transport, which results in a number of internal issues within the customer organisation. Though they can appear as one single actor, there are complex organisations that stand behind this single actor. Top management approved the new concept and supported it throughout development, while the logistics unit was responsible for the overall coordination and management. In the case of Volvo, some internal convincing had to be done to overcome certain scepticism and resistance that was mainly based on the fact that a company whose core business is trucks (and thus road transport) is engaging itself heavily in intermodal transport involving a large share of rail transport. In both cases it was important to communicate properly to the internal actors the change and its implications.

In Volvo the intermodal solution also had to fit the fixed service requirements that existed for the transport between suppliers and assembly plants, while in the case of Coop the change in the transport solution was accompanied by a major change to the warehousing network to enable consolidation. Previously, the Coop suppliers had been responsible for the transport to Coop, but now Coop took control over all inbound transport and started buying Ex works. This included a complex renegotiation of all contracts with product suppliers, who were in general positive to the change of transport solution. However, several larger suppliers expressed interest in being a more visible part of the intermodal solution, but Coop preferred to keep it as an exclusively Coop initiative. Moreover, in the case of Coop, store delivery requirements had been kept unchanged and the logistics department had to make sure the new solution would be able to meet those requirements.

Pillar 3: Infrastructure Management

Intermodal solutions, irrespective of the BM behind them, involve a multitude of actors. Though the number and types of actors may vary from case to case, the typical actors include: shippers; terminal operators; transport operators (different modes – road, rail, sea, intermodal, etc.); equipment-leasing companies; forwarders; government authorities, etc. (Woxenius & Bärthel, 2002).

Both Volvo and Coop rely heavily on partnerships and outsourced operations in their intermodal services, as they do not have the relevant in-house capabilities or resources. They remain as the coordinating actor and channel leader, but all physical transport is outsourced. The terminals in Stockholm and Gothenburg are located close to the Coop warehouse and the Volvo factory, respectively. Coop and Volvo own these terminals, and Volvo also operates their own terminal. The Volvo terminal is the only example in studied cases where the company is directly involved in the physical movement of the goods.

An interesting aspect is the small number of contractual relations compared to the actual number of actors involved in preparations/planning and operations. As mentioned, simplicity is an important aspect of a BM, especially in the case of Volvo where the intermodal solution includes international rail operations. As can be seen from Table 4, Volvo's rail operations are in practice performed by several rail companies: Green Cargo, Schenker Automotive RailNet and DB Schenker Rail Denmark. Despite

Table 4. Actors and Roles (Activities Based on Flodén, 2009b).

Activities and Responsibilities	Coop	Volvo
Overall management and coordination of the intermodal transport	Coop Logistics unit	Volvo Logistics unit
Selection of the business model	Coop Logistics unit	Volvo Logistics unit
Rail transport	Subcontracted to rail operator Green Cargo	Subcontracted to rail operator DB Schenker (+other subcontractors of DB Schenker)
Road transport	Subcontracted to Road Hauliers (Alwex, GDL, Västberga, ACT)	Subcontracted to Road Hauliers (Renus & Ewalds)
Terminal operations	Helsingborg: Existing terminal by Port of Helsingborg Stockholm: Coop terminal. Operations subcontracted to Baneservice AS	Germany: Existing terminal by DB Schenker Gothenburg: Volvo terminal
Marketing and sale of the intermodal transport service	Service only for internal use	Extra capacity on the train is sold exclusively to the hauliers performing pre- and post-haulage

the multiple actors involved, mainly due to the fact that the rail link involves cross-border operations, Volvo has only one contract signed with DB Schenker. However, in practice, non-contractual relations have been built with other rail subcontractors to gain visibility in the rail chain, work with improvement of quality of the service, and when necessary plan for back-up solutions.

In the case of Coop, 1-year contracts were signed with road hauliers and a 3-year contract was signed with the rail operator. In the case of Volvo, the rail contract was signed for 3 years and the contract with road hauliers for 2 years. The contracts signed with rail operators were longer in both cases than the contracts signed with road hauliers and terminal operators. Finding interested road hauliers was slightly problematic, as many perceive intermodal solutions as a competing business. However, the contracts could still be made short due to the stiff competition in the industry with many road hauliers to choose from that easily could be substituted. Price was a key issue in the sense that the intermodal solution was required to provide a certain cost advantage, but it was not the only deciding factor. Rail contracts had to be made longer since there are only a few rail operators to choose from, but also to compensate for the effort it takes to set-up the rail part of the intermodal solution. The decision had to be made as a weighed decision of multiple factors, not just price. The deregulated rail market has allowed Coop and Volvo to have several rail and terminal actors to choose from, which increases the chance to find a suitable partner. This also gives them improved bargaining power compared to only having one potential partner.

Pillar 4: Financial Aspects

An important characteristic of a BM is how financial risk is allocated. In the two empirical cases studied the set-up has been rather different. In the case of Coop, it was mainly the shipper that made the initial required investments: investing in trailers and their own terminal in Bro. In the case of Volvo, the company did invest in their own terminal, as there were no other services available/suitable, but other investments were made by contracted partners: a rail operator has invested in 200 wagons and hauliers have invested in mega trailers against 2-year contracts. One explanation for this is that in the case of Volvo, the rail contract was made with a major rail operator, which did not see the value of the resources required as a risky investment in comparison to their other operations.

The cost structure for Volvo and Coop for the operations is fairly simple, as most operations are outsourced. They are only to a limited extent tied up in long-term investments. The investment risks are born by the subcontractors. It can be assumed that the case companies, as in any outsourcing operation, pay a premium to their subcontractors for them to accept these risks. On the other hand, the subcontractors can balance these risks between different assignments, for example by taking on several clients or moving unused equipment to another client in need of extra capacity, or by employing their own subcontractors. The cost structure is similar to what can be found in a company that has outsourced to a 3PL. For Coop and Volvo to take the role as the intermodal operator also removes a middleman (the external intermodal operator) that would charge a profit margin for their services.

BUSINESS MODELS APPLIED

The case descriptions can be applied to Osterwalder's BM framework. This structures and gives a clearer description of the models, which facilitates further analysis. As can be seen, the BMs are fairly similar. Both have an internal customer and rely heavily on outsourcing. The Volvo system is also open for external customers (but limited to those with whom they have contractual relations already), while the Coop system is strictly internal. See Tables 5 and 6.

Table 5. Volvo Intermodal Business Model.

Pillar	Building Block of Business Model	Description
Product	Value proposition	Intermodal transport of trailers from Germany to Gothenburg, offering lower cost and reduced environmental impact compared to all-road transport. Having their own system, based largely on their own volumes, allows for better control and integration of their specific service requirements into the design of the set-up.
Customer interface	Target customer	Primary – Volvo Trucks and Volvo Cars. Extra capacity is sold to already-subcontracted road hauliers. Other than that, system is closed to external customers.
	Distribution channel	Internal within Volvo.
	Relationship	Close internal within Volvo.
Infrastructure management	Value configuration	Traditional road–rail intermodal transport including road haulage, terminal handling and rail transport.
	Capability	Being able to perform road haulage, terminal handling and rail transport.
	Partnership	Rail operations, terminal operations in Germany, and road haulage – subcontracted. Longer contracts with rail and terminal operator and shorter contracts with other subcontractors. Informal partnership with other rail operators, who act as subcontractors to their contract partner.
Financial aspects	Cost structure	Few fixed costs for Volvo. Main cost on rail haulage, followed by road haulage and last terminal handling.
	Revenue model	Internal payments within Volvo. Payments from sold extra capacity. Cost savings compared to traditional road transport.

Table 6. Coop Intermodal Business Model.

Pillar	Building Block of Business Model	Description
Product	Value proposition	Intermodal transport of trailers from Helsingborg to Stockholm, offering lower cost and reduced environmental impact compared to all-road transport. Having a closed system also allows for better control over the intermodal transport set-up and integration of their specific service requirements into the design of the set-up.
Customer interface	Target customer	Coop.
	Distribution channel	Internal within Coop.
	Relationship	Close internal within Coop.
Infrastructure management	Value configuration	Traditional road–rail intermodal transport including road haulage, terminal handling and rail transport.
	Capability	Being able to perform road haulage, terminal handling and rail transport.
	Partnership	Rail operations, road haulage and terminal operations subcontracted. Longer contract with rail and terminal operator and shorter contracts with other subcontractors.
Financial aspects	Cost structure	Few fixed costs for Coop. Main cost on rail haulage, followed by road haulage and last terminal handling.
	Revenue model	Internal payments within Coop.

These BMs can be compared to the four model types created by Flodén (2009a) and it is clear that both models closely resemble the Own-Account Transport model. The Volvo model has also been combined with the Subcontractor model, as excess rail capacity is sold to external customers. Volvo acts here as a subcontractor to the road hauliers.

ANALYSIS

The models selected by Coop and Volvo position them as channel leaders. The companies perform the role as an intermodal service provider. This means that they control the system and can adapt it to their needs. A common risk in this model is that the companies are not experienced in managing an intermodal solution, which created a lot of internal work. In particular, for Volvo to use an intermodal solution based on mega trailers technology was for them new and not tested. Moreover, both companies have great logistical expertise from conventional road transport, while it must be noted that rail operations are significantly different and require special expertise. The companies have solved this by outsourcing the transport operations to experienced actors and selecting a conventional design of the intermodal system (in load units, rail cars, terminal handling, etc.). The risk that comes with this is that they also become dependent on those actors, which reduces their flexibility. The companies have also experienced problems in changing outsourcing partners when the contracts expired, as old partners refuse to cooperate with the new partners for the transition. The relatively small number of contracted partners used also means that many of them in turn outsource parts of their operations to other companies. This gives rise to a need to establish informal contracts with partners also outside the formal contracts to ensure control over the system and the quality of the transport service.

The high degree of outsourcing also reduces the financial risks for Volvo and Coop as they can end the relatively short contracts, if the intermodal service becomes less successful. Naturally, this risk is transferred to the subcontractors who will charge a premium cost for it. At the same time, Volvo and Coop also absorb the commercial risk of the operations, for example in filling the train. This reduces the risk for the subcontractors, as they are guaranteed payment according to contract. For instance in the case of Volvo, where hauliers invested in the trailers, 2-year contracts were signed compared to the Coop case, where Coop had bought the necessary trailers and hauliers were awarded with 1-year contracts.

The decision by Volvo to also open the system for 'outside customers' highlights the challenge with the large transport volumes required to make intermodal transport cost effective. There are not many companies with large enough volumes to run a full train with only their own goods. This interacts with another challenge for intermodal transport, namely the low flexibility. Volvo actively tried to avoid opening up their system for outsiders to be able to maintain a system tailor-made for their operations and not to complicate it. Outside customers with their own agendas and transport requirements can be challenging to meet in a system tailor-made for one customer. In addition required coordination in operations makes the system more vulnerable to disturbances and creates additional work.

The internal 'politics' within Volvo and Coop also affected the development of the BMs. Although it might appear simple to not have to consider external customers, there are also internal challenges within all large organisations and between departments. These can be just as difficult as those

Table 7. Strength and Weaknesses of the Own-Account Business Model.

Pillar	Building Blocks of Business Model	Weaknesses	Strengths
Product	Value proposition	Intermodal transport is less flexible than traditional road transport and requires large volumes. Potential disturbances in the set-up have major impacts as large volumes of cargo become affected.	Intermodal transport offers cost and environmental benefits.
Customer interface	Target customer	Dependent on one 'customer'. If internal volumes go down, then system must be opened for outside customers or closed. Company has no experience in selling transport services to outside customers.	Allows system to be tailor-made to company's needs. Internal customer guarantees volumes and eliminates marketing costs.
	Distribution channel	Roles as 'buyer' and 'seller' might be blurred, complicating the setting of requirements, etc.	Internal contact gives low costs.
	Relationship	Sensitive for internal 'politics' within the organisation.	Close internal relationship simplifies contacts and planning.
Infrastructure management	Value configuration	Complicated by lack of experience in intermodal transport/management of intermodal transport set-up	The good general logistics experience commonly found in large companies simplifies the design of the system and operations.
	Capability	Dependent on outsourcing and external partners.	A conventional design of the intermodal transport allows transport to be performed according to well-established procedures.
	Partnership	Dependence on external partners and need for close communication. Need to establish informal relations with the subcontractor's subcontractors and other partners to ensure control of the system.	The large degree of outsourcing gives access to important expertise. A limited number of contractual relationships are easy to manage.
Financial aspects	Cost structure	Must accept the business risk for the intermodal system (filling the train, etc.). Transference of financial risk for most investments (rail cars, etc.) to subcontractor gives a higher cost for their services. Remaining initial investments call for long-term commitment to ensure payback.	Low fixed costs and short contracts give large flexibility.
	Revenue model	Internal transfer of money does not allow the intermodal system to charge a profit margin that could be used as a future buffer.	Simple and guaranteed to get paid.

stemming from external customers. Although all departments and employees within the same organisation share a common goal of doing what is best for the company, the notion of what this is can be very different. However, many aspects, such as personal contacts, access to sensitive data and conflict resolution, are simplified within the same organisation. In addition, in both cases the new solutions received approval from the top management of the companies, which meant implementation was highly prioritised and supported throughout the development and implementation.

Strengths and Weaknesses of the Coop/Volvo Own-Account Business Model

The following section summarises the weaknesses and the strengths of the Own-Account Transport Model based on the empirical case studies. The identified weaknesses and strengths point out that the Own-Account Transport Model simplifies the set-up by having an actor responsible for the overall system management and limited amount of contractual relations. Moreover, as a result of direct engagement of the shipper, base volumes are assured. A major weakness is the lost flexibility from the shipper's point of view and the commercial risk for the full train that is completely shifted to the shipper. Finally, the fact that the shipper engages in non-core business activities, that is, transport, creates exposure to new types of risks and limits the options for high utilisation of the intermodal system, unless the shipper somehow engages external users (see Table 7).

CONCLUSIONS

A BM is essential for any sort of commercial undertakings and though the meaning of the concept has to some extent been abused by its popular use, the importance of BMs and research in this context is relevant and necessary. Moreover, even though intermodal transport literature has shown limited interest towards the theme, empirical evidence exists that different and clearly distinguishable BMs are applied in intermodal services. The empirical study conducted highlights the importance of finding the right BM for the viability of the new solutions and calls for further studies in the area.

From the existing research on BMs in the intermodal context, this chapter chose to investigate in-depth the Own-Account Transport Model, which is a rather new and special case of intermodal services. Having applied the

Osterwalder model for describing a BM (as done previously by Flodén), two empirical cases have been analysed to illustrate the viability of this BM in practice, but also to evaluate the weaknesses and strengths related to application of this particular BM. The BM framework has been used to structure and facilitate the analysis.

The empirical research has revealed that the Own-Account Transport Model, as with other intermodal solutions, requires multiple actors to coordinate and sort out roles and responsibilities in order to make the BM functional. An important aspect of the BM is distribution of risks between actors. Though the commercial risk for the full train is born by the shipper, financial risks regarding the necessary initial investments have been more distributed between contractual partners. For the other parties to be willing to "invest into" the new intermodal solution, long-term contracts are required. The shipper controls the channel, but has to rely heavily on the transport operators for their expertise and resources. Thus, a long-term partnership is formed between the main actors, and this is what separates this model from traditional intermodal transport.

Moreover, the Own-Account Transport Model is associated with both strengths and weaknesses that largely differ from other BMs in intermodal transport. The major strength of the model is that the shipper controls the intermodal system and can adapt it to their needs. Heavy outsourcing ensures access to expertise and limits long-term risk. Transport volumes are guaranteed. The major drawbacks are that the shippers must bear the business risk of the intermodal system and are faced with managing a type of system with which they previously have no or limited experience.

A policy maker interested in promoting own-account intermodal transport can aim to simplify the relationships and contracts between the actors, as outsourcing is a key factor. Practitioners should carefully develop their BM and not underestimate the complexity in managing an intermodal transport solution. The studied cases show that the deregulated rail market has been a facilitating factor for these systems to evolve. Although it was possible to buy a full train already before the deregulation, the control of the train was much closer tied to the national rail operator and allowed for limited flexibility in the BM design.

It can be concluded that an actor interested in starting an own-account intermodal system must first ensure the necessary volumes are available to run the system without outside customers, thus keeping it less complicated. An intermodal system using standard technical solutions further simplifies the operations and management. The actor must also decide on the level of outsourcing, where it can be seen that a high level of outsourcing and transference of financial risks increases the flexibility at the expense of higher cost. Most importantly, the complexity in developing and running an intermodal system should not be underestimated.

NOTES

1. Local, for example geographically co-located.
2. PROMIT (Promoting Innovative Intermodal Freight Transport) – a EU project within the Sixth Framework Programme Priority Sustainable Surface Transport, http://www.promit-project.net/

REFERENCES

Afuah, A. (2004). *Business models: A strategic management approach.* New York, NY: McGraw-Hill/Irwin.

Bärthel, F., & Woxenius, J. (2004). Developing intermodal transport for small flows over short distances. *Journal of Transportation Planning and Technology*, *27*(5), 403–424.

Coop. (2012). *Vår historia.* [*Our History.*] Retrieved from http://www.coop.se/Globala-sidor/OmKF/Kooperativ-samverkan/Var-historia1/110-ar-tillsammans/

DIOMIS. (2007). *Report on intermodal rail/road transport in Europe.* UIC, ETF.

Flodén, J. (2009a). Business models for intermodal transport – Intermodal road-rail transport in Europe. *The ninth annual international business research conference*, February 13–14 2009, University of North Florida.

Flodén, J. (2009b). A systems view of the intermodal transport system. WP 2.1, MINT – Model and decision support system for evaluation of intermodal terminal networks. School of Business, Economics and Law at University of Gothenburg.

Flodén, J., Bärthel, F., & Sorkina, E. (2010). Modal choice in Europe – A literature review from the transport customers' perspective. Paper presented at the 12th WCTR Conference, 11–15 July 2010, Lisbon, Portugal.

IFEU, & SGKV. (2002). Comparative Analysis of Energy Consumption and CO2-Emissions of Road Transport and Combined Transport Road/Rail. International Road Transportation Union (IRU), Bundesverband Güterkraftverkehr Logistik und Entsorgung (BGL), Institut für Energie- und Umweltforschung Heidelberg (IFEU), Studiengesellschaft für den kombinierten Verkehr (SGKV).

Kreutzberger, E., Macharis, C., Vereecken, L., & Woxenius, J. (2003). Is intermodal freight transport more environmentally friendly than all-road freight transport? *A Review*, 13–15.

Lammgård, C. (2007). *Environmental marketing of freight transports – The intermodal road-rail case.* Doctoral thesis, School of Business, Economics and Law at University of Gothenburg, Göteborg.

Linder, J. C., & Cantrell, S. (2000). *Changing business models: Surveying the landscape.* Cambridge, MA: Accenture Institute for Strategic Change.

Margretta, J. (2002). Why business models matter. *Harvard Business Review*, *80*(5), 86–92.

Osterwalder, A. (2004). *The business model ontology – A proposition in a design science approach.* Doctoral Thesis, University of Lausanne, Lausanne.

Osterwalder, A., Pigneur, Y., & Tucci, C. L. (2005). Clarifying business models: Origins, present, and future of the concept. *Communications of the Association for Information Systems*, *16*, 1–25.

Pateli, A. (2002). *A domain area report on business models – White paper* (p. 48). Athens: Athens University of Economics and Business.

Pateli, A. G., & Giaglis, G. M. (2004). A research framework for analysing eBusiness models. *European Journal of Information Systems*, *13*(4), 302–314. doi:10.1057/palgrave.ejis.3000513

Porter, M. E. (2001). Strategy and the internet. *Harvard Business Review*, *79*(3), 62–78.

PROMIT. (2007). Promoting Innovative Intermodal Freight Transport. D 4.1 European Benchmarks in Intermodal Freight Transport.

PROMIT. (2009). Promoting Innovative Intermodal Freight Transport. D 3.1–3.3 Best Practice Year 1, 2 and 3.

REORIENT. (2007). Implementing change in the European railway system, Deliverable 6.2, Implications of New Management and Business Models for Rail Operators and IM Companies. Oslo: TOI.

Schweizer, L. (2005). Concept and evolution of business models. *Journal of General Management*, *31*(2), 37–56.

Shafer, S. M., Smith, J. H., & Linder, J. C. (2005). The power of business models. *Business Horizons*, *48*(3), 199–207. doi:10.1016/j.bushor.2004.10.014

Sjöstedt, L., Woxenius, J., & Hulten, L. (1994). Flexibility versus specialisation on the controllability of combined transport systems. *7th IFAC Symposium on Transport Systems*, Tianjin, China.

Spasovic, L. N., & Morlok, E. K. (1993). Using marginal costs to evaluate drayage rates in rail–truck intermodal service. *Transportation Research Record*, *1383*, 8–16.

Tsai, J.-F., Morlok, E. K., & Smith, T. E. (1994). *Optimal pricing of rail intermodal freight: Models and tests*. Department of Systems Engineering, School of Engineering and Applied Science. University of Pennsylvania.

UIC. (2010). *2010 report on combined transport in Europe*. Paris: UIC – International Union of Railway.

Vierth, I. (2012). *Uppföljning av avregleringen av godstrafiken på järnväg*. [*Follow-up of the deregulation of rail freight traffic*.] VTI Report.

Voelpel, S. C., Leibold, M., & Tekie, E. B. (2004). The wheel of business model reinvention: How to reshape your business model to leapfrog competitors. *Journal of Change Management*, *4*(3), 259–276. doi:10.1080/1469701042000212669

Woxenius, J. (1994). *Modelling European combined transport as an industrial system*. Department of Transportation and Logistics. Gothenburg: Chalmers University of Technology.

Woxenius, J., & Bärthel, F. (2002). The organisation of the European intermodal road/rail freight transport industry, FTAM, Delft, 23–24 May.

Woxenius, J., & Bärthel, F. (2008). Intermodal road-rail transport in European Union. In R. Konings, H. Priemus, & P. Nijkamp (Eds.), *The future of intermodal freight transport*. Cheltenham: Edward Elgar Publishing Limited.

Yan, S., Bernstein, D., & Sheffi, Y. (1995). Intermodal pricing using network flow techniques. *Transportation Research B*, *29*(3), 171–180.

Zomer, G., & Krupe, S. (2009). How to materialise synergies in transport logistics. *European Transport Conference 2009*, Noordwijkerhout, The Netherlands.

CHAPTER 9

INTERMODAL BREAK-EVEN DISTANCES: A FETISH OF 300 KILOMETRES?

Dries Meers, Tom Vermeiren and Cathy Macharis

ABSTRACT

Purpose – *In the last two decades, different policy initiatives have been set up to increase the share of intermodal freight transport through a modal shift. In the design of these policies, often critical break-even distances are set, showing the cost or price competitiveness of intermodal transport to delineate transport routes that qualify for such a modal shift. In this chapter, we discuss to which extent such break-even distances can be generalized on a larger scale and how they are calculated.*

Methodology – *We use two price-based models to calculate break-even distances for an intermodal rail and an intermodal barge transport case. General break-even values do not show the price variation in the transport market and vagueness in the calculation of these values adds to this problem.*

Findings – *We find that for the inland waterway case, intermodal barge transport shows potential on shorter distances as well. In addition,*

Sustainable Logistics
Transport and Sustainability, Volume 6, 217–243

ISSN: 2044-9941/doi:10.1108/S2044-994120140000006009

different ways to lower the break-even distance are discussed and a framework for calculating break-even distances is suggested.

Research limitations – *The research elaborates on break-even distances in a European context using price data which are fluctuating over time, location specific and often not publicly available.*

Practical implications – *Policy initiatives promoting intermodal transport should not focus solely on long distance transport. Moreover, evaluating the competitiveness of the intermodal sector solely on a price comparison dishonours its true potential.*

Originality/value – *This chapter challenges the current European policy on intermodal transport by showing the price competitiveness of intermodal transport in two cases.*

Keywords: Intermodal transport; break-even distance; transport prices; modal shift

INTRODUCTION

In the most recent European Commissions' white paper: 'Roadmap to a Single European Transport Area – Towards a competitive and resource efficient transport system', the Commission sets the ambitious goals of increasing mobility by 2050 while significantly decreasing CO_2 emissions by a thorough reform of the current European transport system (European Commission, 2011). The main objective is a reduction of greenhouse gas emissions in the transport sector by 60% in 2050 with respect to 1990. An optimization of the performance of multimodal logistics chains is required, including an increased use of more energy-efficient and sustainable transport modes such as rail and waterborne transport. The promotion of intermodal transport has been on the agenda of the European Commission for some time, to establish a more sustainable transport system (Tsamboulas, Vrenken, & Lekka, 2007), as it is witnessed that in most cases the negative societal impact of intermodal transport is smaller than the impact of road-only freight transport. The Commission aims to shift 30% of road freight over 300 km to these modes by 2030, and more than 50% by 2050. This goal seems very ambitious, but nevertheless, more than half of the goods currently transported on the road are moved over distances below

50 km and more than 75% over distances below 150 km (European Commission, 2011).

Tavasszy and van Meijeren (2011) analysed the numbers stated in this latest European Commissions' white paper and conclude that the policy focus is on road decarbonization for transport distances under 300 km and on a modal shift for the transport segment over 300 km. The modal shift targets of 30% by 2030 and 50% by 2050 relate to this second segment. In addition, they calculated the extent of the required modal shift and found that this 30% modal shift matches with a shift of only 3.4% of road transport for the total freight transport market. In addition, they argue that also promising cases exist for modal shifts below this distance, addressing a bigger market segment and creating a higher net effect.

In this chapter we argue that intermodal transport can indeed be competitive on distances below the 300 km advocated by the European Commission, providing a rationale for a modal shift to these more sustainable transport modes. Therefore, we use the concept of break-even distance, also called critical distance. We start by reviewing the concept of and the existing literature on break-even distances and support the use of price rather than cost data for the calculation of break-even distances. By analysing and comparing previously calculated or empirically tested break-even distances, we aim to address the spatial differences in break-even distances. In the following two sections, we calculate break-even distances for intermodal rail and intermodal barge transport, using Geographic Information Systems (GIS)-based models and conclude that a clear separation is required in policy objectives between inland waterway and rail transport. We also point to different methods in calculating break-even distances, leading to diverging results. In a following section, we provide some possibilities to decrease the break-even distances even further, from a logistics chain perspective. In the final section, we conclude that intermodal transport on shorter distances should also be encouraged by policy makers, when its competitiveness can be shown.

LITERATURE ON BREAK-EVEN DISTANCES

Theory on Break-Even Analysis

The concept of break-even distance is widely used in scientific literature and policy documents to show the cost/price competitiveness of intermodal transport compared to unimodal road transport (Pekin, 2010). The break-even distance can therefore be defined as the physical distance of a transport where the price of intermodal transport equals the price of unimodal road transport. For longer distances, intermodal transport prices will be lower than those of unimodal road transport. Break-even distance analysis supports research in modal choice decisions, in particular when transport price is the major decision variable. The quantification of the break-even distance is thus of great interest for transport researchers, practitioners, policy developers and professionals from the sector.

In general, two alternative approaches exist to estimate the (case-specific) break-even distance. First, empirical data based on real-world transport prices can be used. Alternatively, assuming cost being the basis for price, cost functions can be modelled to calculate the theoretic break-even distance. This second approach demands good insight in the cost structures of the integrated transport activities alongside the intermodal transport chain. Fig. 1 represents a typical intermodal transport chain, including a maritime leg. The black line surrounding the intermodal terminals' market area indicates the locations where the price of intermodal transport and unimodal road transport is equal.

Fig. 2 shows the intermodal cost function for a transport between the sea port and the final destination in the hinterland (a). The diagram has been complemented with the cost function for an alternative continental intermodal transport chain (b) including a pre-haul by truck instead of a maritime sea-leg. The top representation shows that intermodal transport has larger handling costs at the port, compared to unimodal road transport, due to the double use of the expensive ship to shore cranes at the port

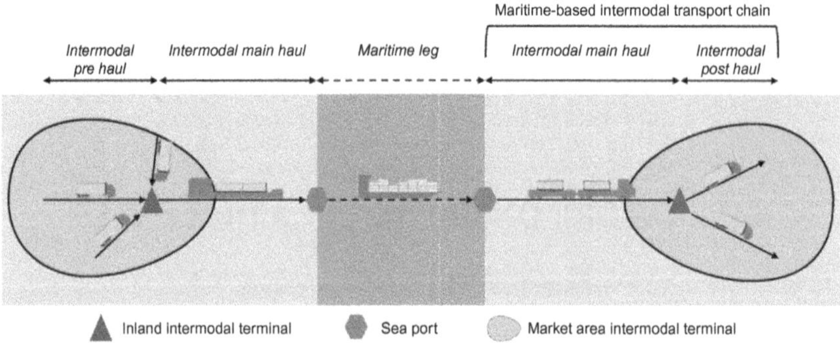

Fig. 1. Intermodal Transport Chain.

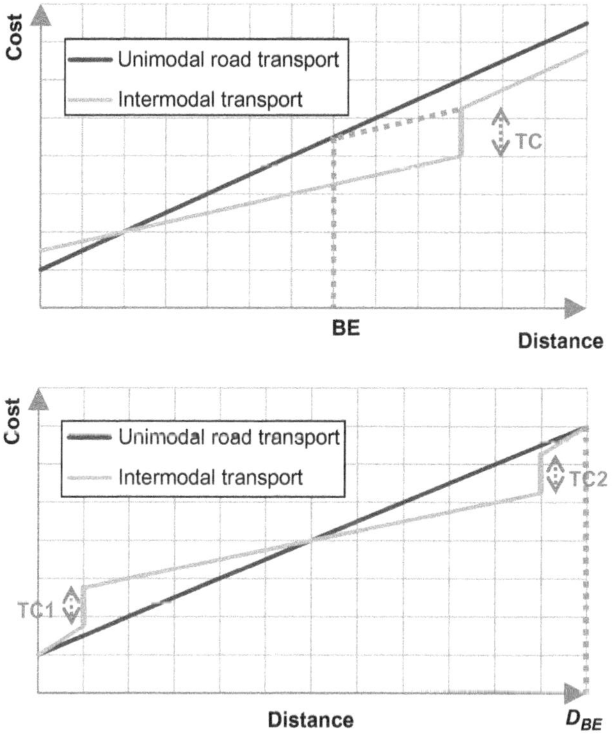

Fig. 2. Cost Functions for Intermodal – and Unimodal Transport: For a Maritime-Based Transport Chain (a, upper) and a Continental Chain Including Pre- and Post-Haulage (b, Lower). (TC = Transhipment Costs). *Source*: Based on Pekin (2010) and Rutten (1995).

terminal. Next, the main haul is performed by rail or barge until the inland intermodal terminal. Due to economies of scale, intermodal transport has lower per unit costs during its main haul, making intermodal transport more competitive over longer distances. Scale advantages in the main haulage of an intermodal transport chain lead to a better performance than unimodal road transport. For example, the capacity of a short sea vessel is equal to at least 200 trucks. For inland navigation vessels, the capacity is 30 trucks or more and trains can carry loading units up to 60 truck equivalents. Scale economies show a decrease in per unit cost of the intermodal loading units transported, although this decrease is not endless. The capacity restrictions will increase the costs per unit above a certain point.

Diseconomies also occur in transhipments, which take place in the intermodal terminals. The second vertical leap (TC) in the cost function represents the extra handling costs at the inland terminal. The post-haul is performed by road until the final destination. This simple cost function representation can be extended by adding for instance non-linear cost functions (e.g. Racunica & Wynter, 2005) or different cost functions for unimodal road transport and post-haul road transport, as is shown in the lower cost function. An extended representation would also involve the return trip, including the handling cost for the return of the empty container which is generally higher for the truck alternative, which returns to the deep-sea terminal (Pekin, Macharis, Meers, & Rietveld, 2012). Contrary, the intermodal transport solution generally returns the empty container to the inland terminal, which can serve as a depot for empty containers with lower handling costs given its lower cost structure outside the port area. The lower cost function representation allows a simple visualization of the break-even distance (D_{BE}), while this is more difficult for the upper case.

Instead of direct transport cost functions, break-even distances could also be derived from logistics cost functions incorporating additional cost benefits of intermodal transport such as a lower contemporary storage cost at the inland terminal compared to the deep-sea terminal and delayed customs at the inland terminal saving on working capital. In particular, broadened logistic cost calculations will account for inventory cost differences between the alternative transport systems.

Cost analyses for calculating break-even distances can be set up in different ways. In the next section, it often remains vague how exactly break-even distances have been calculated. The cost-based break-even distance of an origin-destination couple depends on the transport costs of the available alternatives, which depend on the fixed (e.g. transhipment-) and variable costs. These variable costs on their turn are depended on the time and distance travelled. The cost function of intermodal transport will be leveraged by the assumed economies of scale which is a function of market demand. Other exogenous market conditions such as the available infrastructure network will greatly influence the break-even distance. A high density of transhipment terminals will in general decrease the break-even distance of a region, taking abstraction of the necessary transhipment volumes of a terminal to acquire economies of scale. The varying exogenous conditions will lead to different break-even distances requiring a prudent approach when generalizing the outcomes from different studies.

Another issue when comparing different studies is the applied measurement for distance which can be normalized in different ways using for instance the road distance, intermodal distance or Euclidean distance. Kim and Van Wee (2011) note that the definition of break-even distance can vary between: (i) the door-to-door truck distance, (ii) the hub-to-hub intermodal main haul distance, (iii) the total distance of the intermodal chain, including pre-/post-haulage and (iv) the market distance between two economic activity centres. The underlying definition for distance will clearly have consequences for the comparison of computed break-even distances. There has been no standard set among researchers. In theory, three alternatives can occur for transport between a given origin and destination. The first one is least likely. The distance travelled by road is the same as the distance travelled by the intermodal option (combining barge/rail for the main haul with truck for the post-haul). Second, an intermodal chain can be longer than its road-only alternative. In practice, this will often be the case as the road network is denser and more extensive, especially in comparison to the inland waterway network. A third case is less likely, where the distance travelled by truck is longer than the combined intermodal distance. However, this can be the case for specific trajectories; e.g. in a mountainous region with a rail tunnel, or in the case that high speed road infrastructure demands a big detour to reach a destination. For the dense Belgian network, in only 7% of the studied origin-destination couples (see below), the distance by intermodal barge transport is shorter than by unimodal road transport.

Reported Break-Even Distances

Literature includes numerous studies that have been commissioned to analyse the break-even distances. Kim and Van Wee (2011) provide an overview on the break-even distances of the intermodal freight system. As mentioned before, two approaches can be employed to estimate break-even distances. A first one is based on a direct observation. Researchers address surveys or interviews towards the shippers to portray break-even distances for certain routes. The survey/interview approach also includes market observation from transport routes or market area studies. The Dutch Ministry of Transport, Public Works and Water Management (Ministerie van Verkeer en Waterstaat, 1994) accordingly calculated break-even distances of 100 kilometres for inland waterway transport and 200 kilometres for railway transport when considering maritime containers and 250 and 400 km, respectively when continental chains are considered. Giannopoulos and Aifantopoulou-Klimis (1997) are emphasizing the longer distance for competitive intermodal transport services, quoting 500 km as a break-even distance on the basis of prices. Market perceptions of break-even distances at a European scale are also presented in the works of Bärthel and Woxenius (2004), Bontekoning and Priemus (2004), Macharis (2000), Vrenken, Macharis, and Wolters (2005), and Van Klink and van den Berg (1998). Intermodal rail transport indicates a distinction between domestic and international intermodal services which also affect the break-even distances. According to Bärthel and Woxenius (2004) 550 km is required for domestic intermodal services and 760 km for international intermodal services. Tsamboulas (2008) states distances of 400 km as the minimum for intermodal rail transport in Europe to be competitive, although he finds two major exceptions namely transalpine links and inter-port traffic between Antwerp and Rotterdam. Market observation in North America suggests longer distances of about 500 miles (800 km) as stated in the work of Lim and Thill (2008). Although it is frequently used in the literature due to its simplicity, survey/interview approach cannot directly be applied to other corridors or markets.

Some researchers use cost modelling in case studies to calculate break-even distances. Here, studies use an equation to estimate and compare break-even distances for the current situation and certain scenarios. For example, from Niérats' (1997) market area theory it was possible to calculate break-even distances, with the border of a market area indicating the break-even distance for discrete locations. Literature also includes papers that tackle the drayage operations of intermodal transport in the works of Morlok and Spasovic (1994), Nozick and Morlok (1997), and Spasovic and Morlok (1993). According to their studies, efficient drayage operations can enhance reductions in the break-even distances from 1,300 km to 770 km. Operations and technological development of the railways can also lead to further reductions in the break-even distances, e.g. European Commission (2004b) points out reductions from 738 km to 609 km. ITS (1989) combines higher utilization rates and scale advantages of using 20 foot containers on railways. Compared to the lower degree of utilization of 40 foot containers, this can translate in a reduction from 1,030 km to 520 km. Referring back to the North American railway sector, double-stack technology can reduce the break-even distances from 1,500 km to 645 km (Resor & Blaze, 2004). Rutten (1995) finds clear differences in the break-even distances for intermodal rail transport for maritime-based transport (177 km) and continental transport (417 km). Using cost calculations, he finds that different factors will influence the break-even distance, such as train length, drayage distance, transhipment costs and the train utilization. Janic (2008) models the full costs of an intermodal and road freight transport network and shows the impact of the increasing length of a train. Break-even distances can reduce from 1,100 km to 600–700 km if longer trains are used. A vital segment in intermodal transport operations are the transhipments in intermodal terminals. The European Commission (2004a) highlights that the break-even distance can be reduced from 600 km to 300 km by the introduction of the CargoSpeed system. Finally, pricing is tackled by Wichser, Weidmann, and Nash (2007) through fuel price scenarios. When coupled with higher personnel costs, break-even distances can reduce from 1,400 km to 700 km in Eastern Europe. Janic (2007) showed the effect of internalizing the external costs of transport on the break-even distance. Kreutzberger (2008) examines distance and time in Europe with a generic approach and finds competitive intermodal chains for distances of 600 km or longer. Just like in the literature/survey approach, the mathematical modelling approach is also limited for generalizing the results.

As is clear from the literature review above, break-even distances show a great variation for different regional market conditions and time periods (Table 1). Therefore, it should be noted that most of the reported studies provide break-even distances for specific study areas. Fig. 3 presents the reported break-even distance points or ranges (km) for intermodal container transport in Europe, based on the mentioned studies. It should be noted that most studies calculated break-even distances using cost modelling. The variation in outcome is valid especially for the rail transport which is partly explained by the local market conditions for rail in Europe. Additionally, short run price fluctuations in unimodal road transport may also bias the break-even distance point estimates over time. Fig. 3 also shows the clear difference in break-even distance between intermodal rail and intermodal barge transport. Most studies focus solely on intermodal rail transport, as the network of inland waterways is spatially limited compared to its rail alternative and therefore less studied. In the next sections we stick to this distinction, simulating break-even distances for intermodal rail and intermodal barge transport, using a case-study approach. In addition, we also discuss three other parameters which highly influence the final break-even distance, namely: the direction of the post-haul transport relative to the intermodal terminal, the use of prices versus total logistics costs in calculating the break-even distance and the high network dependency when comparing break-even distances.

Although break-even distance calculations have to be interpreted on a case-by-case basis, we find that in general studies applying a cost modelling

Table 1. Reported Break-Even Distance Points (km) for Intermodal Container Transport in Europe and Evaluation Criteria Explicitly Mentioned (If Applicable).

Study	Break-Even Distance/Range (KM)	Mode	Continental/ Maritime-Based	Length Train/ Capacity	Utilization	Distance PPH/ Terminal Location	Transport Organization/ Network	Cost/Price
Ministerie van Verkeer en Waterstaat (1994)	100–250	Barge	C/M			x		Cost/Price
Macharis and Verbeke (2001)	95	Barge	M			x		Price
Pekin (2010)	57–99	Barge	M			x		Price
Ministerie van Verkeer en Waterstaat (1994)	200–400	Rail	C/M			x		Cost/Price
Rutten (1995)	117–417	Rail	C/M	x	x	x		Cost
Giannopoulos and Aifantopoulou-Klimis (1997)	500	Rail						Price
Janic, Reggiani, and Spicciarelli (1998)	150–500	Rail					x	Cost
Van Klink and Van den Berg (1998)	500	Rail		x				Cost
Kombiverkehr (in Bärthel & Woxenius, 2004)	350	Rail						
Wichser et al. (2007)	1,100–1,400	Rail				x	x	Cost
Janic (2007–2008)	600–1,050	Rail		x	x			Cost
Kreutzberger (2008)	600	Rail				x	x	Cost
Pekin (2010)	173	Rail	M			x		Price
ICF (in Kim & van Wee, 2011)	784–952	Rail						
Sandberg Hanssen, Mathisen, and Jørgensen (2012)	1,858	Rail	C			x		Cost

Source: Based on Vermeiren (2013).

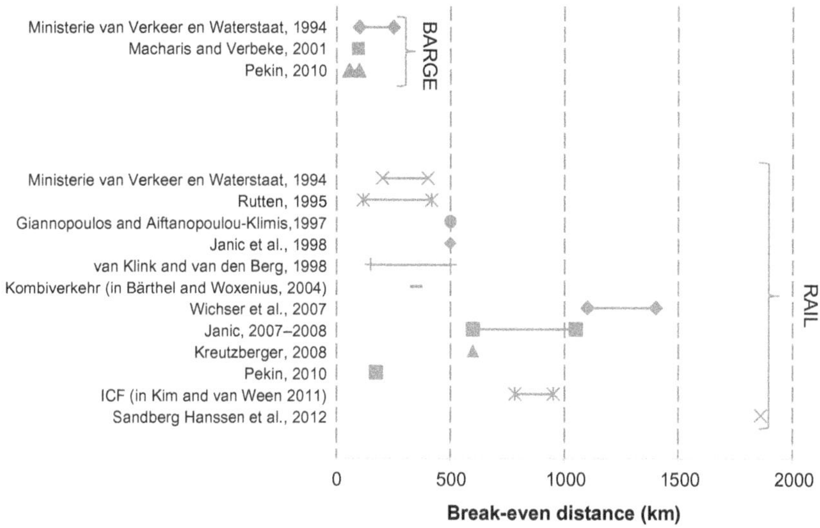

Fig. 3. Visualization of Break-Even Distance Points/Ranges (KM) for Intermodal Container Transport, Mentioned in Table 1.

approach report higher break-even distances compared to those using a data set with prices. We can question if in practice prices are truly set on the basis of cost. Price is a broader marketing concept and price levels can be set on the basis of cost, but also on the basis of competition or even the basis of the perceived value by the customer (Kotler & Armstrong, 2010).

In addition, cost modelling does not reflect the variability in transport prices, accounting for fluctuating short term market conditions. This holds especially for the case of road transport. When analysing break-even distance ranges for modal shift targets, we therefore argue to use price rather than cost. The following two cases for respectively intermodal rail and – barge use GIS-based models to investigate the break-even distance building on a robust set of price data.

BREAK-EVEN DISTANCES OF RAIL TRANSPORT

In this section we calculate break-even distances for intermodal rail transport based on price information derived from the Twin Hub project, a European Union funded INTERREG NWE programme. Twin Hub is an example of a bundling concept that aims to increase the competitiveness of intermodal rail transport by the introduction of innovative train services through a cooperation of competitors (intermodal operators and seaports). This section presents break-even distances from a modal shift analysis for two selected Twin Hub routes. This analysis is carried out by the Euro Terminal Model, which is based on the Location Analysis Model for Belgian Intermodal Terminals (LAMBIT) methodology (Macharis, 2000; Macharis & Pekin, 2009), described below. It should be noted that both LAMBIT and the Euro Terminal Model use price functions, based on real-world market prices, instead of cost functions. This enables to clearly show what the break-even distance means in expenses for shippers. The LAMBIT methodology is developed to analyse the market areas of existing and potential intermodal terminals.

The Euro Terminal Model explores the relative attractiveness of two transportation modes (unimodal road versus rail transport) through a price minimization model. A shortest path algorithm is used in ArcGIS to calculate the routes and the attached transport prices from the Twin hub (Port of Antwerp or Rotterdam) to each NUTS3 region via intermodal terminals and via unimodal road. For each destination, the lowest transport prices for unimodal road and rail/road transport are compared, and the cheapest option is selected. If the Twin hub pilot trains accessing regions have the lowest door-to-door prices, all flows to and from that region will be assigned to rail transport, otherwise to unimodal road transport. The market area of each inland terminal is then highlighted in a map. These visualizations make it possible to see how large the market area of each intermodal terminal is.

Two hinterland cases of the Twin hub pilot network are discussed here: the Slaskie (Poland) and Basel (Switzerland) routes. Fig. 4 presents the results of the model for the two cases with a specific focus on the Poland routes. Intermodal rail transport from the Port of Rotterdam to Poznan is shown together with the unimodal road alternative. The dashed line is the international terminal to terminal rail segment, which is coupled with a solid line representing the post-haulage from the terminal to the centre of the NUTS3 region Leszczyński. Representing the unimodal road route, there is a line that connects the Port of Rotterdam directly to Leszczyński. Results of the Euro Terminal Model highlight that rail services from the Rail Service Center (RSC) in Rotterdam are cheaper compared to the same transport from the Mainhub in Antwerp. Compared to the 1,028 km of rail distance, the road distance is only 943 km following European highways and the major road network. The long distance transport represents opportunities for intermodal rail transport.

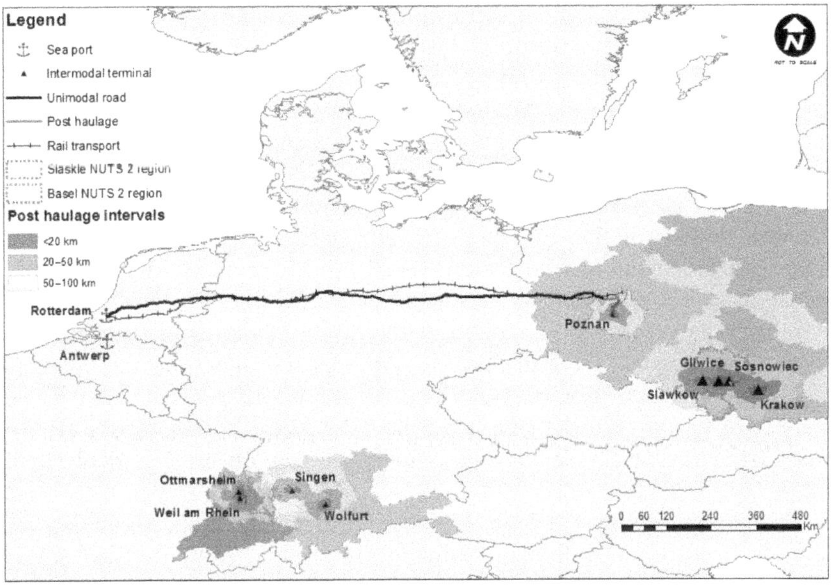

Fig. 4. Slaskie and Basel Twin Hub Cases: The Market Areas of the Intermodal Terminals Are Depicted in Colour Together with Different Post-Haulage Intervals. As an Example the Possible Routes Between Rotterdam and Leszczyński Are Depicted.

The break-even distance for the Slaskie case price function is 464 kilometres if a short post-haulage (20 kilometres) is foreseen, based on the market condition anno 2013. For calculating these break-even distances, it is assumed that the distance of the unimodal haul and the main intermodal haul are equal and that the distance of the post-haul comes in addition. The break-even distance is therefore expressed in door-to-door unimodal road transport distance. When a longer post-haulage is needed, the break-even distance reacts accordingly and increases up to 636 kilometres in the case of 100 kilometres of post-haulage. Fig. 4 also investigates possible connections from RSC to the hinterland terminals in Slaskie province. Considering the hinterland region, the terminal in Sosnowiec is the best option for minimizing the post-haulage.

In this perspective we also calculate the deformation of the market area shape for intermodal rail transport to Slaskie. As we know from Niérats' market area theory (1997), the market area shape of a terminal is not a circle. To illustrate the deformation, we calculate the break-even distance for a fixed intermodal haul distance and with variable post-haulage distances, using Euclidean distances instead of network distances between origin, destination and the terminal. So depending on the destination, both the post-haulage distance and the road-only transport distance will vary, while the intermodal main haul distance remains equal. Fig. 5 shows for the Slaskie price function that for constant main haul distances, the break-even distance will be influenced by the direction of the post-haul transport. If the post-haul is in the direction away from the origin (long), both total trajectories will be equal in length. In all other directions, the intermodal transport chain will be longer than the unimodal one. To break-even, the post-haul distance in the direction of the origin (short) is in this case only 15% of the length of the distance in the opposite direction, for the same main haul distance. For post-hauls in the directions perpendicular to the main haul (up/down), the length will be 31% of the post-haul in direction away from the origin, leaving an egg-shaped market area as depicted in Fig. 1.

Concerning the second Twin Hub case, the model results show that rail services from the Mainhub in Antwerp are cheaper compared to services

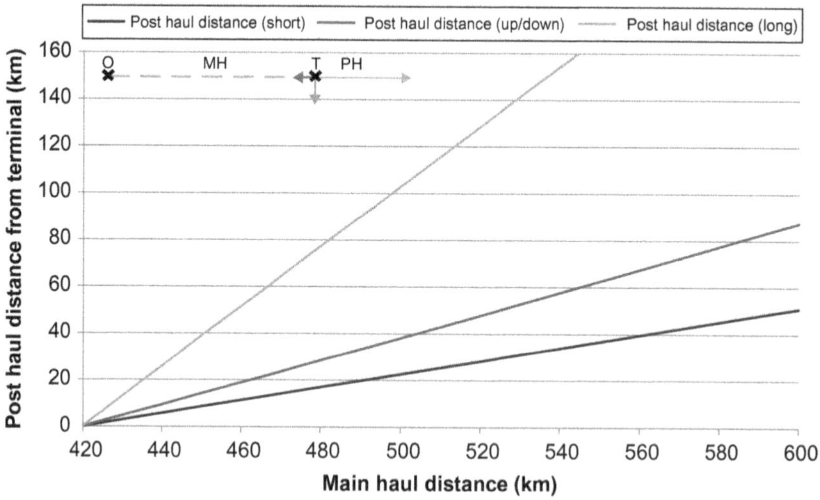

Fig. 5. Break-Even Distances Vary According to the Direction of the Post-Haul Transport. In the Upper Left Corner a Visualization of the Intermodal Transport Chain with the Three Types of Post-Haul Connections (O = Origin, MH = Main Haul, T = Terminal, PH = Post-Haul).

from RSC in Rotterdam, due to the shorter rail distances of the Port of Antwerp to Basel. In the hinterland region, the terminal in Weil am Rhein can capture a major market area in Switzerland including Basel. The break-even distance for that route is 384 kilometres if a shorter post-haulage (20 kilometres) is foreseen, when applying the same preconditions as in the Slaskie case. In the case of 100 kilometres of post-haulage, the break-even distance increases up to 527 kilometres. Compared to the Slaskie case, the break-even distances are lower, which is explained by the higher road transport prices in Switzerland.

Both Twin Hub cases show break-even distances above 300 km if post-haulage distances are over 20 km. For a post-haulage of 20 km, a break-even distance of 464 km was calculated for transport to Slaskie and 384 km for transport to Basel. But, it is also shown that the calculation of the break-even distance is highly influenced by the true and the estimated ratios between the unimodal – the main – and the post-haul and the direction of this post-haul. Infrastructure network characteristics and relative locations are thus crucial in break-even analysis.

BREAK-EVEN DISTANCES OF INLAND WATERWAYS TRANSPORT

The second case deals with the use of inland waterways from the Port of Antwerp accounting for 34% of its total hinterland container transport, which corresponds to 2 million TEU per year. Remarkably, 750,000 TEU are barged between the main ports Antwerp and Rotterdam over a short distance. The strategic objective of the port is even to increase the share of inland waterway container transports to 42% by 2020 (Port of Antwerp, 2013). From the port, a well-developed inland waterway network provides access to the Belgian hinterland, by a relatively dense inland waterway network in the north of the country. Different terminals, who are mostly connected to the port with regular shuttle services, allow the transhipment of containers from road to barges and vice versa (Fig. 6).

To calculate the break-even distances for this specific case – the maritime-based hinterland transport in Belgium – we make use of LAMBIT, developed by Macharis (2000). This is a GIS-based model that is mainly used for policy analysis (e.g. Macharis & Pekin, 2009; Macharis, Van Hoeck, Pekin, & van Lier, 2010). Based on a shortest path algorithm, the model calculates the cheapest routes for all modal alternatives, i.e. road-only transport,

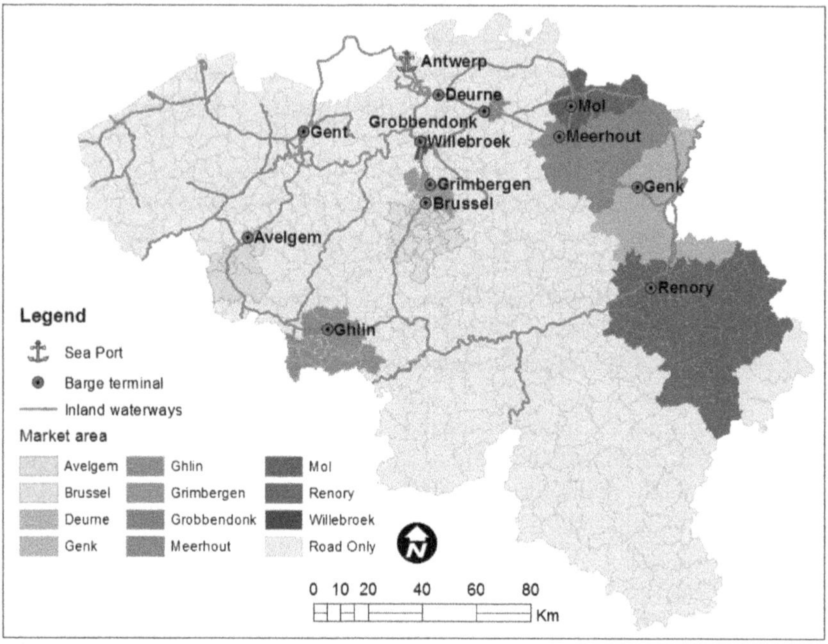

Fig. 6. Locations and Market Areas of Belgian Barge Terminals.

intermodal rail transport and intermodal barge transport. The price information is derived from market surveys and therefore based on real-world market prices. Based on an all-or-nothing assignment procedure, LAMBIT visualizes the cheapest transport alternative for transport between selected origin-destination couples. Taking the Port of Antwerp as origin and all Belgian municipalities as destinations, LAMBIT is able to visualize the market areas of all existing intermodal terminals. Based on this input information, LAMBIT is also able to calculate break-even distances for intermodal barge transport based on the calculated network (and Euclidean) distances. For more elaborate information on the price functions used in the model, we refer to Pekin et al. (2012), where also a comprehensive logistics function is used that includes a value of time component.

A LAMBIT output image (Fig. 6) shows the importance of network distance in break-even analysis. The map shows the market areas of intermodal terminals in Belgium for container transport to and from the Port of Antwerp. Terminals in the east of the country have rather large market areas in comparison to their more western counterparts. One of the main reasons for this phenomenon is the short distance by barge relative to the road-only alternative. This also explains why strategically located terminals in the vicinity of the port can have their own distinct, although small, market areas, as the terminals of Deurne, Grobbendonk and Willebroek prove. The longer inland waterway network distances to the more western terminals lead to smaller market areas.

Based on the LAMBIT price functions, we can calculate the average break-even points for varying post-haulage distances. To calculate the break-even distance to a clear comparison base, we express it as the door-to-door truck distance. To account for network typology, we calculated the ratio between the intermodal main-haul transport distance and the unimodal transport distance for constant post-haul distances. This ratio was calculated as the floating average for a $\{-1;+1\}$ km interval of every possible post-haul distance from 0 up to 50 km, to know how they relate to the total truck door-to-door and the intermodal main haul distance. This ratio allows overcoming the uncertainty on the relative length of both the unimodal and the intermodal transport chain when equalling both price functions.

The resulting Fig. 7 confirms an intermodal barge break-even distance around 85 km for a post-haulage distance of 20 km, assuming the 2013 market conditions. But this figure also shows that even lower break-even distances can be calculated if the post-haul distance is reduced to a few

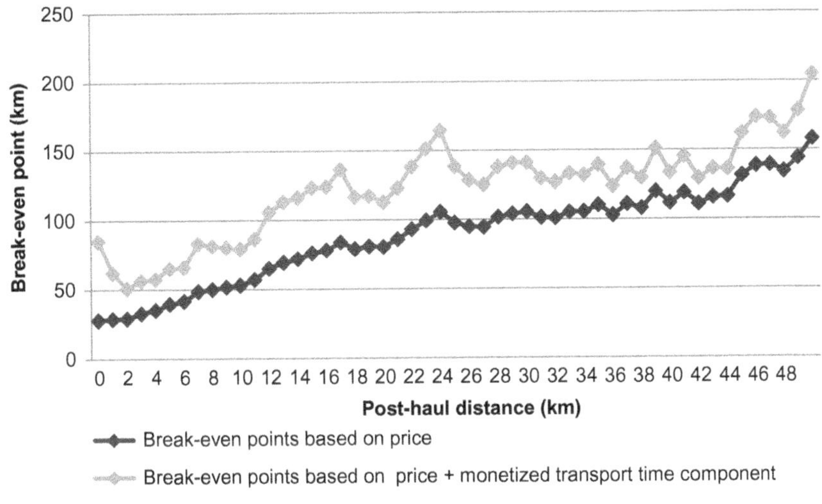

Fig. 7. Average Break-Even Points Barge Transport in Belgium.

kilometres. Multiple reasons contribute to these relatively short break-even distances for intermodal barge transport in Belgium. First, due to the use of shuttle services between the terminals and the sea port, the performance can be improved if transport volumes are big enough and a reasonable frequency can be achieved (Konings, Kreutzberger, & Maraš, 2013). Second, different support measures aim to (in)directly reduce the transport prices of barge transport in Belgium. Also a fierce price competition between operators leads to prices with limited or absent profit margins. Nevertheless should these results be interpreted with some caution, as LAMBIT solely uses linear price functions, possibly underestimating the intermodal price for short distance transport. In addition, similar break-even points were calculated using a total logistics price function that accounts for the value of time (Pekin et al., 2012). As intermodal transport will be slower in any case, the break-even points will raise, with an approximate value of 120 km for a post-haulage distance of 20 km. The trend breaks in both functions are a consequence of the varying distance ratio, explaining the relatively long break-even distance for the second price function under a post-haulage of 0 km, being a consequence of the specific transport networks characteristics.

This Belgian example, together with the break-even distance studies on intermodal barge transport mentioned in Fig. 3, show that break-even distances in intermodal barge transport can be much shorter, compared to intermodal rail transport services. A case-specific break-even distance of 85 km was calculated when a post-haulage distance of 20 km was considered. Therefore, the modal shift targets of the European Commission, focusing solely on transport distances over 300 km, seem inappropriate. Moreover, this example shows that a case-specific (or corridor-specific) approach is more relevant for determining modal shift targets. Especially, intermodal barge transport seems to be competitive on much shorter distances, than the ones stated for intermodal rail transport. Taking into account other relevant modal choice aspects such as transport time might however increase the break-even distance.

WAYS TO FURTHER LOWER THE BREAK-EVEN DISTANCE

In the previous sections, we discussed current break-even distances for two real-world examples. In this section, we elaborate on possibilities to further decrease these current break-even distances. Therefore, we return to the cost functions described above and discuss which parameters can change these functions which on their turn form the basis for long-term sustainable pricing and calibrate the applicable break-even distance. This cost modelling approach remains useful to grasp the sensitivity of the cost determinants.

Fig. 8 shows how the break-even distance can be lowered, by directly impacting the cost function of intermodal transport. A first option (A) to decrease the break-even distance, is lowering the transhipment costs in the inland terminal (Konings, 2009). For instance ICT applications can make the transhipment operations more reliable, reducing the overall transhipment costs. Faster transhipment cranes lower the non-productive berthing time of the barges and can therefore lower the transhipment cost per unit and increase the service level. Macharis and Verbeke (2004) calculated that

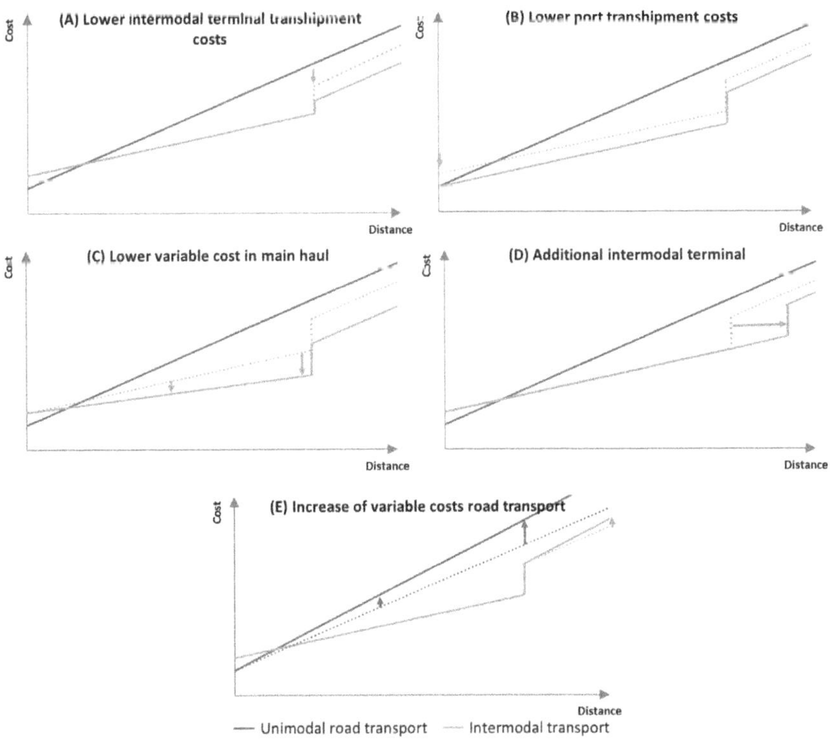

Fig. 8. Changes in the Cost Function Can Lower the Break-Even Distance. *Source*: Partly based on Macharis and Verbeke (2004) and Nemoto, Browne, Visser, and Castro (2006).

approximately 30% of the total costs of hinterland intermodal barge transport relate to transhipments. Alternatively, the transhipment costs in the port could be lowered as well (B). Another option is a decrease in the per unit variable transport costs of the intermodal main haul (C). The loading degree and the transported volume are important determinants of this cost function, but are commonly not explicitly studied in the context of break-even distances. Konings (2009) indicates that barge operators estimate a break-even loading degree on 75%, although this depends on the circulation times. Also the frequency of the service and the type of network connections (e.g. hub-and-spoke versus shuttles) are crucial. Economies of scale can be exploited through the use of bigger barges or longer trains, or through a higher capacity utilization. But the types and length of barges and ships is highly dependent on the characteristics of the infrastructure network, in first instance the granularity. For the inland waterways for instance, the ship type that can be used, depends on the capacity of the waterways, water levels, locks, height of bridges etc. A fourth possibility (D) is the set-up of a new terminal. A new terminal can decrease the post-haul distance, while at the same time the main haul distance increases. As the per km transport cost during the main haul is lower, the total intermodal transport chain will become cheaper, lowering the break-even distance.

A second series of possibilities that directly impact the break-even distance, relate to the cost function of road transport. A first option is to increase the variable costs of road transport (E). This also means that the variable cost of the post-haul (and if applicable the pre-haul) are influenced. For instance an increase in the fuel prices can lead to increased variable costs. Also increasing congestion costs impact the cost function. Especially the areas in the vicinity of the major ports suffer from increased road congestion, not only impacting the total transport cost, but also the transport time and the on-time delivery reliability. Increasing the fixed costs of road transport, will also impact the cost of the intermodal post-haul transport. This increase has no effect on the competitiveness of intermodal transport, unless the impact is different. As indicated above, also the share of empty trips will strongly influence the final road transport cost. Next, the handling cost for the returned empty container could be included in these calculations. When the intermodal terminal can serve as a depot for empty containers, the return trip will be shorter in comparison to the unimodal alternative. The handling cost in the deep-sea terminal will also be higher than the one in the inland terminal, pointing to the importance of the balance of transport flows. On the other hand can the number of empty trips be reduced when innovative IT systems are used for better alignment of inbound and outbound flows, leading to lower road transport prices. Some measures can combine different effects described above. By internalizing the external transport costs, the break-even distance will also decrease, as the unimodal transport costs will witness a higher increase than the intermodal alternative.

In order to gain further insight in the relative importance of the various parameters, we have modelled the actual cost structures of trucking and barging operations transporting containers between the port and the hinterland. The managerial cost comparison between the two modes accounts for the so-called running costs such as fuel, labour, repairs and maintenance, as well as fixed overheads and capital outlays. Our cost model for truck yields an average total cost of €0.96 per km. on the basis of 216 operating days per year with an average daily road distance of 585 km. The cost calculations for the barge indicate a much lower inland shipping cost, ranging between €0.18 and €0.30 per km depending on the actual load factor. The advantageous cost per km of the barge, however, will be eroded when adding the cost of the additional handling at the inland terminal and the cost of pre-/on carriage by truck to the final destination. Needing to follow the network of canals and rivers, the main haul distance of the barge may also be longer and therefore increase its total cost. Using the managerial cost model, we now explore the relative sensitivity of the break-even distance towards varying assumptions about the intermodal transport solution, namely the applicable load factor of the barge, the drayage distance, round whether single trip of the barge and the effect of a detour in the main haul due to the limited inland waterways network, compared to the road distance. In the theoretic case where the trucking distance would precisely match the barging distance and the inland shipping alternative would not require any pre- and on carriage by truck, the break-even distance (computed on the basis of the cost model) appears to be very low, namely 27 kilometres. The savings from the lower operating cost of the barge per km. are than just enough to off-set the extra handling cost at the inland terminal. In practice, the intermodal solution generally requires pre- and on-carriage by truck which adds costs to the combined transport alternative and therefore increases its break-even distance against unimodal truck. Also potential longer main hauls on the river or canal network compared to the denser road network translate into extra costs and increase the break-even distance to beat truck. The cost calculation, however, shows that the drayage distance seems to be the main driver behind the outcome of the applicable break-even distance (Fig. 9). The impact of varying load factors or the application of a round trip barging model instead of a single trip is much

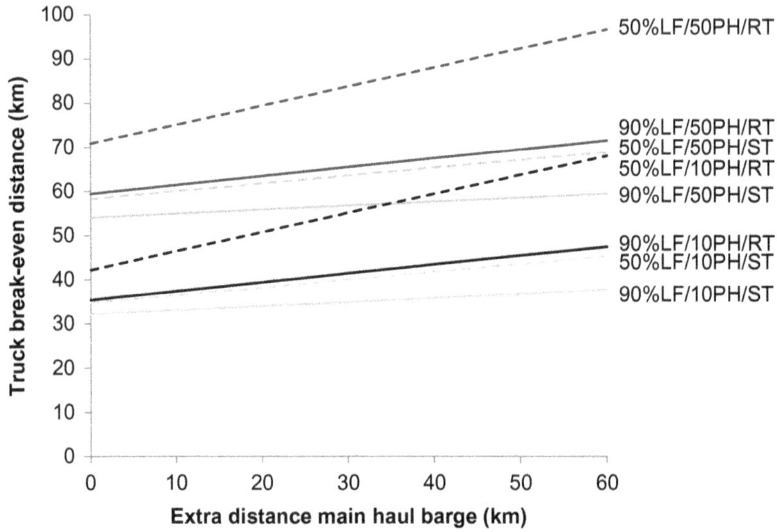

Fig. 9. Changes in Truck Break-Even Distance When Influenced by a Detour of the Barge Transport (*X*-axis), the Load Factor (LF), the Post-Haulage Distance (PH) and a Single Trip (ST) Versus a Round Trip (RT).

less pronounced. The cost assessment implies that the potential for a cost competitive intermodal transport solution on the short distance is great, but only if the last miles by truck can be kept short.

So roughly, the variables influencing the break-even distance can be divided into two interrelated subgroups. The first group includes network-related variables such as: unimodal road distance, drayage truck distance and intermodal main haul distance which depend on the locations of origin, destination and the intermodal terminal and in addition the possibility of a round trip. The second group consists of price-related variables, including transhipment rates and prices of the intermodal and the road-only transport. Obviously these rates can depend on a multitude of factors such as fuel price, tax regimes and subsidies and the balance of traffic. In this perspective, Woxenius (2012) discusses the concept of directness in freight transport chains with a focus on detours.

As noted in the literature review section, different ways exist to shorten the break-even distance of a transport. As Fig. 8 showed, the cost function can be directly impacted to make intermodal transport more competitive. From a total logistics chain perspective other factors such as value added services can also ease the transport service. Empty container depots or the use of intermodal services for the return of empty containers can impact the total logistics price (Pekin et al., 2012). Another possibility would be a terminal located in the vicinity of the sea port, which can serve as an extended gate (Konings et al., 2013) decongesting the port area.

Based on the previous, we propose a framework for estimating and comparing break-even distances on a case-study basis, where the following parameters should be included (Fig. 10). The actual break-even distance is directly influenced by the prices of road-only and intermodal transport, but a variety of factors should be monitored when translating break-even distances to concrete modal shift policies.

On the other end of the spectrum, also possibilities remain to increase the break-even distance of intermodal transport. Flodén (2007) argues that the competitiveness of road transport will also depend on the type of lorry and the type of intermodal loading unit used. One notable example being the use of longer and heavier vehicles (LHVs) that can reduce the per unit transport costs of door-to-door truck transport, as the Dutch example proves (Ministry of Infrastructure and the Environment, 2011). Although, LHVs can also be used for the pre- and post-haulage of intermodal transport,

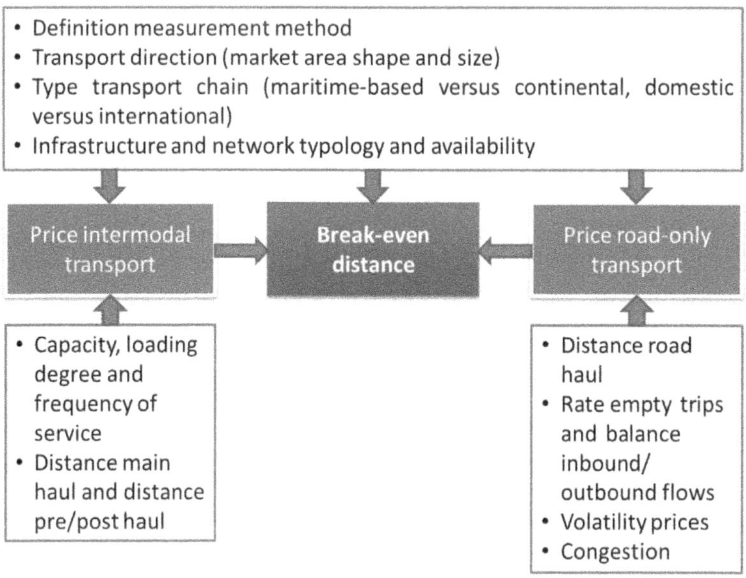

Fig. 10. Framework for the Estimation of Case-Specific Break-Even Distances.

leading to cost reductions of 5–10% (Bergqvist & Behrends, 2011). Finally, it is a misconception that transport mode decisions are solely made, based on a transport price comparison. Other mode choice criteria such as speed and reliability clearly influence the preference for one mode over another.

CONCLUSIONS

In this chapter we showed that, based on a literature review and model simulations, a modal shift policy focussing solely on distances over 300 km in European hinterland transport is a too one-sided approach that might originate from the perception that truck has its domestic market in this distance segment. Applying a 300 km threshold can also be understood given that the European Commission aims triggering transports across different member states. The majority of maritime containers, however, are shipped inland over considerably shorter distances and are thus not on the radar screen of the European Commissions' roadmap. However, these transports of containerized goods form the backbone of the global economic model and directly impact the economic wellbeing.

By linking the competitiveness of intermodal transport to the break-even distance, we found that intermodal transport can be competitive on shorter distances as well, which is especially interesting for maritime-based container transport in Western Europe where transport distances are short and volumes large. First, a literature review showed that break-even distances in Europe are estimated within a wide range of distances. Vermeiren (2013), in his review, reports break-even distances ranging from below 100 km up to 1,300 km. This range can be explained by local market conditions and/or assumptions made by the researchers. Also the definition of break-even distance and its calculation is crucial in comparing reported distances. Second, simulations, based on market price data, show that intermodal transport in Belgium can be cheaper than its road-only alternative on distances below 100 km, but this largely depends on the current market prices and the spatial characteristics of the alternative routes. Especially short post-haul distances can reduce the break-even distances of intermodal transport drastically.

Based on our findings, we advocate a more case-specific modal shift approach, based on local, spatial and price-related variables, resulting in a broad break-even framework (Fig. 10) that should be used when designing local modal shift policies. Policy makers shouldn't solely focus on the long distance transport market, but investigate the local competitiveness of the intermodal transport sector. We showed that the modes available and their locational network characteristics partly determine the competitiveness of intermodal transport. In addition, different policy levels can act to decrease the break-even distance in a certain region. To promote the use of intermodal transport, these characteristics should be taken into account when designing modal shift policies, as different measures exist that can impact the cost function of intermodal transport (e.g. set-up of new intermodal terminals) or of both intermodal and unimodal transport (e.g. internalization of external transport costs, road pricing).

ACKNOWLEDGEMENTS

This research was conducted in the framework of two projects. The research on the Euro Terminal Model was conducted in the framework of the Twin Hub network project, financially supported by the European Union funded INTERREG NWE programme. The further development of the LAMBIT model is conducted within the framework of the Policy Research Centre MOBILO, financially supported by the Flemish Government.

REFERENCES

Bärthel, F., & Woxenius, J. (2004). Developing intermodal transport for small flows over short distances. *Transportation Planning and Technology*, *27*(5), 403–424. doi:10.1080/0308106042000287586

Bergqvist, R., & Behrends, S. (2011). Assessing the effects of longer vehicles: The case of pre- and post-haulage in intermodal transport chains. *Transport Reviews*, *31*(5), 591–602. doi:10.1080/01441647.2011.584980

Bontekoning, Y. M., & Priemus, H. (2004). Breakthrough innovations in intermodal freight transport. *Transportation Planning and Technology*, *27*(5), 335–345. doi:10.1080/0308106042000273031

European Commission. (2004a). Final report: Cargospeed (Cargo Rail Road Interchange at Speed).

European Commission. (2004b). REALISE (Regional Action for Logistical Integration of Shipping across Europe).

European Commission. (2011). *White paper – Roadmap to a single European transport area – Towards a competitive and resource efficient transport system* (p. 30).

Flodén, J. (2007). *Modelling intermodal freight transport. The potential of combined transport in Sweden*. Göteborg: Göteborg University.

Giannopoulos, G. A., & Aifantopoulou-Klimis, G. (1997). Intermodal freight transport networks in Europe and policies for their development. Transportation systems, pp. 485–495.

ITS (Institute for Transport Studies, U. of L.). (1989). In what circumstances are inter-modal systems competitive?

Janic, M. (2007). Modelling the full costs of an intermodal and road freight transport network. *Transportation Research Part D: Transport and Environment*, *12*(1), 33–44. doi:10.1016/j.trd.2006.10.004

Janic, M. (2008). An assessment of the performance of the European long intermodal freight trains (LIFTS). *Transportation Research Part A: Policy and Practice*, *42*(10), 1326–1339. doi:10.1016/j.tra.2008.06.008

Janic, M., Reggiani, A., & Spicciarelli, T. (1998). The European freight transport system: theoretical background of the new generation of bundling networks. In 8th world conference on transportation research Antwerp, Antwerp, Belgium.

Kim, N. S., & Van Wee, B. (2011). The relative importance of factors that influence the break-even distance of intermodal freight transport systems. *Journal of Transport Geography*, *19*(4), 859–875. doi:10.1016/j.jtrangeo.2010.11.001

Konings, R. (2009). *Intermodal barge transport: Network design, nodes and competitiveness*. Delft: Delft University of Technology.

Konings, R., Kreutzberger, E., & Maraš, V. (2013). Major considerations in developing a hub-and-spoke network to improve the cost performance of container barge transport in the hinterland: The case of the port of Rotterdam. *Journal of Transport Geography*, *29*, 63–73. doi:10.1016/j.jtrangeo.2012.12.015

Kotler, P., & Armstrong, G. (2010). *Principles of marketing* (13th ed., p. 630). Saddle River, NJ: Pearson Prentice Hall.

Kreutzberger, E. D. (2008). Distance and time in intermodal goods transport networks in Europe: A generic approach. *Transportation Research Part A: Policy and Practice*, *42*(7), 973–993. doi:10.1016/j.tra.2008.01.012

Lim, H., & Thill, J.-C. (2008). Intermodal freight transportation and regional accessibility in the United States. *Environment and Planning A*, *40*(8), 2006–2025. doi:10.1068/a38336

Macharis, C. (2000). *Strategische modellering voor intermodale terminals, Socio-economische evaluatie van de locatie van binnenvaart/weg terminals in Vlaanderen*. Vrije Universiteit Brussel.

Macharis, C., & Pekin, E. (2009). Assessing policy measures for the stimulation of intermodal transport: A GIS-based policy analysis. *Journal of Transport Geography*, *17*(6), 500–508. doi:10.1016/j.jtrangeo.2008.10.004

Macharis, C., Van Hoeck, E., Pekin, E., & van Lier, T. (2010). A decision analysis framework for intermodal transport: Comparing fuel price increases and the internalisation of external costs. *Transportation Research Part A: Policy and Practice*, *44*(7), 550–561. doi:10.1016/j.tra.2010.04.006

Macharis, C., & Verbeke, A. (2001). Het intermodale transportsysteem vergeleken met het unimodale wegvervoer. *Tijdschrift Voor Economie En Management*, *46*, 39–63.

Macharis, C., & Verbeke, A. (2004). *Intermodaal binnenvaartvervoer: Economische en strategische aspecten van het intermodaal binnenvaartvervoer in Vlaanderen* (p. 176). Antwerpen, The Netherlands: Garant.

Ministerie van Verkeer en Waterstaat – Projectgroep Intermodaal vervoer. (1994). *Intermodalitijd – stimulering intermodaal vervoer – plan van Aanpak* (p. 37). Ministerie van Verkeer en Waterstaat.

Ministry of Infrastructure and the Environment. (2011). *Monitoring modal shift: Longer and heavier vehicles The follow-up measurement (2011)* (p. 67).

Morlok, E. K., & Spasovic, L. N. (1994). Redesigning rail-truck intermodal drayage operations for enhanced service and cost performance. *Transportation Research Forum, 34*, 16–31.

Nemoto, T., Browne, M., Visser, J., & Castro, J. T. (2006). Intermodal transport and city logistics policies. In E. Taniguchi & R. Thompson (Eds.), *Recent advances in city logistics* (pp. 15–30). Oxford: Elsevier.

Niérat, P. (1997). Market area of rail-truck terminals: Pertinence of the spatial theory. *Transportation Research Part A: Policy and Practice, 31*(2), 109–127.

Nozick, L. K., & Morlok, E. K. (1997). A model for medium-term operations planning in an intermodal rail-truck service. *Transportation Research Part A: Policy and Practice, 31*(2), 91–107. doi:10.1016/S0965-8564(96)00016-X

Pekin, E. (2010). *Intermodal transport policy: A GIS-based intermodal transport policy evaluation model*. Vrije Universiteit Brussel.

Pekin, E., Macharis, C., Meers, D., & Rietveld, P. (2012). Location analysis model for belgian intermodal terminals: Importance of the value of time in the intermodal transport chain. *Computers in Industry, 64*, 113–120. doi:10.1016/j.compind.2012.06.001

Port of Antwerp. (2013). Annual report 2012 (p. 102). Antwerp, Belgium.

Racunica, I., & Wynter, L. (2005). Optimal location of intermodal freight hubs. *Transportation Research Part B: Methodological, 39*(5), 453–477. doi:10.1016/j.trb.2004.07.001

Resor, R. R., & Blaze, J. R. (2004). Short-haul rail intermodal; Can it compete with trucks? *Transportation Research Record, 1873*, 45–52.

Rutten, B. J. C. M. (1995). On medium distance intermodal rail transport: A design method for a road and rail inland terminal network and the Dutch situation of strong inland shipping and road transport modes.

Sandberg Hanssen, T.-E., Mathisen, T. A., & Jørgensen, F. (2012). Generalized transport costs in intermodal freight transport. *Procedia – Social and Behavioral Sciences, 54*, 189–200. doi:10.1016/j.sbspro.2012.09.738

Spasovic, L. N., & Morlok, E. K. (1993). Using marginal costs to evaluate drayage rates in rail-truck intermodal service. *Transportation Research Record, 1383*, 147–166.

Tavasszy, L. A., & van Meijeren, J. (2011). Modal shift target for freight transport above 300 Km: An assessment (p. 20).

Tsamboulas, D. (2008). Development strategies for intermodal transport in Europe. In R. Konings, H. Priemus, & P. Nijkamp (Eds.), The future of intermodal freight transport – Operations, design and policy (pp. 271–301). Transport Economics, Management and Policy Series. Cheltenham: Edward Elgar Publishing Limited.

Tsamboulas, D., Vrenken, H., & Lekka, A.-M. (2007). Assessment of a transport policy potential for intermodal mode shift on a European scale. *Transportation Research Part A: Policy and Practice, 41*(8), 715–733. doi:10.1016/j.tra.2006.12.003

Van Klink, H. A., & van den Berg, G. C. (1998). Gateways and intermodalism. *Journal of Transport Geography, 6*(1), 1–9. doi:10.1016/S0966-6923(97)00035-5

Vermeiren, T. (2013). *Intermodal transport – The delta in the Delta*. Vrije Universiteit Brussel.

Vrenken, H., Macharis, C., & Wolters, P. (2005). In H. Vrenken, C. Macharis, & P. Wolters (Eds.), *Intermodal transport in Europe*. Brussels: European Intermodal Association (EIA).

Wichser, J., Weidmann, U., & Nash, A. (2007). Strategies for increasing intermodal freight transport between Eastern and Western Europe. European transport conference, Association for European Transport (pp. 1–18).

Woxenius, J. (2012). Directness as a key performance indicator for freight transport chains. *Research in Transportation Economics, 36*(1), 63–72. doi:10.1016/j.retrec.2012.03.007

CHAPTER 10

THE CARGO TRAM: CURRENT STATUS AND PERSPECTIVES, THE EXAMPLE OF BRUSSELS

Mathieu Strale

ABSTRACT

Purpose – *The purpose of this chapter is first to establish a state of the art about cargo tram, in order to understand its opportunities and constraints. In a second step, the aim is to add an often overlooked spatial dimension in the analysis, which may offer a renewed insight into the issue and extend the understanding of the potential use of urban light rail for freight.*

Design/methodology/approach – *The first step of the analysis is carried out through a study of the literature. In particular, the potential use of cargo tram for the various urban supply chains is evaluated. The aim is to establish a European state of the art on urban light rail use for freight. Then, to concretise these elements and to add an empirical geographical approach, the Brussels situation is studied in detail.*

Findings – *This combination of a theoretical analysis and a field approach of the case of Brussels allows us to lead an original study. Obstacles, opportunities and interest for the implementation of cargo*

Sustainable Logistics
Transport and Sustainability, Volume 6, 245–263

ISSN: 2044-9941/doi:10.1108/S2044-994120140000006010

trams are specified. In addition, the use of geographical approach gives a new point of view since most of the publications in this field are based on economic or technical approaches.

Practical implications – *This study may help public and private actors involved in urban freight matters, as well as public transport companies, to better understand the issues related to cargo trams. In a context where this transport mode is promoted and discussed in a growing number of cities and urban supply chains, it is useful to get an objective synthesis and a prospective analysis on this topic. This research may also have social implications in the way it helps to define a more sustainable urban logistics.*

Originality/value – *The originality of this research is related to the combination of a theoretical and empirical approach, to the link that is made between urban supply chains and capabilities of cargo trams and finally to the prospective study on the Brussels case. Indeed, because of the novelty of the topic, there are very few studies, either ex-post or ex-ante.*

Keywords: Light rail transport; freight; European situation; Brussels; retail deliveries

INTRODUCTION

It is well known that urban transport generates nuisances at local, regional or global levels. However, most of the scientific literature concentrated on passenger issues, whereas freight-related matters have been widely ignored (McKinnon, Cullinane, Browne, & Whiteing, 2010).

Yet urban goods transport generates significant environmental, social and economic harm. It is estimated that this activity corresponds to 10–20% of urban road traffic and of about 30% of polluting emissions (Dablanc, 2009; MDS Transmodal Limited & Centro di recerca per il transporto e La Logistica, 2012). These polluting emissions are more harmful in urban areas because of the denser population, which increases the amount of harm to the people. In addition, most freight vehicles are not adapted to urban areas, creating congestion, noise and higher risk of accident. Finally, most urban vehicles are diesel powered, generating severe polluting emissions (e.g. particulate matter emissions). This situation has led to a reduction in life expectancy in urban areas (McKinnon et al., 2010).

Nevertheless, urban transport and logistics are essential services for the economy and the inhabitants of the cities. The recent evolutions of the economy, that is its higher competition and internationalisation, have led to a growth of the volume of freight exchanges and of the related logistics services (Carroué, 2002; Rodrigue, Hesse, Comtois, & Slack, 2009). In urban areas, these evolutions are quite visible because many clients of logistics services are located in urban areas: inhabitants, retailers, as well as business or industrial activities (Hesse & Rodrigue, 2004; McKinnon et al., 2010; Regan & Golob, 2005). In addition, new logistics services, such as express deliveries or e-commerce, that generate an additional transport demand, are particularly used by urban clients (Browne, Allen, Nemoto, Patier, & Visser, 2012; Browne, Allen, Nemoto, & Visser, 2010; SETRA, 2009). Finally, urban agglomerations generally operate as transport nodes, concentrating freight transport and logistics activities operating at wider geographical scale (Rodrigue et al., 2009; Savy, 2006; Wemelbeke & Mariotte, 2007).

Recent reflexions have been led both by public bodies and by scientific researchers on these issues (APUR, 2011; Dablanc, 2007; ERZ, 2005; McKinnon et al., 2010). One of the potential solutions is the use of alternative transport modes for urban transport. In this aim, the use of urban light rail, underground rail services as well as cargo tram is the main part of the attention (Dinwoodie, 2006; Patier & Routhier, 2009a). It seems to be a good solution: this transport mode is mostly electric powered, reducing polluting emissions and noise (McKinnon et al., 2010), at least in urban spaces. However, multimodal freight transport is mostly studied at a wider scale, regional or national level (McKinnon et al., 2010; Rodrigue et al., 2009), so scientific literature remains quite scarce. This situation is probably the consequence of the lack of data and visibility of the urban freight sector. Another element is the fact that successful examples of the use of urban light rail for freight are rare (Robinson & Mortimer, 2005). This lack of success is mostly interpreted as the result of the inefficiency, the poor profitability or the need for adapted infrastructures required by the use of urban light rail. Yet, the use of this transport mode for goods is a quite ancient practice, both inside or outside cities (Regué & Bristow, 2013).

In this context, the purpose of this chapter is twofold. Firstly, it is the critical presentation of the use of urban light rail for freight transport and search of the related literature. This is very much needed since this literature is very scarce and there is almost no synthesis of experiments in this field. For treating this topic, the confrontation of recent and old schemes should be relevant. Light rail was used as a freight transport means before the development of road transport. Secondly, by studying the Brussels (Belgium) case, we want to determine the limits and constraints of the use of urban light rail, technically, economically and geographically. Indeed, we intend to prove that this often neglected geographical approach is needed in order to understand the opportunities and constraints related to the use of a rail transport for urban freight.

This chapter is organised as follows: in the next section, the ancient and current experiences of the use of urban light rail for freight transport and related literature are analysed. Following this section, Brussels situation is treated as an example of the potentialities and difficulties related to the use of light rail service for urban freight transport. Finally, we conclude by establishing a critical synthesis on the topic.

THE USE OF URBAN LIGHT RAIL FOR FREIGHT: A CRITICAL REVIEW

Urban light rail use for freight transport raises various issues, regarding environment, urban planning, transport management and economy. For now, there are very few examples of freight schemes using tramlines because of various reasons. However, this has not always been the case.

Indeed, the use of light rail infrastructures for freight transport, both in intra- and suburban exchanges, is an ancient practice. Before the development of an efficient road transport, rail and waterway transports were the only ways for consolidating freight flows (De Block & Polasky, 2011). In this context, light rail infrastructures have been used for urban and non-urban freight transport. For example, there were underground schemes in Chicago and London, and mainly standard tramlines were used for freight in other cities (Arvidsson, 2010; Visser, 2006). However, as in other rail freight businesses, the competition with road transport has been very unfavourable to rail services (Rodrigue et al., 2009). The situation was worst in urban areas because freight flows were transported on short distances and less consolidated and because of the high impact of just-in-time strategies in final customer and retail deliveries that represent the majority of goods exchanges in urban spaces. So, during the second half of the twentieth century, almost all urban light rail freight services disappeared (Arvidsson, 2010).

Nevertheless, a recent interest has grown on the (re)use of light rail services for urban freight (Arvidsson, 2010). Indeed, if the current urban logistics organisation seems quite economically efficient, its social and environmental impacts have been the subject of growing concerns. In addition, the growing congestion of urban areas and the bad image of road transport and of its users raise new issues. So public bodies, scientific researchers and some private actors seek innovative solutions (APUR, 2011; Regué & Bristow, 2013). In this context, a few new rail use experiences have been launched in Europe with various successes. These services are synthesised in Table 1. In addition to the four 'classic' examples of European cargo tram projects (Arvidsson, 2010; Robinson & Mortimer, 2005) – Dresden, Zurich, Vienna and Amsterdam – the Paris case is also integrated in the analysis. This case is special, because it is not a cargo tram, but a regional express network service that links Paris city centre and a distribution centre of the Monoprix supermarket enterprise, located in the neighbourhood (APUR, 2011). Even if the regional express network is rather a heavy rail infrastructure, Monoprix example fits with the other freight light rail services, in terms of use and objectives.

The first result is the fact that only small-scale projects, in terms of actors involved or of geographical coverage, survive. The cargo tram in Dresden is operated by a single actor, created and exploited by Volkswagen for carrying spare parts between its distribution centre and a factory in Dresden city centre, using only one tramline (Arvidsson, 2010). Monoprix train, in Paris, follows a quite similar strategy; the train is used by a single private operator that sends a full train every evening to Paris city centre for delivering its most central shops (APUR, 2011). On the public side, Zurich cargo tram is used as a mobile waste collection point, accepting particular waste: electronic or bulky ones (ERZ, 2005; Regué & Bristow, 2013). It works on a regular basis, covering the whole city with a dozen collection points. The current Vienna project intends to follow the same scheme (Regué & Bristow, 2013). On the opposite, all projects involving large-scale city distribution failed. Amsterdam's cargo tram project intended to create an entire city distribution network by light rail. However, the difficulties for funding this project, the cost of infrastructure investments and the difficulties for sharing the tram network with the passenger operator led in 2008 to the bankruptcy of cargo tram in Amsterdam, before its actual entry into service (Regué & Bristow, 2013). Similarly, prior to the Vienna project, another project was aimed at creating a city distribution network too, but preliminary studies and consultation concluded that this would not be a profitable system, leading to the reduction in scope of the cargo tram. In this context, another common element is the difficulty of creating sustainable services based on the consolidation of goods flows or on the

Table 1. The Current European Situation on Freight Use of Light Rail Networks.

	Dresden	Zurich	Vienna	Amsterdam	Paris
Type	Private/industrial tram	Waste collection	Waste collection (city distribution)	City distribution	Internal logistics of Monoprix
Geographical coverage	1 line	Whole city (nine collection points)	Whole city	Whole city	1 line
Frequency	16 tram/day	100 trips/year		40–50 trams	1–2 trains per day
Current situation	In operation	In operation	On hold/partial failure	Failure	In operation
Date of creation	2000	2003	2005	2007	2007
Type of goods	Spare parts	Electronic and bulky waste	Waste	Consumer goods	Consumer goods
Operator/investor	Private (Volkswagen)	Public (municipality)	Public (municipality)	Private (City Cargo)	Private (Monoprix)

Sources: Arvidsson (2010) and Regué and Bristow (2013).

cooperation of multiple actors (Verlinde, Macharis, & Witlox, 2012). The Vienna and Amsterdam projects of consumer goods distribution failed and, on the opposite, projects at Dresden, Paris and Zurich, which are mono-product or mono-actor, are in operation.

Scientific analyses of these projects are quite new. The privileged way of studying this issue is the economic approach; indeed, the current urban logistics organisation seems, at least, economically efficient (McKinnon et al., 2010; Regué & Bristow, 2013), hindering the development of alternative organisations that could be more environmentally or socially efficient. Arvidsson (2010) studied the potential use of trams and electric distribution vehicles for urban distribution by analysing the various European case studies in these fields and Amsterdam situation in particular. His conclusions are that light rail transport costs are higher than truck for the same distance; consequently the use of light rail must be coupled to attractive or exclusive services such as evening delivery, final delivery, truck access restriction and temporary storage. Another element is the fact that only small-scale projects, based on one actor or on one kind of goods, seem to be economically viable. Finally, the infrastructure issues, that is the sharing of tracks with passenger services, are highlighted (Trentini & Malhene, 2012). Regué and Bristow (2013) emphasise the key role of urban logistics for productivity and efficiency of urban economy and the risk that the social and environmental issues get higher because of the development of just-in-time strategies and e-commerce and of the high urbanisation rate in general. The main observation about light rail use for urban goods transport is the lack of concrete achievements in comparison with numerous theoretical studies about 'green logistics'. In this context, Regué and Bristow (2013) follow a cost–benefits analysis of the use of urban light rail in the case of Barcelona. Alessandrini, Delle Site, Filippi, and Salucci (2012) use a similar methodology, but in a more theoretical approach since there are no concrete applications of their findings. The main benefits of the use of light rail for freight seem to be the reduction of noise, air pollution, congestion and the potential integration in a consolidation scheme of urban goods. Indeed, as Institut National de Recherche sur les Transports et leur Sécurité (INRETS) research (2006) highlights, tramways consume less space than motorised vehicles, generate less polluting emissions in urban spaces because of the electric propulsion and create a smaller amount of noise than trucks or vans. Nevertheless, INRETS (2006) stresses that light rail vehicles may be very noisy at a local scale, where traffic is dense or close to particular infrastructures such as crossroads or curves. This may complicate the use of these infrastructures in dense urban areas and during the night in particular. In addition, according to Alessandrini et al. (2012), main difficulties are the lack of flexibility, the additional transhipment and transport costs and the competition for the use of infrastructures. So, necessary elements should be empty capacities on infrastructures and the concentration of flow generators around these tramlines. In this context, retail distribution would be unprofitable since it would need the creation of distribution centres and of new infrastructures, generating a more expensive solution than truck distribution. Robinson and Mortimer (2005) agree with this hypothesis, reporting the difficulties for adapting logistics chains and tram distribution, in particular for retail and other final distribution activities. Indeed, urban actors are seeking for just-in-time logistics, which are hard to conciliate with the use of light rail (Foyer, 2001). In addition, the knowledge of urban freight flows is weak since they are generally not recorded, which hardens the development of alternative solutions (Arvidsson, 2010). Thus, light rail use would remain limited to waste transportation or to other goods that do not need consolidation. However, the rising cost of the energy should be a good incentive for the use of urban light rail (Robinson & Mortimer, 2004). Following a more technical approach, Rien and Roggenkamp (1995) analyse the way light rail may be integrated in local distribution centre strategies or in the use of new technologies. Author highlights the fact that a wide geographical coverage by rail infrastructures is a need for an efficient use of light rail in a goods distribution strategy.

The organisation of urban deliveries is fundamental for understanding the opportunities of urban light rail for freight transport development. Supply chain of retail activities is divided between independent retailers, supermarkets and chain stores (Fernie & Sparks, 2009). Logistics of independent retailers are mainly based on intermediation via wholesalers and wholesale market. This organisation has several consequences. Firstly, wholesalers and wholesale markets remain located in and around cities. Thus, last delivery takes place in the city. Secondly, these deliveries are generally not consolidated between different shops and wholesalers, in favour of a non-environmentally optimal organisation, based on private passenger cars and small vans. Regarding the use of light rail, this situation raises challenges. Since the last mile part of deliveries is not consolidated and represents an important part of urban freight traffic, it should be integrated in the scope of the development of a light rail scheme for freight. However, the dispersion of actors, both retailers and wholesalers, and the fact that last miles are provided inside the urban area make the use of a higher capacity transport mode difficult. The need for many loading and unloading stops, and the short transport distances challenge the profitability and the efficiency of the use of light rail for these clients. Indeed, this explains the lack of successful experiences.

In chain stores and supermarkets, final distribution is centralised on distribution centres. These large warehouses are located out of the urban agglomeration for accessibility and real estate cost or land availability reasons (Hesse, 2002). Logistics is organised at the regional, national or continental level. Thus, goods flows are grouped, facilitating the use of bigger transport vehicles. In this context, using of light rail may be an opportunity for the final transportation between distribution centres and shops located in the cities. Yet, these warehouses may be located far away from urban limits and also from urban light rail infrastructures. Linking them to the network by building new infrastructures is expensive, reducing the profitability of such initiatives. In consequence, there are very few effective initiatives of the use of urban light rail and they are limited to some retail chains that have warehouses located near their shops in the city centre and consistent with the tram network (APUR, 2011). E-commerce logistics is organised in a very similar way. It is based on distribution centres (Fernie, Sparks, & McKinnon, 2010). However, final distribution is very different because it is not based on shops. Thus, it can be handled by the final consumer, a parcel distributor or by the e-commerce provider itself. Since the main issue is the rapidity and the flexibility of the service, consolidation initiatives are quite rare, reducing the opportunities for the use of freight trams. Nevertheless, some express distributors are able to consolidate their deliveries and to use alternative transport modes. But it is only on an individual initiative base, without sharing their infrastructures, because of the large volume of goods they deliver in urban agglomerations (APUR, 2011). Tertiary activities are another major source of urban goods exchanges. These are mixed fluxes, made of express deliveries and of other deliveries for office equipment. So, final distribution issues are a mix of the various deliveries of the organisations as shown in Fig. 1.

This review of the literature reveals the lack of knowledge about the way light rail could be integrated in urban logistics chains. Indeed, on the one hand, the organisation of these logistics chains has long been unknown because it was considered as an issue for private actors. On the other hand, these chains have been profoundly transformed by new economic organisation. Moreover, reflexions on urban light rail use for goods transport are very new. This reduces the concrete knowledge and the successful experiences on this topic. It seems that the main issues are the difficulties for developing consolidation initiatives and the sharing of infrastructures with

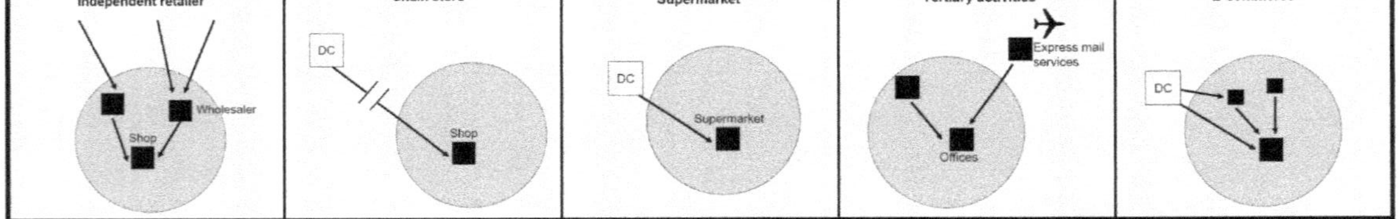

Independent retailer	Chain store	Supermarket	Tertiary activities	E-commerce
Final distribution of independent retailer is generally provided by wholesalers or organised via wholesale market. Thus, last miles, before the delivery of the shop, take place inside the urban area. There is generally no consolidation of this kind of fluxes. So, private car or small vans carry the vast majority of these goods, generating a important amount of intra-urban traffic. Delivery hours are included in shop opening hours because they are managed by the shop owners.	Final distribution of chain stores is organised via national or continental distribution centres. Final delivery route may be long because distribution center locations are optimised at a national or international level. Goods fluxes are consolidated. Deliveries are organised on delivery rounds, at the city, regional or national level. Trucks are the privileged transport mode. Delivery hours may be flexible but they are generally included in shop opening hours because of the lack of storage space and of delivery regulation into cities.	Final distribution of supermarkets is organised via national or regional distribution centres. Supermarket distribution centres are generally located nearby cities. So final distribution takes place between suburbs and cities. Goods fluxes are consolidated. Deliveries are organised on delivery rounds, at the city or regional level. Trucks are the privileged transport mode. Delivery hours are more flexible because all supermarkets are not located in the dense urban area and because of the possibility to store parts of the goods inside shops. However, delivery times may be regulated by public authorities.	Goods distribution of tertiary activities are twofold: Express mail services and other deliveries. Express mail services are mainly provided by global integrators. Their locations are mainly concentrated in suburbs, near major transport nodes such as airport or, at least, highways. Final distribution is provided by vans or, sometimes, alternatives modes like bikes or electric vehicles. These deliveries are organised by rounds. Other deliveries, furniture, food etc. are handled by 'classic' wholesalers, that may be located into cities. These deliveries aren't necessarily consolidated.	E-commerce distribution may follow different ways. Fluxes managed by biggest actors pass through distribution centres. So, they are consolidated and delivered via delivering rounds to final clients. However, if it is asked by the client, these goods may be sent to intermediate point (pick up point, post offices, temporary storage areas). Thus final clients take the last mile in charge. Other e-commerce may be originated from other actors, like food distributors. In these cases, delivering circuits may transit via stores or dedicated distribution centres. Delivering vehicles are various, depending on the provider and the circuit choice.

Fig. 1. The Organisation of Deliveries of Retail Actors. *Sources*: Fernie and Sparks (2009); Fernie et al. (2010); Strale (2013) and Wayens (2006).

passenger services. The successes are related to exclusively private or public projects which involve a limited number of actors and are based on goods that are already consolidated. Another important element is the absence of a geographical analysis of this topic; the only spatial elements that are mobilised in the literature are the fact that an extensive light rail network is more efficient than a small one and the need for a spatial coverage of the key freight generators by this network.

By analysing the Brussels case, this chapter aims at answering this need for new concrete facts in two domains: the organisation of urban logistics chains and the way light rail services may be integrated in these chains. The methodology is based on a geographical approach, analysing the location of actors and fluxes inside Brussels urban area.

LIMITS AND CONSTRAINTS: THE BRUSSELS CASE

Brussels regional administrative area is small (161 km^2). It covers half of the morphological urban area and has around one million inhabitants. Like most of the European cities, Brussels economy is mainly based on tertiary activities and retail, whereas industrial activities have declined steadily. In consequence, urban deliveries are mostly related to retail and tertiary activities (Fig. 2). Since there is no statistical data on deliveries in Brussels,

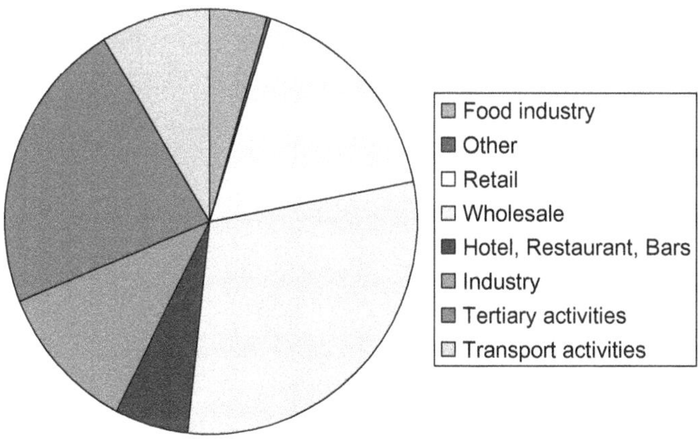

Fig. 2. An Estimation of the Deliveries in Brussels. *Source*: Patier and Routhier (2009b) and author's own calculations.

Fig. 2 is based on the employment in Brussels region and on delivery ratio (Patier & Routhier, 2009b). Nearly 50% of deliveries come from wholesale and retail activities. Around 25% are related to other tertiary activities. The other deliveries result from industries and transport activities.

Tram network in Brussels is quite dense for a West European city. Brussels has kept a large part of its historic tram network, even if part of it were deleted or replaced by bus or metro lines. As such, some current tramlines in the city centre are under the ground – they are called 'premetro' lines – because, initially, they had to be transformed in full underground lines. Consequently, the network length was nearly halved between 1945 and 2000. Yet, nowadays, the tram network is expanding again, thanks to new interest for high capacity public transport.

Regarding retail geography in Brussels in relation to this tram network (Fig. 3), one can conclude that major retail areas are connected to the tram network, which seems also to be true for shopping malls. This could be

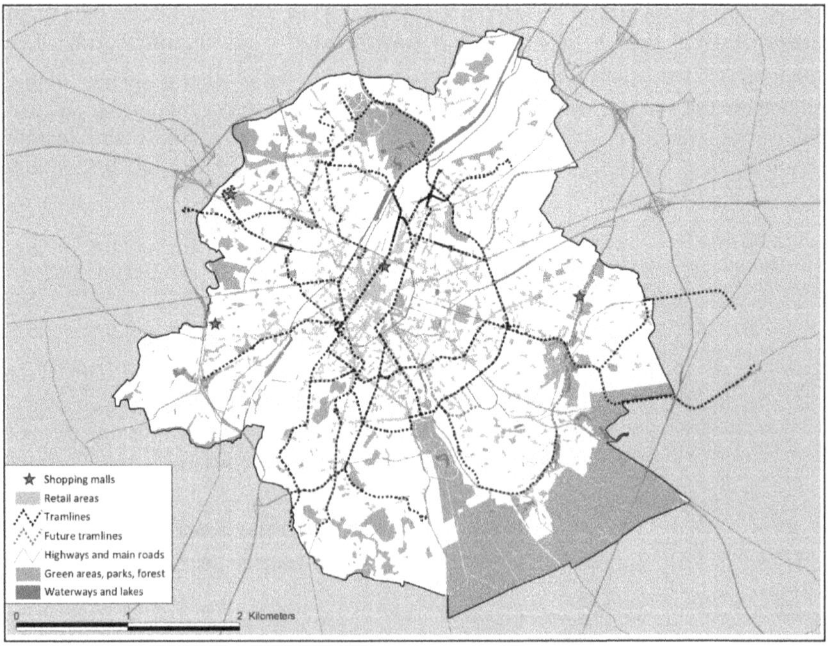

Fig. 3. The Brussels Tram Network and the Location of Major Retail Areas. *Source*: Wayens (2006).

favourable for the use of the light rail network for urban goods distribution. However, the situation is far from so easy. Firstly, as we said, part of the tram network is underground. It complicates its use for retail delivering, because of the need for handling capacities and new infrastructures linking the surface and the tramlines. Secondly, because the Brussels tram network is old, many lines run in narrow streets in the dense urban area, not benefiting from exclusive lanes. Thus, the availability of space for handling goods or creating new infrastructure dedicated to freight trams is very low. Moreover, traffic jams reduce the reliability of tram services, which is already a problem for passenger services (Lebrun, Hubert, Huynen, De Witte, & Macharis, 2013), but may be more severe for deliveries of time-sensitive goods. In relation to this last element, it is important to know that the current tram network is already saturated by passenger traffic. This reduces the capacities for the development of freight services, both in time and scale. It also signifies that an extensive goods delivery service would rapidly demand large investment in new capacities for not disturbing the passenger network. It is also a risk in terms of reliability of hypothetical freight services, because passengers vehicles will surely have priority on freight ones.

Moreover, the geography of deliveries is conditioned by the location of distribution centres, wholesalers or wholesale markets. In Brussels, this geography creates various constraints (Fig. 4). On the one hand, wholesalers and wholesale markets have very central locations; most of them are located inside the regional limits. In consequence, many final deliveries of Brussels' retailers take place inside the agglomeration. So, these trips are in the operating range of freight trams. Yet, such freight tram service would need lots of loading and unloading stops and very short trips, which increase costs and reduce the efficiency. An element that could change the situation is the medium-term project of Brussels' authorities to regroup wholesale activities in the northeast of the agglomeration, thereby creating a cluster that may be better suitable for freight tram services (Région de Bruxelles-Capitale, 2013, 2014). However, at this time no new tramlines are projected in this area. At the opposite, distribution centres of large retailers and supermarkets are located outside the city limits. Indeed, truck access is better in the surrounding of the city, because of the geography of highways, and there is a better land availability. In consequence, supermarket warehouses are outside the limits of the Brussels tram network. To link these warehouses with this network would be very expensive, but one can say that it could be useful for the suburbs population. In the Belgian federal state, public transport and land planning are regional matters. So, an

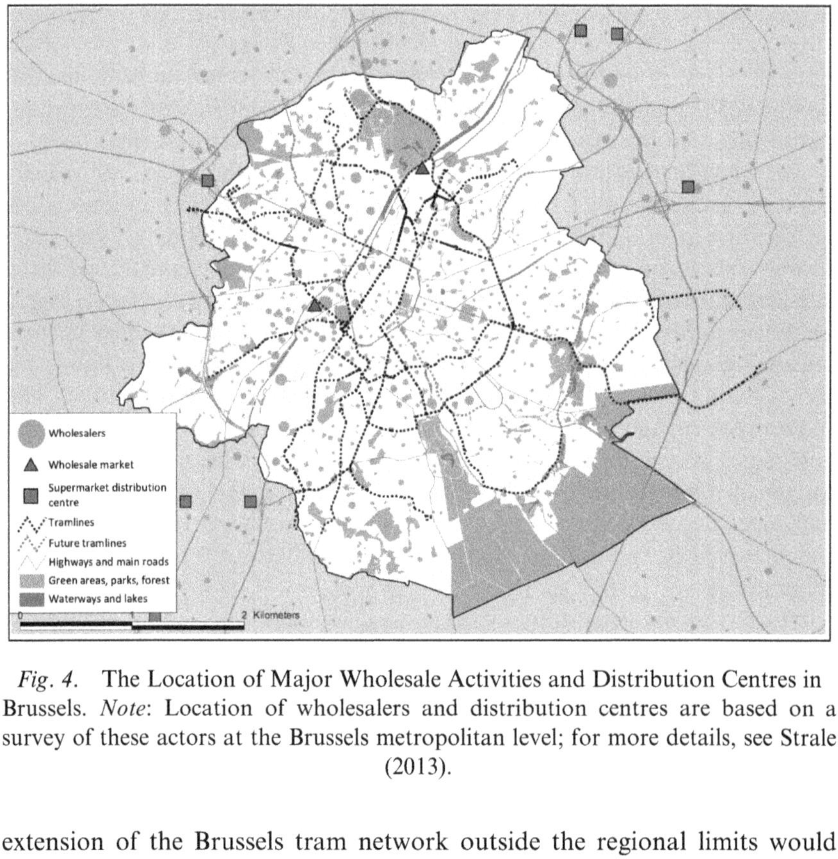

Fig. 4. The Location of Major Wholesale Activities and Distribution Centres in Brussels. *Note*: Location of wholesalers and distribution centres are based on a survey of these actors at the Brussels metropolitan level; for more details, see Strale (2013).

extension of the Brussels tram network outside the regional limits would require consultations between Brussels and Flemish authorities, duplicate administrative procedures, etc. This is one of the reasons why there are so few interregional tramlines. Nevertheless, the Flemish public transport operator is interested in creating new tramlines in Brussels' suburbs that would be linked with the metropolitan network (Lebrun et al., 2013). If new lines are close to the location of distribution centres, this could be an opportunity for developing freight services between them and the shops located in the city. However, the dispersion of warehouses in the suburbs complicates the creation of an efficient service. So it seems impossible to create common deliveries gathering supermarket operators.

Since freight tram services for retail distribution seem to be complicated, it may be useful to look at other potential clients. These are express mail services. Indeed, this sector has various strengths. Actors are geographically concentrated around the airport, northwest of the city. Secondly, their goods are mainly parcels, which are suitable for light rail distribution. Finally, an efficient and rapid link between the tertiary spaces of the city centre and the main warehouses of express mail services should be very attractive, because of the congestion of roads (Macharis, Dobruszkes, & Hubert, 2014). For now, there is no tramline linking these places and the city centre, but a new line is projected in the future. However, this raises the same issues about interregional collaboration. Nevertheless, it seems to be one of the best ways for implementing a freight service on tramlines in Brussels (Fig. 5).

Other potential goods to be carried by freight trams are wastes. There are reflexions about the use of alternative transport modes for this activity. However, the focus is on the use of waterway between the south of the agglomeration and the waste incinerator in the north. Since this incinerator is close to existing tramlines, the development of a *waste tram* may be an interesting line of thought. It could be based on night or evening services, not competing with passenger trams. However, it would imply the

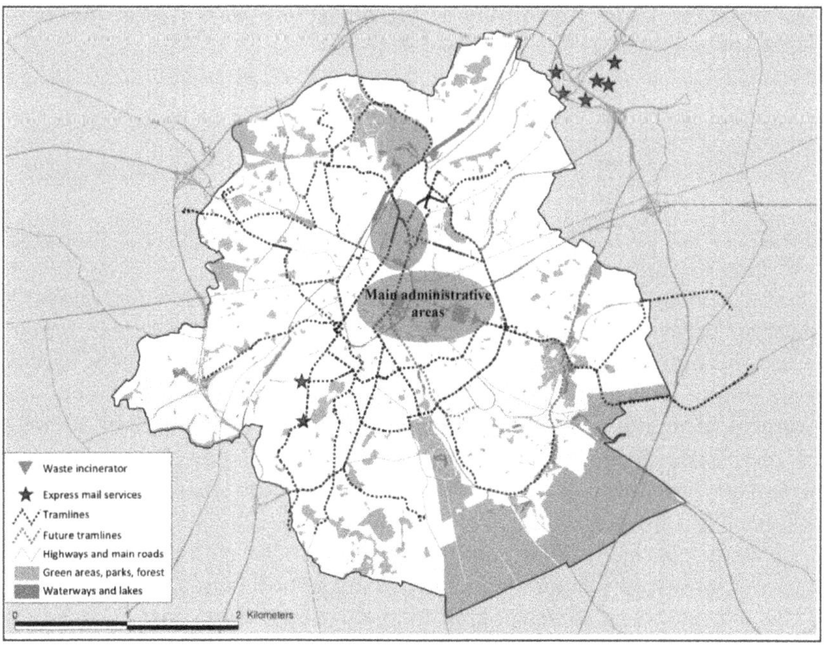

Fig. 5. The Other Freight Issues in Brussels. *Source*: Strale (2013).

development of new infrastructures such as collecting points or the use of silent vehicles.

In this context, reflexions on the development of a freight tram in Brussels are at their very beginning. The new Brussels strategic plan for freight transport (Région de Bruxelles-Capitale, 2013), which specifies the future public policy in freight issues, integrates freight tram in various ways. Brussels' authorities intend to use tram network for shop deliveries in general and as a tool for an integrated freight transport approach, which would include an urban distribution centre. In these projects, the whole tram network should be used. The integration of the light rail network with the underground network and the heavy rail network for freight distribution is also considered. Yet, these are long-term projects, with no concrete implementation strategy.

Regarding our results, it seems very complicated to complete these ambitions, because of the Brussels context – policy issues, dispersion of the actors, inherent difficulties of the light railway network – and the whole European situation about the freight tram – a very complicated adaptation of tram schemes to urban freight needs. However, on some particular markets, such as waste transport, or for clients that need a large amount of goods, there may be some opportunities.

CONCLUSIONS

The use of urban light rail networks for freight transport is a new issue, both for scientific researchers and for policy makers. It meets various issues: making urban logistics more sustainable, developing innovative transport solutions and renewing light rail for urban transport.

In this context, the development of freight tram is very limited, because of the many difficulties. These are the consolidation of urban goods, the divergent demand of urban economic actor regarding freight services, the poor knowledge about urban freight flows and the competition with passenger services on light rail networks. Hence, only small-scale or private initiatives can survive.

Nevertheless, many urban authorities want to develop new freight tram services, because of the good and modern image of this solution and because of the redevelopment of urban light rail networks in European cities.

The Brussels situation illustrates these ambitions and the various difficulties that are linked to the concrete use of light rail network for freight.

The density of the urban fabric and the configuration of the network, the location of the various key spaces for urban freight and the operational and political issues must be taken into account. For now, the development of a new cargo tram in Brussels seems very complicated.

An effective development of freight tram services requires essential elements: the good knowledge of urban freight realities of the city, the identification of free capacities on light rail network and of the needed infrastructure investments and finally the definition of the targeted clients and goods. In this context, some sectors, such as express mail services, wastes or large retailers, may be interested in the development of freight trams in Brussels. But it would imply the building of new tramlines and other accompanying infrastructures for moving goods.

As in other innovative urban freight initiatives, the willingness is not sufficient to get results. But, with good preliminary analyses and a careful approach, freight trams may participate in the renewing of the organisation of urban goods exchanges.

REFERENCES

Alessandrini, A., Delle Site, P., Filippi, F., & Salucci, M. V. (2012). Using rail to make urban freight distribution more sustainable. *European Transport/Trasporti Europei*, *50*(5), 17.

APUR – Atelier parisien d'urbanisme. (2011). Le projet tram-fret – Mise en place d'une expérimentation de transport de marchandises par le tramway. Point d'avancement, p. 38.

Arvidsson, N. (2010). New perspectives on sustainable urban freight distribution: A potential zero emission concept using electric vehicles on trams. Article présenté à la 12th WCTR, July 11–15. Lisbon, Portugal.

Browne, M., Allen, J., Nemoto, T., Patier, D., & Visser, J. (2012). Reducing social and environmental impacts of urban freight transport: A review of some major cities. *Procedia Social and Behavioral Science*, *39*, 19–33.

Browne, M., Allen, J., Nemoto, T., & Visser, J. (2010). Light goods vehicles in urban areas. *Procedia Social and Behavioral Sciences*, *2*, 5911–5919.

Carroué, L. (2002). *Géographie de la mondialisation*. Paris: Armand Colin.

Dablanc, L. (2007). La notion de développement urbain durable appliquée au transport de marchandises. *Les Cahiers Scientifiques du Transport*, *51*, 97–126.

Dablanc, L. (2009). *Freight transport for development toolkit: Urban freight*. Department for International Development, The World Bank, Washington, DC, 57p.

De Block, G., & Polasky, J. (2011). Light railways and the rural/urban continuum: Technology, space and society in late nineteenth-century Belgium. *Journal of Historical Geography*, *37*, 312–328.

Dinwoodie, J. (2006). Rail freight and sustainable urban distribution: Potential and practice. *Journal of Transport Geography*, *14*, 309–320.

ERZ Entsorgung + Recycling Zurich. (2005). Cargo-Tram Zurich – The environmental savings of using other modes by Gottfried Neuhold, Bestufs 2005 conference, Amsterdam.

Fernie, J., & Sparks, L. (2009). Retail logistics, changes and challenges. In J. Fernie & L. Sparks (Eds.), *Logistics and retail management: Emerging issues and new challenges in the retail supply chain* (pp. 3–37). London: Kogan Page.

Fernie, J., Sparks, L., & McKinnon, A. C. (2010). Retail logistics in the UK: Past, present and future. *International Journal of Retail & Distribution Management*, *38*(11–12), 894–914.

Foyer, P. (2001). Just-in-time by rail: Enabling rail freight to return to the city. BESTUFS Workshop 5, Rail based transport: A disappearing opportunity or a challenge for urban areas? August, 30–31, DVB-Dresden, Germany.

Hesse, M. (2002). Shipping news: The implication of electronic commerce for logistics and freight transport. *Resources, Conservation and Recycling*, *36*, 211–240.

Hesse, M., & Rodrigue, J.-P. (2004). The transport geography of logistics and freight distribution. *Journal of Transport Geography*, *12*, 171–184.

INRETS – Institut National de Recherche sur les Transports et leur Sécurité. (2006). *Bruit et vibration des tramways: Caractérisation des sources et évaluation de la perception.* Enquête auprès des riverains: méthodologie et premiers résultats. Rapport INRETS-LTE No. 0614, 28p.

Lebrun, K., Hubert, M., Huynen, P., De Witte, A., & Macharis, C. (2013). *Les pratiques de déplacement à Bruxelles* (Vol. 2, p. 108). Bruxelles: Cahiers de l'Observatoire de la mobilité de la Région de Bruxelles-Capitale.

Macharis, C., Dobruszkes, F., & Hubert, M. (2014). *Mobilité et logistique à Bruxelles/ Mobiliteit and logistiek in Brussels* (p. 217). Brussels: VUBPRESS Brussels University Press.

McKinnon, A. C., Cullinane, S., Browne, M., & Whiteing, A. (2010). *Green logistics. Improving the environmental sustainability of logistics.* London: Kogan Page.

MDS Transmodal Limited, & Centro di recerca per il transporto e La Logistica. (2012). European commission: Study on urban freight transport, final report, DG Move.

Patier, D., & Routhier, J.-L. (2009a). La logistique urbaine, acquis et perspectives. Introduction au dossier. *Les Cahiers Scientifiques du Transport*, *55*, 5–10.

Patier, D., & Routhier, J.-L. (2009b). Une méthode d'enquête du transport de marchandises en ville pour un diagnostic en politique urbaine. *Les Cahiers Scientifiques du Transport*, *55*, 11–38.

Regan, A. C., & Golob, T. F. (2005). Trucking industry demand for urban shared use freight terminals. *Transportation*, *32*(1), 23–36.

Région de Bruxelles-Capitale. (2013). Plan stratégique pour le transport de marchandises en Région de Bruxelles-Capitale, adopté par le Gouvernement de la Région de Bruxelles-Capitale le 11 juillet 2013.

Région de Bruxelles-Capitale. (2014). Projet de Plan Régional de Développement Durable. Retrieved from http://www.prdd.be/

Regué, R., & Bristow, A. L. (2013). Appraising freight tram schemes: A case study of Barcelona. *European Journal of Transport and Infrastructure Research*, *2013*(1), 56–78.

Rien, W., & Roggenkamp, M. (1995). Can trams carry cargo? New logistics for urban areas. *World Transport Policy and Practices*, *1*, 36–40.

Robinson, M., & Mortimer, P. (2004). Urban freight and rail. The state of the art. *Logistics and Transport Focus*, *6*(1), 46–51.

Robinson, M., & Mortimer, P. (2005). Rail in urban freight: What future, if any? *Logistics and Transport Focus*, *6*(2), 33–37.

Rodrigue, J.-P., Hesse, M., Comtois, C., & Slack, B. (2009). *The geography of transport systems* (2nd ed.). New York, NY: Routledge.

Savy, M. (2006). *Logistique et territoire, Délégation interministérielle à l'aménagement et à la compétitivité des territoires (DIACT)*. Paris: Collection Travaux.

Sétra – Service d'étude sur les transports, les routes et leurs aménagements. (2009). Les bâtiments logistiques – Fonction et impacts sur les territoires. *Synthèse des connaissances*, 64.

Strale, M. (2013). La logistique: localisation des activités et impacts territoriaux, thèse de doctorat en Sciences, présentée à l'Université Libre de Bruxelles, sous la direction de C. Vandermotten et M. F. Godart.

Trentini, A., & Malhene, N. (2012). Flow management of passengers and goods coexisting in the urban environment: Conceptual and operational point of view. *Procedia Social and Behavioral Sciences*, *39*, 807–817.

Verlinde, S., Macharis, C., & Witlox, F. (2012). How to consolidate goods without setting up an urban consolidation center? *Procedia Social and Behavioral Science*, *39*, 687–701.

Visser, J. (2006). Underground freight transport systems. In E. Taniguchi & R. G. Thomson (Eds.), *Innovations in freight transport* (pp. 125–152). Boston, MA: WIT Press.

Wayens, B. (2006). Structure et dynamique du commerce de détail bruxellois. Thèse de doctorat en Sciences à l'Université Libre de Bruxelles.

Wemelbeke, G., & Mariotte, H. (2007). L'essor de grandes zones logistiques accompagne la massification des flux routiers. *SESP En bref*, 4.

ANTICIPATION

CHAPTER 11

TOWARDS ZERO EMISSION URBAN LOGISTICS: CHALLENGES AND ISSUES FOR IMPLEMENTATION OF ELECTRIC FREIGHT VEHICLES IN CITY LOGISTICS

Hans Quak and Nina Nesterova

ABSTRACT

Purpose – *Electric freight vehicles (EFVs) are one of the solutions to improve city logistics' sustainability. EFVs, that are electric powered light and heavy vehicles with a number plate, have the potential to make zero emission city logistics possible within the urban area. However, although trials have been undertaken for the last years, large-scale usage of EFVs in city logistics does not occur yet. EFVs are technically possible, but the implementation of EFVs in practice is relatively limited.*

Sustainable Logistics
Transport and Sustainability, Volume 6, 265–294

ISSN: 2044-9941/doi:10.1108/S2044-994120140000006011

Design – *This chapter examines by reviewing current and past EFV implementations, what are the challenges, barriers and success factors for EFVs in city logistics operations. EFVs have especially positive environmental effects, but are overall usually more expensive (especially in procurement) than conventional vehicles. Besides, other technical and operational issues remain to be solved, and many uncertainties still exist on long-term usage.*

Findings – *Three main barriers for large-scale EFV uptake are identified. The current logistics concepts are developed for conventional vehicles and should be redesigned to fit EFVs better. Local authorities' support is essential in order to find a positive (or not too negative) business case. And EFV implementation requires companies that want to be sustainable. This contribution presents examples of how some companies or authorities deal with these barriers.*

Value – *This chapter concludes by identifying elements that are necessary for acceleration of EFV uptake in city logistics operations.*

Keywords: Electric freight vehicles; city logistics

ELECTRIC FREIGHT VEHICLES IN CITY LOGISTICS

City logistics operations are usually recognized for their unsustainable impacts. Even though the modern urbanized civilization requires an efficient city logistics system in order to sustain the demands for goods and services and to remove waste, decreasing the negative impacts of the system takes more effort. City logistics operations have a negative impact on all sustainability indicators in several ways, that is:

- People (consequences of accidents with large vehicles, nuisance (noise) and the effects of local emissions on public health);
- Profit (logistics inefficiencies due to last mile issues and local regulations, but also by adding to the often severe urban congestion);
- Planet (adding to the CO_2 emissions and the consequences of local emissions and nuisance on organisms in the urban area).

One of the difficulties in city logistics is that many different actors with different (and sometimes even conflicting) interests are involved, which makes it a challenge to come up with sustainable solutions in this field. Van

Rooijen and Quak (2013, p. 505) illustrate it on the CIVITAS example: 'compared to the other thematic clusters of CIVITAS, the urban freight logistics cluster has not reached the results that were hoped for. A large part of the measures in this cluster is not continued after the project or not completely implemented at all during the project. The main reason is that for freight transport the participation of private companies is needed because these measures take place in a competitive market'.

This conclusion highlights one of the problems in city logistics: even though local authorities and residents are the most affected by the sustainability problems in the cities (e.g. emissions and nuisance), solutions to improve sustainability are usually lying within the scope of influence of private companies (especially carriers, and indirectly on their principals such as the shippers). Carriers often already organize their operations as efficient as possible within their ability, since this is their core business. The carriers' solution space is limited due to local authorities' restrictions (e.g. time access windows), infrastructure in inner cities or commercial requirements (e.g. delivery times requested by (opening hours of) receivers). However, from a city perspective the situation is not as optimal as it is from most carriers' perspectives. Different carriers visit the same streets and shops with only a limited volume of goods, optimizing their own routes and operational planning. From a city perspective it would be far more optimal to bundle goods based on the destination, thus decreasing the number of vehicles entering the city. This is opposite of how most supply chains are designed; usually goods are bundled from the origin (e.g. a shipper's or carrier's depot). As a result, it appears to be, from a city perspective, a poorly organized urban logistics system (see, e.g. Dablanc, 2007).

Although the described city context seems very complex, we can distinguish three main directions for solutions to improve the sustainability of city logistics operations (Quak, 2011):

- Technology (including vehicle technology and ICT solutions)
- Policy
- Logistics

Very often, technology, policy and logistics solutions are used separately and are initiated by different actors, whereas a combination of efforts in each of the directions is required in order to make city logistics really sustainable. One reason to focus more on making city logistics operations more sustainable is the EU objective that aims at zero emission city logistics (see EC, 2011): 'Halve the use of "conventionally-fuelled" cars in urban transport by 2030; phase them out in cities by 2050; achieve essentially

CO_2-free city logistics in major urban centres by 2030'. In order to do so, technical solutions are necessary; electric freight vehicles (EFVs) are able to answer these requirements.

Implementation of EFVs in city logistics is not a new phenomenon. The characteristics of many city logistics operations, including a relatively limited number of vehicle-kilometres per trip and multi-drops (and collections) per trip, make these operations an apparent, perfect match for electric vehicles (EVs). The first (modern) trials/demonstrators were undertaken more than 20 years ago (e.g. within EU-funded projects CIVITAS VIVALDI, MIRACLES, TELLUS, EVD POST project). The main challenges faced then by operators in implementation of EFVs in city logistics were high procurement costs, limited range of vehicles, little or no after-sale support, long waiting time for spare parts, low performance of the lead–acid, nickel–cadmium and ZEBRA battery technologies, low vehicle speed and limited payload. In short, these early EFVs were far from perfect and were not a serious alternative for conventional vehicles (i.e. internal combustion engine, ICE) in city logistics operations. The development of more reliable and better-performing batteries was considered crucial for all types of EVs to become (more) competitive to conventional vehicles.

Today, despite the significant progress being made in improving EV performance and after addressing many of the issues from the early trials and demonstrations, the implementation of EFVs in city logistics is still in its infancy. EFVs are defined as freight vehicles with a licence plate, so excluding electric bicycles and tricycles. The load capacity of these EFVs runs from relatively small vehicles, for example Goupils and Renault Kangoo ZE (about 700 kg cargo capacity), to larger vans and small trucks (e.g. Smith). Large-scale uptake, which is necessary to achieve considerable impact on the sustainability of the city logistics operations, does not take place yet. The new EFVs have a greater range and an improved loading capacity, but the actual implementation of EFVs in practice shows other new and emerging challenges.

This chapter examines the state of the art of the EFVs implementation in city logistics. Using EFVs for city logistics operations is only one of the solutions to organize city logistics in a more sustainable way. Other ways are equally important but are not in the scope of this chapter. Nevertheless, when investigating this technological solution, it was also considered how and in which way the other solution directions – policy and logistics – should be incorporated in order to more successfully implement this new technology in practice. This contribution is based on the state-of-the-art study on EFVs implementation in city logistics that was undertaken as a start of the FREVUE project (see FREVUE, 2013, for full report and all references used for this state-of-the-art study). FREVUE is a project co-funded by the European Commission under the Seventh Framework Programme, Theme 7, Sustainable Surface Transport.

METHODOLOGY

The objective of the state-of-the-art study is to identify what the main strengths and weaknesses are in the current implementation of the EFVs in the city logistics and what lessons can be learnt from these first implementation initiatives. The research approach includes two interrelated steps:

1. Existing projects in which EFV implementation was evaluated (EU, national, regional and local projects as well as initiatives from the private sector) were reviewed in order to define the main challenges and success factors from past and current implementation of EFVs in city logistics.
2. An in-depth analysis of a limited number of selected cases was carried out, in order to identify lessons to be learnt and to provide inspiring examples of how EFV implementation is possible.

This two-step method is used, so that this state of the art shows the general issues concerning EFV implementation, but can also be used as a more inspiring document in which specific cases show where EFV implementation was possible and positive (however, from these cases generalizability is more difficult). This chapter presents the overall results and conclusions from both steps in the following way: the results of the first step are presented in this section (i.e. the projects) and the overall project results in the section 'Challenges and Factors of Success for EFV Implementation'. Next, the results of the second step, the in-depth cases, meant as more inspiring examples of how EFV implementation is possible, are presented in the section 'Dealing with EFV Implementation Issues: Good Examples'.

This study does not attempt to be exhaustive in providing an overview of all past and ongoing projects and initiatives concerning EFVs in city logistics operations. In the selection of the projects for review, we only focused on those that meet a number of criteria (based on the scope as well as for practical reasons):

- Projects/initiatives that took place in Europe;
- Projects that include actual physical (operational) running of the EFVs and that present results of the EFV demonstration and implementation, in terms of technological, economic, systemic or environmental impacts;

- Projects that include implementation of light or heavy delivery vehicles, but not any other electromobility solution for the urban logistics (e.g. electric bicycles, tricycles or trams);
- Projects/initiatives reported in English, French, Dutch or Norwegian.

These projects/initiatives were identified through consultation and engagement with industry and other stakeholder groups such as FREVUE project partners and via desk-based research, including Internet and literature searches of scientific articles. The first step resulted in the review of 9 EU projects running EFVs demonstrations, 17 national initiatives on electromobility and 29 private operators' experiences. The overview of the literature review is summarized in Table 1. For the full list of projects and short description of results, see Appendix.

The analysis of the results from demonstrators was organized along five main topics:

1. Technical performance of the vehicles, the batteries and the charging systems;
2. Economic performance comprising the business case and analysis of the total cost of ownership (TCO);

Table 1. Overview of Reviewed Projects and Initiatives.

Category	Reviewed Projects/Initiatives
EU projects	ENCLOSE, SMARTFUSION, CIVITAS, CO_2NeuTrAlp, ELCIDIS, TURBLOG, EVD Post, SELECT
National initiatives	MOVELE (ES), Plugged-In Places programme (UK), Source London (UK), Plugged-in Fleets (UK), LuccaPort (IT), ElektrofahrzeugGroßversuch (DE), Electric Mobility Pilot Regions (DE), Frankfurtemobil (DE), MOBI.E (PT), Elektrisch Vervoer Centrum Rotterdam (NL), E-Laad (NL), Flemish Living Lab Electric Vehicles (BE), EVE – Electric Vehicle Systems Programme (FI), EMIL, EMKEP (DE), Cologne-Mobil (DE), SendSmart (NO), Stadsleveransen (NO)
Initiatives from private companies	STEF–TFE (FR), Deret (FR), Peter Appel (NL), CityShopper (NL), UPS (DE), EROSKI (ES), TNT Express (UK), Correos (ES), SEUR (ES), 020 stadsdistributie (NL), Eco-Logis (IT), PostenNorge (NO), La Poste (FR), Seymour Green (UK), Tesco (UK), Cargohopper (NL), Abby Couriers (UK), Brewers (UK), EntrepriseMouchel (UK), Gnewt Cargo Ltd (UK), Melrose and Morgan (UK), Nappy Ever After (UK), Office Depot (UK), Sainsbury (UK), Speedy (UK), Go-Ahead (UK), L'Oreal (FR), Label Route (FR), Post Nord (DN)

3. Systemic impacts of EFV usage in cities on the transport network and the environmental effects, including impacts on congestion, air quality and CO_2 emissions;
4. Wider attitudinal and social impacts of EFVs, such as customers' and drivers' perspective, as well as traffic safety issues and
5. The effects of policies, procurement and governance on the uptake and performance of EFVs.

In the second step of this study, six demonstration cases were selected for a more in-depth analysis. These cases were chosen based on the following criteria: availability and depth of information, duration and continuation (ongoing or not) of the case, type of vehicles employed, existence of quantified results, operational and regulatory environments and project sponsors (private and/or public initiatives).

CHALLENGES AND FACTORS OF SUCCESS FOR EFV IMPLEMENTATION

Electromobility is slowly gaining its place in current practices and in city logistics in particular. There are multiple major initiatives at the EU, national, regional and local levels supporting the development of EV infrastructure, demonstrating the personal and business usage of EVs, focusing on the creation of a supportive regulatory environment for EV implementation, raising awareness and promoting electromobility. These initiatives and demonstrations have different initial motivations, but ultimately all contribute towards a common goal: decreasing the use of fossil energy by transport and therefore reducing CO_2 emissions as well as local emissions to improve air quality. The EU supports sustainability measures in transport in order to achieve the goals stated in the White Paper on Transport (2011) and, on a more global level, to fulfil its commitments to the Kyoto Protocol. National and local stakeholders seek to implement sustainability measures on the urban level to meet EU requirements and to improve the quality of life for their inhabitants. Private companies are implementing these measures because, along with a good image and innovative behaviour, it might bring them cost savings due to reduced energy costs and improved operational efficiency.

The demonstrators, trials and initiatives that include actual running of the EFVs were reviewed with the aim of understanding the current challenges and factors of success. The assessment criteria included their technical, economic, operational, environmental, social and wider systemic performances of the EFVs in city logistics and what lessons can be learnt from them. The state-of-the-art review resulted in identification of the challenges and success factors for EFV implementation and uptake in daily city logistics operations, resulting in the SWOT analysis presented in Table 2.

Technical Performance EFVs

The range of EFVs is usually not more than 100–150 km at most. Next, the climatology and topography of a city have a great impact on the actual vehicle range. The range promised by the manufacturer is often not reached, although new vehicles have a higher real range. Whether the range is a limiting factor depends on the logistics operations, since many typically city logistics operations start and finish close to a city and have many stops (e.g. parcel deliveries), an actual range of 100 km might be sufficient. Technical issues observed include: failing batteries (and limited or late) support, equipment availability issues, relatively long charging time and the necessity to adapt charging infrastructure for fleet needs. The rapid improvement in the technology is mentioned as a reason for waiting to acquire EFVs, as (some) carriers do not want to buy vehicles knowing that the newer generation will perform better. This is often an argument to wait with actually purchasing EFVs, whereas the price of EFVs is expected to decrease in case of large implementation. (This can be seen as a chicken-and-egg problem.) The limited availability of standard vehicles and vehicle types is also a factor that is seen as a barrier for EFV implementation, and this is especially the case for larger vans and trucks. Some operators retrofit their ICE to EFVs, because the desired EFVs are not available from a manufacturer (or are too expensive).

Operational Performance EFVs

EFVs demonstrate both positive and negative operational performance characteristics compared to conventional vehicles. Because of their environmental performance and reduced noise level they are sometimes permitted in larger geographical areas and during larger time window periods in cases where any of those restrictions exist. Some technological features, like an acute turning range, steering circle and improved visibility facilitate the

Table 2. SWOT Analysis of the EFVs Compared to ICE Vehicles (in City Logistics Operations).

Strengths EFVs Compared to ICE Vehicles	Weaknesses EFVs Compared to ICE Vehicles
Economics/operational performance: • Lower maintenance and fuel cost • In case of government support (e.g. larger time windows, environmental zones and free parking) allow more efficiency in operation Social and environmental impacts: • Good environmental performance and improved air quality • Reduced noise nuisance since vehicles are silent (some noise is introduced for safety reasons though), but only for (un)loading equipment and drivers behaviour • Drivers are happy with vehicles and acute turning range, which is helpful on city's streets • Positive general acceptance from public • Contributed to the positive image of transport operator and shipper	Technical performance: • The range promised by the manufacturer is not reached and is limited in comparison to ICE vehicles • Failing batteries • Limited or late technical support • Relatively long charging times • Necessity to adapt charging infrastructure for large fleet needs • Limited availability and types of vehicles on the market Economics/operational performance: • High procurement costs • High cost of battery • Long downtimes (due to malfunction) and limited availability of spare parts • Limited loading capacity • Slow maintenance services and vehicles • Extra transhipment costs might occur (transhipment from ICE vehicles to EFVs)
Opportunities EFVs Compared to ICE Vehicles	**Threats EFVs Compared to ICE Vehicles**
Technical performance: • New(er) vehicles have higher real range • The availability of public charging points is seen as a confidence boosting measure • New vehicles and batteries are available in near future (from one side it is positive, from another – reason to wait (at the moment) to buy). Increasing battery performance also follows from uptake of passenger EVs Economics/operational performance: • If regulative environment supports the introduction of EVs the benefits from EVs implementation are higher • Innovative vehicle/battery leasing schemes • For vehicles on the fixed routes, charging in the company may be sufficient; for vehicles on variable distance routes it is necessary to consider complimentary charging methods • Political support, procurement demands, public opinion and awareness, private companies and surrounding municipalities are factors of success of demonstrators	Different uncertainties and risks: • Logistics concepts might require to be adapted for the usage of EFVs (i.e. an extra transhipment point is necessary to tranship goods from ICE vehicles to EFVs) • Equipment availability issues • Uncertain safety level • Other uncertainties and risks involved in the implementation of the new technology

manoeuvring of the vehicles in dense city areas. At the same time, charging, load capacity, maintenance and the need to adapt logistic concepts for the usage of EFVs (i.e. sometimes extra transhipment is necessary close to a city, from ICE vehicles that transported the goods from the depots to the city to EFVs making the final urban deliveries and pickups) are seen by operators as the main existing operational challenges. Not all freight operations are currently suitable for using EFVs, which is particularly the case for the long-haul operations and operations that require vehicles with a large loading capacity (in volume, or in weight). However, in terms of the range, the payload and overnight charging, current EFVs performance levels are good enough for many less than truckload city distribution operations.

Economical Performance EFVs

Currently, the purchase price and TCO for (large) EFVs are significantly higher than for conventional vehicles. That is largely due to the high battery cost and limited production volumes of these vehicles. Theoretically (see, e.g. Lee, Thomas, & Brown, 2013) the TCO of EFVs could be similar to ICE (for a time period of about a decade), but there are many uncertainties in actual costs and advantages of EFVs, since there is hardly experience with running these vehicles for 10 years. In the longer term it is expected that EFVs will become more competitive, incorporating savings from improved operational performance, reduction in purchase prices due to the mass production and associated environmental benefits. Currently, as operators are usually more focused on short-term benefits, the wider uptake of EFVs is difficult. The fact that the second-hand market and residual value of EFVs are not yet known, or expected to be non-existent due to technical improvements of new EFVs and limited life time of batteries, holds back some of the operators in their purchase decision. Leasing and financing companies are also reluctant to invest due to these uncertainties. Battery leasing or swapping options are regarded as possibility to reduce vehicle purchase and operational costs.

Environmental Performance EFVs

Undoubtedly EFVs' environmental performance is better than that of ICE vehicles, manifested in reduced CO_2 emissions and reduced local emissions. For the full picture, well-to-wheel emissions need to be considered and therefore certification of the electricity supply becomes important. Many projects report the improved environmental performance (e.g. FREVUE, 2013): PostenNorge (NO) reports the savings of 1.4–2.1 tonnes of CO_2 per vehicle per year by using EVs instead of fossil fuel vehicle. Tesco (UK) operates 15 Modec EVs and states that each EV saves between 13 and 15 tonnes of CO_2 each year. Deret (FR) has 38 times less CO_2 emission per truck for the EV compared to a conventional truck of the same capacity. UPS reports about 20% well-to-wheel carbon saving. Next EFVs show considerable gains in the local emissions; no local emissions (e.g. NO_x) are emitted by EVs, so air quality does not deteriorate due to these logistics activities, and the noise nuisance is minimized due to the low noise emissions.

Social and Attitudinal Performance EFVs

EFVs are less noisy and more environmentally friendly than conventional vehicles. EFVs are often very visible (in contrast to other solutions to improve urban freight transport's sustainability), and (partly therefore) very well perceived by the general public. EFVs receive positive feedback from drivers in most of the initiatives. One key point is that drivers first have to get over the 'range anxiety'. Trials show that after a few weeks of driving an EFV, and after getting familiar with the range estimator in the vehicle, as well as the reliability of this estimator and the vehicle itself, the driver's confidence increases and with that the attitude towards the EFV. Training and experience is necessary in order to familiarize drivers and general transport operators with the technical and operational particularities of the vehicles in order to achieve better results from the vehicle performance. The low noise generated was sometimes reported as a concern for the EFVs operations in the populated areas. Drivers considered the positive comments of passers-by as a pleasant change in comparison to more negative comments or attitudes towards ICEs.

Impact of Local Policy and Governance Structure

At the current stage of the EFV market development, a supportive government policy regime is considered necessary in order to achieve the wider uptake of the EVs. Measures both supporting the usage of EFVs and discouraging the usage of ICEs are required and are already being successfully implemented by several European municipalities. Another way to

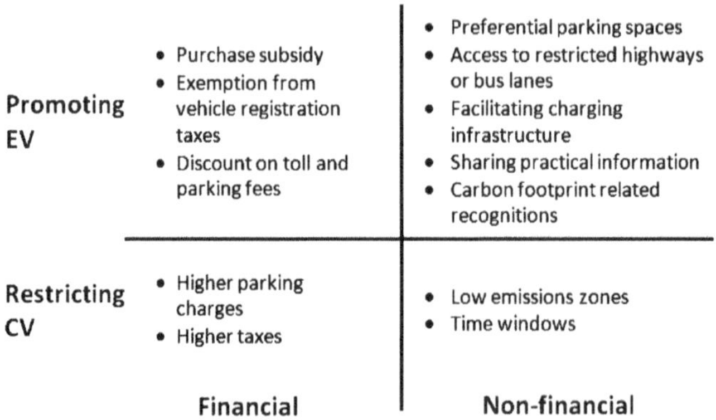

Fig. 1. Government Policy Supporting EVs or Policies to Limit ICE Usage. *Source*: FREVUE (2013), deliverable D1.3.

stimulate the wider uptake of EFVs is by using them in the authorities' fleets. Fig. 1 shows different ways (financial vs. non-financial) for local authorities to promote EFVs in city logistics. These policies are often used in combination.

DEALING WITH EFV IMPLEMENTATION ISSUES: GOOD EXAMPLES

The SWOT analysis shows that currently the main strengths of the EFVs are of environmental and social value. The main weaknesses have an economic and/or operational character, and basically come down to the negative business case of EFVs (i.e. higher costs) compared to ICE vehicles. Opportunities lie in the improvement of the vehicles' technical performance, supporting governmental regulations (which makes the business case for EFVs more positive, compared to ICE vehicles) and increasing scale advantages in production of, for example, batteries (which might reduce costs). Finally, main threats are related to the fact that there is a high uncertainty level about availability of the equipment, safety level, residual value and other risks.

Based on the overview of EV initiatives in city logistics, key challenges are identified as well as cases on how these are addressed in different demonstrators/initiatives. Based on the challenges and factors of success mentioned in the previous section, we identified three (high-level) issues (containing several of the challenges and factors) that should be taking in account to increase EFV implementation:

1. The need for an adapted logistics concept that enables the use of EFVs in city logistics operations to overcome range and load concerns. This is an issue from both economic (extra costs for transhipment area and handling) and operational (extra handling and planning) performance as well as the driving range.
2. The need (or desirability) of authorities' support to increase EFV uptake in city logistics activities. The examples in many of the discussed cases show that a positive business case using EFVs is difficult without supporting actions from authorities.
3. The opportunities that EFVs offer for private logistics companies to demonstrate their commitment to improving their environmental performance, that is their green image, and their visibility in cities.

In the following section some examples are discussed of how to deal with these issues in real-time implementations. In this section, we take a closer look at the approaches taken in six examples and the lessons learned. This section does not present the cases in full details (for further details see FREVUE, 2013). The aim of this section is to provide illustrative examples of how to deal with the three main issues that were identified based on the review of initiatives in the previous section.

Based on the selection criteria for step 2, these six cases of the EFV implementation in city logistics were selected (which are described in more detail in FREVUE, 2013, but in this chapter mainly the cases' lessons are discussed): Chronopost and Distripolis in Paris (France), ELCIDIS in La Rochelle (France) and the mobile depot of TNT (Belgium) in Brussels (part of the STRAIGHTSOL project) show examples of adapted logistics concepts; the EV project of Transport for London (United Kingdom) and to a lesser extent the Cargohopper (the Netherlands) in Utrecht show examples of authorities support; and the Cargohopper (the Netherlands) illustrates the opportunities EFVs offer logistics companies for a corporate sustainability, marketing and green image reasons.

Logistics Facilities Enabling Efficient Use of Freight EVs

Not all logistics operations are directly suitable for EFVs as existing operations were designed for ICE vehicles rather than EFVs which could be a barrier for EFV take up. So sometimes, more is required than only replacing an ICE vehicle with an EFV. However, instead of seeing this as a reason not to use EFVs, it is also possible to redesigning the logistics concepts so that these fit the characteristics of EFVs (i.e. smaller range and loading capacity) in order to enable the usage of EFVs in city logistics operations. Several options are available. On a high conceptual level most solutions are similar; an extra transhipment point is added in the transport chain. This point is usually necessary because often the origin of the goods (e.g. the carriers' or shippers' depot) is located too far from the city to be able to make the final deliveries in the cities and the transport from the depot to the city as well as the return from the city to the depot within the range of the EFV. How this will be done in practice depends on many factors, such as the type of city (and the corresponding infrastructural and policy restrictions), the type of deliveries, the volume and the depot locations of carriers (or shippers).

The cases of Chronopost, Distripolis, Elcidis and TNT's mobile depot illustrate how the logistics concept can be redesigned (see FREVUE, 2013, for full details). A common factor involves the introduction of a physical hub (transhipment point) that enables two-phase delivery and facilitates the use of environmental friendly transport in the urban areas. This is especially the case for those trips that contain many drops in a small area (i.e. high drop density). For other city logistics activities, for example more FTL (full truckload) deliveries, two-phase deliveries are less suitable because it is likely that additional value-added activities (such as consolidation and sorting) are required to further support the investment in a physical hub. When transferring a FTL to another truck, there is – from logistics point of view – little advantage to be gained while incurring additional cost. The location of urban facilities close to their customer base enables efficient EFV use. The four above-mentioned cases show some lessons that are of importance for transhipment concepts, that is:

- *Weight of goods.* The weight of goods should not be too high; two-phase delivery is only suitable for goods that can be transferred from one truck to another without too much effort (or other systems should be in place to transfer heavier loads, but that is not easy to implement in these urban facilities). The weight to be transported is a particular determinant on the actual range of EFVs; high volume, low weight operations are well suited for EFVs. The Distripolis case shows for example that a distinction is made between deliveries over and under 200 kg. Heavy goods are directly transferred to the receiver by a conventional truck in this case.

- *Adding value on logistics facilities.* Many urban locations are quite expensive, and have more productive economic uses (higher value) than for logistics activities. Considering the cost of land in urban areas, the logistics facility should be part of a system that adds value to the logistics operations in order to be financially viable. However, when the facility only serves as a location where the goods are simply transferred from the large truck to the EFV, it will not likely be financially viable to use a secure and accessible logistics facility at a prime location. The Chronopost, Elcidis and TNT cases show that value could be added at the new facility with sorting and consolidation activities. More consolidation benefits are possible, when the facility is accessible for multiple operators, like in the Distripolis case. Another option to add value to the facility (but not shown in the cases) can be to provide a customer collection point.
- *Bundling.* Bundling of cargo could be done for several carriers in such a facility. This does create the issue of parcel tracking and liability, which could be handled with a common IT system between the carriers. Bundling also increases drop density for the EV deliveries in the city, and can also reduce costs for the last mile logistics. The chance that value-added services, as described above, can be offered at a hub is much higher when there are many different consignees served with one vehicle (i.e. groupage delivery). This is why the concept works well for postal services (see all cases).
- *Property requirements.* Property requirements for logistics facilities are a difficult area to address. If (local) authorities support the idea and concept of the logistics facilities, they may help in finding the right location for the right price. It is often a big challenge to find an available, suitable and affordable property for the activities. The case of the mobile depot is included to show an alternative for physical buildings to facilitate transhipment. TNT's mobile depot could serve a similar purpose, but requires less space. However, parking space (public) is still required to make use of such a solution. Therefore, the support of the municipality and/or another governmental body in finding a site is crucial, both financially and regulatory (e.g. giving permissions) in this transition period.
- *Architecture.* The architecture of the logistics facility should be attractive to avoid visual nuisance in the city centre. Especially when it concerns public space. Either, by not being visible (e.g. underground like in the Chronopost case), or by being innovative (e.g. TNT mobile depot). The risks of vandalism and crime and possible security measures should also be considered in the development of the concept.

- *Long-term vision.* It should be considered whether and how the municipality can give support or grant certain permissions to other parties in the future (i.e. be part of long-term vision). There will be a limit to the number of mobile depots that can be placed in a public park (TNT case), without receiving complaints from citizens. Not all transport operators can use such a depot and park it in public space, because then there will simply be too little space available (especially in denser city centres). Another reason is that municipalities should support market competition. This is why municipalities are often reluctant to support an individual initiative. However, initial public (financial) support is often required to provide confidence to private sector investors.

Local Government Support to Accelerate the Introduction of the Freight EVs in a Large City

The adoption and implementation of EFVs has not been smooth. Where the business case for the logistic operators is often uncertain and usually negative, the benefits to society are evident. Authorities therefore have a clear interest in the uptake of EFVs, and hence, are in general very encouraging. This is required because the development and procurement of EFVs is costly and the introduction requires adaption (e.g. charging points), additional logistics activities and facilities. Without the encouragement and support of the authority, the logistic market is often not able or willing to implement EFVs massively. Supportive public policy or exemptions from restrictive policies often make the business case for EFVs better (see also Fig. 1): congestion charge discounts or exemptions, access to priority lanes, for example bus lanes, or access to pedestrian areas, no or extended time-window restriction, free parking and priority in public charging points.

In these cases, the use of EFVs for freight transport becomes more efficient than similar operations with conventional trucks or vans. Therefore, well-considered, coherent and consistent government strategy can support and facilitate the introduction and implementation of EFVs in city logistics, as is illustrated in the cases of London and Utrecht (see FREVUE, 2013).

The London example shows that even if government' support and strategy provide a good framework to initiate the EFVs purchase, this is not sufficient to take away the existing barriers for operators at this moment. Even though the actual EFVs uptake in London is still low, it is a good example to illustrate how the committed policy of municipality can initiate the EFVs uptake.

Although the business case for logistics operators improves with a supportive regulatory framework, reducing uncertainties on performance and costs seem to be necessary to convince more operators and increase EFV uptake, important lessons on local governmental support are:

- *Set up a comprehensive package of measures.* The lesson is to develop policies supporting a wide range of measures that do not only focus on facilitation and promotion of EFVs, such as building charging infrastructure and giving financial incentives, but also restricting the use of conventional vehicles, for example with low emissions zones and time windows.
- *Look at the needs of the logistic service providers in the city.* Collaboration with the private sector and the provision of transparent, practical information on how to implement EFVs will help the uptake of EFVs as well.
- *Enabling or restricting measures* work best depends on the initial situation. For example, in the old city of Amsterdam, there are a lot of (infrastructural) restrictions in place for freight vehicles. Logistics service providers would be (more) interested in driving EFVs if EFVs are exempted from these restrictions. This does not work in a modern city like Rotterdam for example as the city does not have many restrictions in the first place.
- *Availability of public space.* When the logistics concept need to be adjusted and new facilities are needed (see pervious section) then it often concerns public space. It is a challenge for logistics operators to find an affordable space in congested, urban areas. This is one of the key aspects for a positive business case for operators to start using EFVs. The local authorities can play a great supportive role by giving permissions to build or use a logistic facility, see for example the Chronopost case.

Private EFV Initiative to Increase the Company's Clean Image

EFVs are (at this moment) still a striking appearance in cities; therefore this is one of the (few) sustainability measures private carriers can take that are actually noticed by the public and the local authorities, in contrast to for example better bundling or cooperation. This opportunity gives motive to some carriers to choose for EFVs (even though it is more expansive).

The extra costs for using EFVs instead of conventional vehicles give a company a (more) sustainable image; so the EFVs are considered as a sign board for the company's sustainability policy. An example from this perspective is Coca-Cola (USA), that decided to use (parts of) its marketing budget for paying the extra costs of EFVs, as using EFVs gives this company the sustainability image it wants to show to the public.

Next to the economic arguments, including the supporting authorities' measures as discussed in the previous section, companies might have other arguments or reasons to start using EFVs for their urban freight operations. The state-of-the-art review studied more in detail the Cargohopper case (operating in Utrecht), illustrating what a private company can do in order to increase the EFV uptake. Major lessons to be learnt from this case are (see FREVUE, 2013, for fuller case description):

- *Strong commitment of the private entrepreneur.* The Cargohopper case demonstrates that if a private entrepreneur, in this case the managing director at Hoek Transport, really believes in the concept of electric freight transport for city logistics, progress is possible. The entrepreneur developed an electric-powered road train to do the city logistics operations in Utrecht (Cargohopper I). The Cargohopper activity has generated a lot of positive media coverage and subsequent awards. The use of the Cargohopper did not directly result in an increase in profits. Relatively few new clients joined Hoek Transport or TransMission (the network collaboration of local carriers in which Hoek Transport is active). The customers, that is shippers acting on a national or even European level, are not always aware of the Cargohopper (local actors are, but these are often not clients of transport operators). Many shippers are also not able to move instantly to another carrier due to contractual issues. At the local level some small customers were welcomed as new clients, but that was not enough to compensate for the extra costs of the Cargohopper and its operations. After the success of Cargohopper I, Cargohopper II has been developed (i.e. an electric truck able to pull a little trailer (more information is available on www.cargohopper.nl) and TransMission operates the Cargohopper II in two other Dutch cities (Amsterdam and Enschede), in order to make their operations more sustainable.
- *High visibility.* The vehicle itself made the transport company visible and distinguishable from other carriers, to local shop owners in Utrecht, residents and (local) authorities. The striking appearance of the vehicle, together with the PR campaign, resulted in a change of attitude towards

the carrier. The technical innovation (i.e. the actual development of the Cargohoppers by order of the carrier Hoek and TransMission) makes the logistics concept of TransMission visible, that is regional bundling of cargo and cooperation between carriers. Without this non-visible logistics concept, the amount of volume for the Cargohopper would be far too low to actually make it economically feasible to use. Both Hock Transport and TransMission were invited to discuss issues with local authorities, whereas in the past it was hard for the carrier to be in contact with authorities; mostly local authorities were not willing to discuss carriers' issues. Other local authorities asked the carrier whether it would be possible to operate a Cargohopper in their city as well.

- *Knowledge of logistics.* Many local authorities in the past started city logistics initiatives or freight EV initiatives from the wrong perspective due to relatively limited logistics knowledge (see Quak, 2012; Van Rooijen & Quak, 2013). As noted above, the Cargohopper initiative started from an entrepreneur that believed in the concept; this is different than an initiative started by authorities and in which no transport companies (or shippers or receivers) are actually involved. Therefore, Hoek and TransMission explained to several authorities that this solution could only work when the right logistics concept was in place. For the Cargohopper this means that (among other things) it is necessary that facilities are in place close to the city centre and that there is sufficient demand (drops and collections) to fill the vehicle. A top-down approach can make the business case for EFVs feasible on a shorter term, but local entrepreneurship seems to be necessary to get the EFVs actually running in freight operations.
- *High drop density and the availability of logistics facilities.* Local facilities need to be available and high drop (and collection) density is required.
- *Innovative vehicle.* Other lessons learnt from the Cargohopper case include the vehicle itself; the trailer is equipped with solar-power panels that provide electricity to the vehicle. Subsidies are considered as a short-term market stimulus and funding programs often provide financial support only for the truck. Many additional investments are consequently not subsidized, whereas they are part of the concept. The parties that provide subsidies would often like to be able to provide support for all additional costs of the Cargohopper, but are not allowed to do so by the regulations (often self-imposed). Discussing the possibilities to subsidize the entire concept also slows down the actual implementation of EFVs. This is also the case for when the vehicle is not a standard design, as for example, in a case of the Cargohopper I and II. This may result in significant difficulties

to actually get the vehicle approved for use on public roads; it took a long time to finally receive a licence plate for the Cargohopper II.

The Cargohopper is now used in three Dutch cities (Utrecht, Enschede and Amsterdam). These cities all have regulations and restrictions for freight transport. These regulations result in an advantage for the electric-powered Cargohopper in comparison to the conventional vans and trucks that were used; access to pedestrian areas and longer or no time windows, are making the use of the Cargohopper economically more feasible (see also Van Duin, Tavasszy, & Quak, 2013).

DISCUSSION

The review has indicated that there is already a lot of information available on EVs in general, but the majority focuses on promotion, support and introduction of the EVs for passenger usage as opposed to freight. The scope and number of the projects/initiatives focussing on freight EVs is limited, but slowly growing. From the available freight EV literature and information, only a few sources actually present real evaluation results; the majority focus on pre-trial or context descriptive information only.

The challenges and success factors from the previous and ongoing initiatives and demonstrators of freight EVs illustrate that back in the 1990s the main challenges with implementation of the EFVs were related to high procurement costs, limited product diversity and low performance of the vehicles. Nowadays a lot of technological progress has been made and technological reliability as well as efficiency of EFVs is much higher. Trials and demonstrators are moving on with larger/heavier vehicles that have greater range and capacity. Nevertheless, further improvement of EFVs' technical characteristics is necessary as well as their practical and commercial viability. The high procurement price and the uncertainty on the residual value of the vehicle remains one of the main challenges.

Although (theoretical) studies show that the EFVs' TCO can be lower than for conventional trucks, the novelty, uncertainty, low availability (especially for larger vans and trucks) and ongoing technical improvements are barriers for transport companies in purchasing EFVs (next to the high initial purchase costs). More large-scale demonstrations using EFVs that show the operational feasibility, for example technically reliable, and that provide more certainty on the residual value (which should make it easier to lease or finance EFVs) are necessary to convince more companies to invest in EFVs for (city) logistics activities.

This review on city logistics and EFVs shows four main points, which in essence show why the actual use and implementation of EFVs is limited in (city) logistics at this moment:

1. Logistics operators defer purchasing and using EFVs due to concerns over a mismatch between characteristics of EFVs and the vehicle requirements to perform efficient operations. The limited range of the vehicles, the relatively long charging time and the limited loading capacity are barriers for EFV usage in logistics. However, this review also showed that for some operations this mismatch is basically due to the fact that the current logistics concepts are designed for conventional vehicles. An adapted logistics concept can enable the use of EFVs in city logistics while it deals with these issues.
2. The high initial purchasing costs for EFVs. The review shows that for some operations the TCO can be at least comparable to (or lower than) conventional trucks. This turns out to be particularly the case for areas in which supporting measures are in place, such as fewer restrictions for EFVs than for conventional vehicles, lower taxes or charges and obviously purchasing subsidies. These supporting measures do not only result in direct financial benefits but also in increased operational efficiency (in urban areas).
3. The early adoption of EFVs in logistics operations can also have another advantage for carriers, improved visibility in a city for residents and local authorities, which could result in more business and in more influence on, and a better relationship with these stakeholders. It does not automatically result in more customers, since for many shippers (the paying customers of many carriers) these last mile EFVs are not visible.
4. Finally, there are still issues and uncertainties to be dealt with. Rapid technical development is an issue for some operators that are afraid their vehicles will become obsolete and as a result have no residual value at all after just a few years. The limited availability of (larger) EFVs in the market is also an issue, but this is improving as more manufacturers start to offer EFVs (especially for smaller vans). Questions remain on which operations are suitable for EFV use, how the operational and technical performance of the vehicles is influenced by other elements, such as climate and geographical characteristics and city characteristics (hills, pavements, etc.).

CONCLUSION

A number of practical recommendations can be given to transport operators and demonstrations showing EFVs. Operational lessons to be learnt from previous and ongoing trials and initiatives:

- A detailed planning of a demonstration process is very important;
- For municipalities it is important to be coherent and consistent in their policy approach following a step-by-step method: building infrastructure, promoting, supporting EFVs and restricting conventional vehicles;
- Private–public cooperation is important especially during the initial trials that involve EFVs;
- From an operational point of view, driver training is important alongside the establishment of the correct vehicle charging routine;
- Sharing the results with others outside a trial is important, in order to encourage far wider uptake of EFVs, since this might reduce uncertainties for companies not familiar with EFVs.

The state-of-the-art review shows that within EU projects, EFVs are not seen as a stand-alone measure but are usually dealt with in combination with the implementation of an urban consolidation centre, time window regulations or an environmental (low emission) zone. In a lot of these cases, EFVs are bought with a support of (local) authorities or in the framework of EU or national demonstration projects. In cases where the purchase of an EFV is done directly by a private company, it concerns mostly light duty commercial vehicles (< 3.5t GW), with a very limited number of the medium and large-size vehicles. Most freight EVs in city logistics are used in the postal and parcel sector.

Nowadays there are still many uncertainties about EFVs implementation and therefore many risks. The main strengths of the EFVs in city logistics are related to its low CO_2 emissions (depending on power generation), no local emissions and its limited noise nuisance. The main weakness comes, next to some uncertainties in the technical and operational performance, from the economics side; that is it is at this moment it is really difficult to find a viable business case. Next, the SWOT analysis shows that a supporting governmental scheme is very helpful to improve the business case for EFV implementation, in order to become more financially sustainable as well. Challenges with implementation of the EFVs in city logistics operations during the first trials differ from today's challenges. The SWOT analysis shows that today some private operators (e.g. the front runners in the first EFV implementations) show some scepticism in purchasing EFVs:

they are waiting for further technology development and a viable business case. The use of EFVs for city logistics still has to gain benefits from practical demonstrations in order to be recognized notably by fleet managers.

The following three elements that are necessary for acceleration or uptake of EFVs are identified:

- Adaptation of logistics concepts that enable (large-scale) EFV use in urban logistics;
- Local government support to accelerate EFV introduction;
- Companies that want to be sustainable.

The state-of-the-art review illustrates that there are many gaps in knowledge about the freight EVs technological and operation performance, which constrain private operators in their decision to buy an EFV. If we want to achieve the higher uptake of the EFVs in city logistics operations, these frequently asked questions regarding EFVs should be answered in future research. The review conducted indicates the most important questions for future research are:

- What is the residual value of the freight EV and what are the factors that determine the residual value of the freight EVs?
- What kind of operations are suitable for freight EVs and how should logistics concepts be designed in order to make EFV usage financially viable for operators?
- What are the wider systemic impacts from the freight EVs implementation?

ACKNOWLEDGEMENTS

The authors thank the FREVUE partners for their comments on the state-of-the-art study, as well as the two reviewers of the report that contributed to FREVUE's final deliverable 1.3 state-of-the-art city logistics and EV. Special thanks go to the co-authors of the deliverable: Susanne Balm (TNO), Isabelle Roche-Cerasi and TerjeTretvik (both SINTEF).

REFERENCES

Dablanc, L. (2007). Goods transport in large European cities: Difficult to organize, difficult to modernize. *Transportation Research Part A*, *41*(3), 280–285.

EC. (2011). White Paper Roadmap to a Single European Transport Area – Towards a competitive and resource efficient transport system, COM/2011/0144 final.

FREVUE. (2013). Deliverable 1.3 State of the art of the electric freight vehicles implementation in city logistics, December 2013.

Lee, D. Y., Thomas, V. M., & Brown, M. A. (2013). Electric urban delivery trucks: Energy use, greenhouse gas emissions and cost-effectiveness. *Environmental Science and Technology*, *47*, 8022–8030.

Quak, H. J. (2011). Urban freight transport: The challenge of sustainability. In C. Macharis & S. Melo (Eds.), *City distribution and urban freight transport – Multiple perspectives* (pp. 37–55). Cheltenham: Edward Elgar.

Quak, H. J. (2012). Improving urban freight transport sustainability by carriers. In E. Taniguchi & R. G. Thompson (Eds.), *Procedia – Social and Behavioral Sciences* (Vol. 39, pp. 158–171). Amsterdam: Elsevier Ltd.

Van Duin, J. H. R., Tavasszy, L. A., & Quak, H. J. (2013). Towards E(lectric)-urban freight: First promising steps in the electric vehicle revolution, *European Transport/Trasporti Europei*, *54*, 9.

Van Rooijen, T., & Quak, H. J. (2013). City logistics in the European CIVITAS initiative. In E. Taniguchi & R. G. Thompson (Eds.), *City logistics* (Vol. 8, pp. 494–508). Amsterdam: Elsevier Ltd.

APPENDIX: OVERVIEW OF EU, NATIONAL AND PRIVATE PROJECTS/INITIATIVES INCLUDED IN THE STATE-OF-THE-ART REVIEW

A.1. Overview of the Reviewed EU Projects.

Project Acronym/ Duration	Project Full Title	Demonstration Case
ENCLOSE (2012–2015)	Energy efficiency in City Logistics Services for small and mid-sized European Historic Towns	LuccaPort: up to date 6 EVs and purchase of 3 more during the project duration (Italy) Trondheim city logistics: 15 EV in Norway Post and planned to increase till 37 by 2015 (Norway)
SMARTFUSION (2012–2015)	Smart Urban Freight Solutions	Testing of EV and HV for distribution of perishable goods; testing of EV equipped with metering devices (Italy)
STRAIGHSTOL (2011–2014)	Strategies and measures for smarter urban freight solutions	Usage of 2 EVs in DHL demonstration (Spain)
CIVITAS (2002–ongoing)	Cleaner and better transport in cities	TRENDSETTER. Deployment of EVs for the department store in Graz (Austria) VIVALDI. Bristol logistics platform demonstrator, using Smith EVs operated by DHL (UK) TELLUS: Demonstration of EVs in Rotterdam (the NL) RENAISSANCE: Usage of 2 Smith EVs for the Urban freight Logistics in Bath (UK) MODERN: Utilization of freight EVs in Brescia (IT) MIMOSA: Running up Cargohopper
CO_2NeuTrAlp (2009–2012)	CO_2 – Neutral Transport for the Alpine Space	Inteporto: testing of 1 EV that includes refrigerated units for perishable goods distribution (Italy)
ELCIDIS (1998–2002)	EVs City distribution systems	Rotterdam and Stockholm deployed large electric vans (payload 1000–1500 kg, Rotterdam) and hybrid electric trucks (payload max. 11 tonnes, Stockholm); La Rochelle

A.1. (*Continued*)

Project Acronym/ Duration	Project Full Title	Demonstration Case
		focused on the deployment of EVs with a payload of approximately 500 kg; Stavanger, Milan and Erlangen focused on the deployment of (hybrid) EVs for in-house goods and mail distribution for companies
TURBLOG (2010–2013)	Transferability of urban logistics concepts and practice from a worldwide perspective	EV demonstration by Chronopost: 14 small electric vans and two wheeled containers (France); Cargohopper (the NL)
EVD Post (1998–2000)	EVs deliveries in postal services	Demonstrators of implementation of EVs in the postal services in Germany, Finland, Sweden, France and Belgium
SELECT (2012–2015)	Suitable electromobility for commercial transport	The project's central objective is to understand the technical and practical user requirements for using EVs in commercial transport and to develop a set of methods for the fleet management of electric and mixed fleets

A.2. Overview of the Reviewed National Initiatives.

Initiative	Ownership	Country	Relevance
MOVELE (ongoing)	National	Spain	Government is planning to install an electric car infrastructure in several cities and subsidize private purchase of a small fleet of electric cars (as well as buying some for itself)
Plugged-In Places programme (ongoing)	National	United Kingdom	Programme offers match-funding to consortia of businesses and public sector partners to install EV charging points. Does not include demonstration of freight EVs
Source London (ongoing)	Municipal	United Kingdom	Source London is a city-wide charge point network key to increasing the uptake and usage of EVs in London. Does not include demonstration of freight EVs
Plugged-in Fleets (ongoing)	Municipal	United Kingdom	Within this initiative, Energy Saving Trust, funded by the Office for Low Emission Vehicles offers free analysis and tailored review on how to successfully use plug-in vehicles in their business
LuccaPort (ongoing)	Municipal	Italy	LuccaPort is an operative division of METRO Srl, participatory society of the Municipality of Lucca. They have 6 different EV, plus 5 to be purchased in 2013–2015.
ElektrofahrzeugGroßversuch (Rügentest) (1992–1996)	National	Germany	A large-scale EV test on Rügen with the aim to test the potential of electric mobility
Electric Mobility Pilot Regions (2009–2011)	National	Germany	8 Demonstrator projects with some on freight EVs
Frankfurtemobil (since 2010)	Municipal	Germany	Provides information on important electric mobility projects and current events in Frankfurt
MOBI.E	National	Portugal	Nationwide programme for the implementation of an intelligent charging network for the deployment of electric mobility

A.2. (*Continued*)

Initiative	Ownership	Country	Relevance
Elektrisch Vervoer Centrum Rotterdam (since 2011)	Municipal	The Netherlands	Information and trial centre on electromobility in the Netherlands
E-Laad	Private, on national level	The Netherlands	Association of Dutch energy net owners to provide public charging points
Flemish Living Lab Electric Vehicles	Regional	Belgium	5 platforms supporting different electromobility initiatives. EVTecLab focusing on new technologies for electric heavy – duty vans
EVE – Electric Vehicle Systems Programme (2011–2015)	National	Finland	Programme that aims to create a community of EV and support system developers. Test environment ECV support electric commercial vehicles
EMIL, EMKEP	National	Germany	Promote the use of EV to reduce the petrol dependence and pollution
Cologne-Mobil	Regional	Germany	Promotes the electromobility together with a new GPS system
SendSmart (2012–2014)	National	Norway	Objective is to develop sustainable freight transport in urban areas. Running of 1 Renault Maxity with lithium battery for one week (parcel delivery)
Stadsleveransen (2012–2015)	Municipal	Goteborg, Norway	Last mile delivery in the city centre of Goteborg Deployment of Melex 391 vehicle with special-designed trailer

A.3. Overview of the Private Initiatives.

Company	City/Country	Dates	Type of Business	EV's Fleet
STEF–TFE	Lyon/France	2011–2012	Distribution services of refrigerated products for Carrefour	1 refrigerated Renault Midlumtruck, 16 tonne
Deret	Different locations/ France	Ongoing	Last mile delivery to the city centres	50 electric vans
Peter Appel	Amsterdam, the Netherlands	Was planned in 2012	Distribution services for AlbertNL	Renault Maxity EV. Re-build from conventional vehicle. Range is 100 km; max speed is 90 km/h
CityShopper	Nijmegen, the Netherlands	Ongoing	Distribution services for AlbertNL and Cornelissen	First tests with Smith, which appeared not reliable in their case. Now service with Renault Maxity Electric. Range is 100 km; max speed is 90 km/h
UPS	UK, Germany	Since 2009	Parcel and post delivery	12 Modec vehicles (6 in UK, 6 in Germany). Range 100 miles
EROSKI	Bilbao, San Sebastian, Vitoria/ Spain	Since 2011	Home delivery from supermarkets	5 Mercedes Vito E-Cell vehicles. Range 130 km; max speed 80 km/h
TNT Express	UK, Scotland	Ongoing	Business to business parcel delivery	Newton vehicle; 7.5 tonne
Correos	Spain	Since 2008	Parcel and post delivery	15 Comarth vehicles; range 100 km; max speed 50 km/h
SEUR	Valencia, Madrid/ Spain	Ongoing	Home delivery services for organic supermarket	Comarth Cross Rider vehicle
020 stadsdistributie	Amsterdam/the Netherlands	Since 2009	Distribution of refrigerated goods	3 EVs
Eco-Logis	Brescia/Italy	Since 2012	Serving last mile delivery to the city centre from the UCC	3 EVs
PostenNorge	Norway	Ongoing	Postal service	40 Comarth Cross Rider vehicles; another 500 vehicles are ordered in 2012/2013
La Poste	France	Ongoing	Postal service	250 Comarth vehicles

A.3. *(Continued)*

Company	City/Country	Dates	Type of Business	EV's Fleet
Seymour Green	London, Westminster/UK	Ongoing	Sustainable marketing company	10 Aixam Mega Vans
Tesco	London/UK	Since 2007	Online shopping	15 Modec vehicles; range around 90 miles
Cargohopper	Utrecht, Enschede/the Netherlands	Since 2010	Last mile delivery in a city centre	1 Cargohopper I, several Cargohopper II
Abby Couriers	London/UK	Ongoing	Furniture and antique items delivery service	3 Goupil vehicles
Brewers	London/UK	Since 2008	Decorating materials store and delivery	1 Modec van
EntrepriseMouchel	London/UK	Since 2008	Integrated highways maintenance provider	10 Modec vehicles
Gnewt Cargo Ltd	London/UK	Ongoing	Parcels delivery	
Melrose and Morgan	London/UK	Ongoing	Grocery shop	1 second-hand Axiam Mega van
Nappy Ever After	London/UK	Since 2005	Weekly cotton nappy laundry services	1 brandshaw taxi travel
Office Depot	London/UK	Since 2009	Supplier of office solutions	4 electric vans
Sainsbury	London/UK	Since 2005	Supermarket chain	19 EVs; in a process of procuring 50 new Smith EVs
Speedy	Manchester, Glasgow, London/UK	Since 2007	Provider of equipment rental and support services	10 Modec vehicles
Go-Ahead	London/UK	Since 2008	Provider of passenger transport services	15+ Smith EVs
L'Oreal	Paris/France	2003	Serving L'Oreal deliveries	10 tonne EFV to deliver goods in centre of Paris
Label Route	Montpellier/France	Since 2009	Provider of city centre deliveries (groceries, courier services, distribution)	2 roues et petits utilitaires électriques (Two scooters and small electric vans)
Post Nord	Sweden, Denmark		Mail delivery	5,050 EVs (out of 20,000 vehicle fleet), mainly bicycles, mopeds and club cars

CHAPTER 12

THE COST AND EFFECTIVENESS OF SUSTAINABLE CITY LOGISTICS POLICIES USING SMALL ELECTRIC VEHICLES

Sandra Melo, Patrícia Baptista and Álvaro Costa

ABSTRACT

Purpose – *In the recent decades, research and industry on city logistics have tried to seek for environment-friendly solutions that are efficient enough to satisfy both society and suppliers' needs. One of the potential solutions is the use of small-size electric vehicles (SEVs), due to their improved energy efficiency, local zero emissions, and lower traffic disturbance.*

In spite of all the benefits of SEV for society, advertised through experimental trials focused on social and environmental benefits, research on these vehicles' impacts seems to overlook the effects on private stakeholders operations, namely, disregarding the replacement rate needed to assure the same delivery patterns and their purchasing and battery charging implications.

Sustainable Logistics
Transport and Sustainability, Volume 6, 295–314

ISSN: 2044-9941/doi:10.1108/S2044-994120140000006012

Design – *In this chapter, the authors contribute in filling this research gap by considering private interests, related to operation costs levels (running and driving costs), service levels, and efficiency in the promotion of SEV. Simultaneously, its balance with public interests, related with sustainability, quality of life, mobility, and environmental issues are also addressed.*

Findings – *The authors aim to evaluate the usage of SEV in this research and to estimate the effects of replacing conventional vans by SEV on city logistics operations. The results of this quantitative analysis enlighten if SEVs are indeed a viable solution to satisfy public and private stakeholders, when operational and external costs are fully accounted.*

The chapter presents a case study that addresses the effects of replacing vans by SEV on city logistics operations in the city of Oporto (Portugal), considering public and private stakeholders' interests. The study compares four scenarios of 5%, 10%, 30%, and 70% of SEVs replacing diesel vans used in transport and unloading operations. The four scenarios are tested on different geographical scales: street and city levels. First, the authors estimate how the use of SEV in city logistics affects traffic, energy consumption, and emissions. Second, the respective operating and external costs are quantified and the acquisition and battery issues are discussed.

Originality/value – *When considering the goal of promoting SEV as a sustainable city logistics policy, under a methodology focused on mobility, operational performance, and environmental externalities, the authors concluded (a) the replacement rate SEV:van is determinant to make a decision on whether or not to use SEVs replacing vans, (b) SEVs are economically competitive with conventional vans if the replacement rate is 1:1, (c) SEVs have a better performance at the street level rather than at the city level, (d) SEVs can be used with normal traffic as a niche of market (lower than 5%), and (e) SEVs benefits exist, but they are not significant enough to drive suppliers for their adoption.*

Keywords: City logistics; small electric vehicles; electric mobility; sustainable city logistics policies

INTRODUCTION

The increasing urbanization, population growth, and the changes in the freight demand patterns, favoring just-in-time solutions, added to the reduced stock in stores, have led to the increase in freight movements in cities. The increase in freight movements has been followed by a decrease in the freight and passenger mobility on transportation systems due to the lack of associated increases in road capacity. As a result, despite the relevant role of urban goods distribution in the sustainable development of cities, freight movements have also been related to negative impacts on the economic power, accessibility, quality of life, and on the attractiveness of urban areas (Barredo & Demicheli, 2003; Wolpert & Reuter, 2012). The most common examples of those impacts are mainly the effect on traffic congestion created by commercial vehicles, the road capacity reduction caused by frequent stops for loading or unloading operations, and the resulting higher energy consumption and emissions (Dablanc, 2007, 2011; Giuliano, O'Brien, Dablanc, & Holliday, 2013; Russo & Comi, 2012).

Under such context, cities and its local public administrators' are facing a difficult challenge. Public administrators must maintain and promote city's sustainability, mobility, and quality of life, while ensuring that urban goods distribution systems efficiently serve their city's needs. As a result, urban goods distribution solutions that are environment friendly and are, at the same time, efficient enough to satisfy, both society and distribution companies, should be promoted (Melo, 2011).

Many measures and initiatives are being encouraged with this goal, namely, the introduction of small-size electric vehicles (SEVs). A SEV is a vehicle with smaller dimensions than conventional commercial vehicles, powered by an electric motor instead of a conventional internal combustion engine. In terms of operations, SEVs are easier to park than vans or trucks and are viewed as less intimidating and safer by the public (Browne, Allen, & Leonardi, 2011). In addition, SEVs allow the reduction in the total curbside space occupied by vehicles making on-street deliveries, further reducing the impact of unloading operations on traffic congestion. SEV also presents social and environmental advantages that can include noise reductions, less conflicts with other road users, and greater pedestrian safety (Browne et al., 2011; Leonardi, Browne, & Allen, 2012).

In spite of all the benefits of SEV for society, advertised through experimental trials focused on social and environmental benefits, research on these vehicles' impacts seems to overlook the effects on private stakeholders operations (Melo, Baptista, & Costa, 2014). In this chapter, the authors contribute to fill this research gap by considering private interests, related to operation costs levels (running and driving), service levels, and efficiency in the promotion of SEV and simultaneously, its balance with public interests, related with sustainability, quality of life, mobility, and environmental externalities. The effect of replacing vans on city logistics operations by SEV, equivalent to downsizing city logistics vehicles without penalty costs of unsatisfied demand, is analyzed, considering public and private stakeholders' interests in the city of Oporto (Portugal).

The impacts of this replacement on mobility and environment (public good) as well as on operational efficiency costs (private concerns) are discussed in order to identify the potential market share for SEV and in which operational conditions can it succeed.

STATE OF THE ART

City logistics takes place in areas with high density of buildings and population and a high demand for goods and services. The current urbanization levels added to important phenomena and transformations on the market such as the high pressure to reduce costs and to improve the type and level of service to customers contribute to a higher number of movements of goods and services to urban areas. Simultaneously, such context is being followed by an intense pressure to reduce emissions and congestion induced by commercial movements.

In Europe, the major urban centers should be CO_2-free in city logistics by 2030. To achieve this ambitious goal, motivated by the predictable reduction in oil resources, research and industry started to investigate solutions for the electrification of commercial vehicles, namely, electric trucks (Birkner, 2012). The argument behind this approach is that the electrification of commercial vehicles could reduce emissions and, thus, contribute to a better quality of life in cities. Such reasoning makes it an appealing approach for public stakeholders' sustainability goals.

Nonetheless, in spite of its advertised social and environmental benefits, the use of electric trucks still raises some apprehension on private stakeholders, namely on suppliers. The operational competitiveness of electric trucks in comparison with conventional diesel ones is still not convincing suppliers to adopt it in a generalized way (Duin, van Tavasszy, & Quak, 2013). The running costs of electric vehicles (EVs) are actually lower, but these do not stand out sufficiently on a total operational cost basis (Thiel, Perujo, & Mercier, 2010). The running costs, directly influenced by the price of electricity, are in fact predictably competitive in the near future. Electricity presents lower and more stable prices in comparison to fossil fuel energy sources' volatility, and it is unlikely that electricity price will increase due to the EV phenomenon. Moreover, the current scarcity and slow penetration rate of EV ensure that the effects of EV can be anticipated by public electric companies and, therefore, the supply of electricity is not a constraint to the deployment of electric trucks (Davis & Figliozzi, 2013). However, the purchase costs of electric trucks are still much higher, which negatively influences its operational competitiveness. Several authors even identify the high purchase cost of EVs as the major obstacle to widespread implementation of electric trucks at the moment (Kley, Lerch, & Dallinger, 2011; van Rooijen & Quak, 2010; Vermie, 2002). Similar conclusions were reached in other studies as well (Feng & Figliozzi, 2013), concluding that electric trucks are only competitive if EV purchase prices fall at least 30% and if its utilization level is high. The higher price of EV added to its technical limitations (namely, range, weight, speed, reliability) are still far from convincing suppliers to replace their conventional fleets by EV (Duin et al., 2013). Environmental awareness, social perception, and green image of the company can promote the adoption of EV, but they are not enough to boost the EV market within suppliers (Kley et al., 2011). Moreover, despite the growing interest in EV technology and the fact that these vehicles can indeed contribute to mitigate the environmental impacts of city logistics, they are not more effective in reducing traffic disturbance caused by commercial vehicles during unloading operations. Their size is similar to conventional diesel trucks and, consequently, so is their respective traffic disturbance. As traffic disturbance is a major issue when it comes to city logistics operations, solutions with lower dimensions have recently been considered. The downsizing of delivery vehicles without "penalty costs" of unsatisfied demand can be obtained by using SEV, such as SEVs with only four wheels, electric cargo bikes, and electric tricycles. These vehicles can respond to environmental and traffic disturbance concerns over transportation energy use (Bartern et al., 2012). SEVs are able to access to narrow streets in historical centers and avoid the weight and size restrictions imposed on many European cities. In fact, in European urban areas, as more than 80% of the urban road freight trips are performed in distances below 80 km (Ruesch & Petz, 2008), the use of SEV for city logistics can potentially be a feasible operational solution.

Altogether, this context has promoted the use of SEV on city logistics operations within European cities. In Paris, 272 km of bicycle lanes were recently created, making it possible for 60 electric cargocycles in "Petite Reine" to enhance the productivity of 1 million parcels in 2008. These SEVs, running from a consolidation center with 600 m^2, avoided 203 tons of CO_2 and 84 kg of particles of atmospheric pollution and reduced noise pollution in a twelve-month period (SEStran, 2010).

A similar trial experiment was carried out in London between 2009 and 2010 and, results of the test of 6 cycles, 3 electric vans, and 1 truck revealed a total savings of 62% of CO_2 emissions (in kilograms per parcel) and the total distance traveled was reduced by 20% (Browne et al., 2011). Brussels presents the example of Ecopostale, using 4 bicycles, 7 cycles, and 1 electric van, reaching a savings of 30 tons of CO_2 within its first two years of operations (Etienne, 2012). Barcelona estimated savings of 912 kg CO_2 emissions/year/trike with the implementation of such system.

CASE STUDY

Overview of the Area

This chapter analyzes the effects of the use of SEVs replacing vans and working as a complement to trucks in city logistics movements and operations and analyzing its effects in mobility, environment, and operational efficiency. The case study is carried out in the city of Oporto (Portugal), using the delivery patterns information collected with detailed windshield surveys performed in March 2009. Two geographical levels of study were chosen: street and city levels. Fig. 1 illustrates those geographical levels of analysis and identifies the critical areas for city logistics operations. The two levels of analysis allow in the definition of what should be the geographical coverage for the use of SEV within the city, its significance in terms of replacement share that maximizes the benefits of all (public and private) interests involved, and the comparison of its traffic performance, energy efficiency, resulting emissions, operating, and external costs with the ones of diesel vans.

The streets marked in dark gray represent areas with significant traffic disturbance due to city logistics operations. The areas of pressure due to

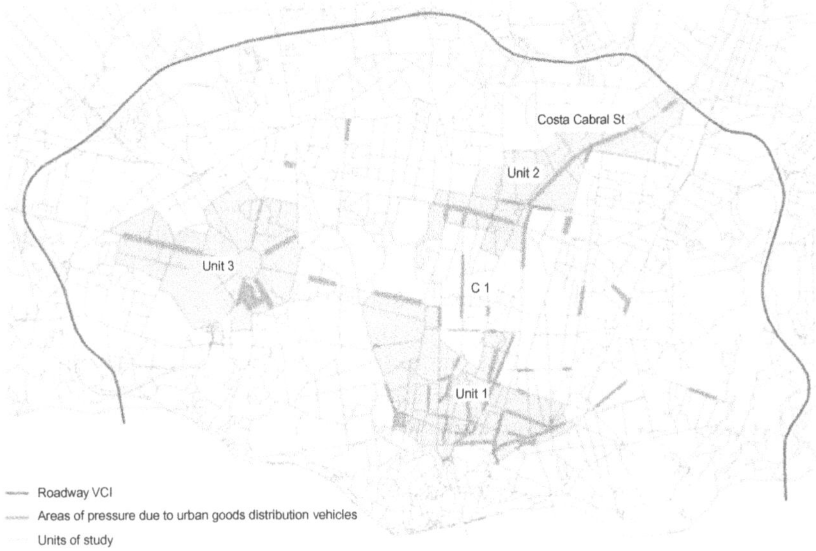

Fig. 1. Illustration of the Geographical Levels of Study. *Source*: Melo (2010).

urban goods distribution vehicles were determined based on the daily observations collected by the windshield surveys. Considering the homogeneity of activities, landscape and commercial activities identified in Melo (2010), three green "units" were defined: Unit 1 represents Oporto downtown; Unit 2 is an area with residential and commercial mixed uses; and Unit 3 is an area of services. The area inside the first ring (VCI roadway) is considered the "City."

The case study analyzes the impact of introducing SEV at city level and street level (in this case, the main commercial street of Unit 2: Costa Cabral Street). Unit 2 comprises 12 blocks and is an area with ancient and historical buildings around narrow streets. The urban design of its streets neither allows to increase the capacity of the streets nor to consider solutions that imply to change their layout. Moreover, the delivery patterns within Unit 2 and the small and one-box type of parcel being delivered on their commercial activities are the most suitable to a potential shift to the use of SEV. The selection of the Costa Cabral Street within Unit 2 was based on the fact that it is the one with a higher number of commercial activities at this unit and, therefore, the most appropriate to test the impacts of the shift to SEV in terms of external and operational costs. These impacts will be assessed by specific traffic, environment, and efficiency variables for different vehicles typology in the fleet, in each of the geographical levels analyzed within the case study.

The Model: Base Conditions, Assumptions, and Outputs

The estimation of traffic (mobility) impacts was carried out using the microscopic traffic commercial software AIMSUN, through specific vehicle parameter adaptations (vehicle dimensions, speed, and acceleration profiles) to include SEV within the default vehicle typologies.

For the purpose of designing the network, 231 intersections and 559 sections were defined, and their respective geometry and movements were later confirmed in detail in loco. The simulation exercise covers the daily peak of deliveries. The traffic model assumes that the traffic demand for public, passenger, and freight transport remains the same of the business-as-usual (BAU) scenario. Due to the fact that trucks are targeted for different demands of the ones of SEV, namely, on the geographical coverage, location of receivers on peri-urban areas, type and size of packages, truck demand remains the same on the O/D traffic matrix along the four scenarios. It is assumed that SEV replaces diesel vans moving goods and thus the sum of both O/D traffic matrices (vans and SEV) for each of the considered four scenarios corresponds to the values of the O/D vans traffic matrix of the BAU scenario calibrated with the respective replacement rate between vans and SEV to assure there are no penalty costs of unsatisfied demand. The range of parcel sizes and weights makes it viable for all parcels transported by vans to be fully delivered by SEV, despite these vehicles having a smaller capacity. The four replacement share scenarios to be tested are presented in Table 1.

Table 1. Scenarios of SEV Share on Light Commercial Vehicles.

	Share of SEV Replacing Diesel Vans (%)	Assumptions
Scenario 1	5	• Same O/D traffic demand matrix from the BAU scenario
Scenario 2	10	• Same incidents
Scenario 3	30	• SEVs use the same paths used by vans
Scenario 4	70	• Same supply capacity
		• Increasing market penetration of SEV with scenarios
		• Same delivery patterns

Seven categories of vehicles were considered: SEV, car, bus, truck, van, taxi, and other public transport. Vans have average external dimensions of 5 m long and 2 m wide. Trucks have external dimensions of 7.5 m long and 2.3 m wide. SEVs have external dimensions of 2 m long and 1 m wide. The considered SEV is based on the characteristics of a Toyota prototype (Toyota COMS, 4.8 kW electric motor, 2.4 kWh battery, and maximum speed of 45 km/h), with an empty weight of 400 kg, including the two batteries (without the driver and load weight) and a load space of 1 m^3. The considered SEV has an average speed of 30 km/h in free flow conditions. The SEV requires 8 hour charging overnight. The considered vans are based on the characteristics of Opel Combo (Combo Van L1H1 1.6 CDTI 105cv; Combo Van L1H2 1.6 CDTI 105cv; Combo Van L2H1 1.6 CDTI 105cv), Renault Kangoo (Kangoo Express Compact 1.5 dCi 75cv Confort; Kangoo Express Compact 1.5 dCi 75cv Bussiness; Kangoo Express Compact 1.5 dCi 75cv Societé), and Fiat Fiorino (FioninoQubo 1.3 Multijet 16v95cv; Fiorino Cargo 1.3 Multijet 16v75cv).

The delivery patterns adopted on the exercise are the ones obtained during delivery windshield surveys in the area. Roughly 90% of the stores of the area receive a single parcel. At some stops, the driver makes deliveries to more than one store due to their close proximity to one another. The flow unit is vehicles, not passengers or cargo. Such fact is critical as the replacement rate of 1 SEV per 1 van cannot assure the same delivery pattern in the four scenarios (as is presented in Table 2).

Incidents representing illegal parking were defined in a detailed way. Sections where incidents occur, position on the lane, and length of the event were defined according to real data obtained on windshield surveys. On the BAU scenario, incidents were mostly caused by vans rather than trucks. Vans incidents were replaced by SEV incidents on the proportion of each of the four scenarios: 5%, 10%, 30%, and 70%, respectively. Outputs of the simulation exercise refer to the hourly average of the daily peak period of deliveries. AIMSUN was run for 15 iterations per scenario guaranteeing the convergence of the model.

The analysis of the stakeholders' effects was carried out through the quantification of relevant indicators by the type of vehicle: SEV, vans,

Table 2. Replacement Rate SEV:Van for the Four Scenarios.

	Sc. 1 (5%)	Sc. 2 (10%)	Sc. 3 (30%)	Sc. 4 (70%)
City	1:1	1:1	2:1	2:1
Street	1:1	1:1	1:1	2:1

trucks (transporters/suppliers), passenger vehicles (citizens and city users), buses and taxis (public transport operators), and on the total system (sum of all users of the road system).

The following indicators were quantified: mobility (distance traveled, average speed excluding stops to make deliveries, travel time, delay time, density of traffic, flow), energy and environment (fuel consumption and global CO_2 in a fuel life-cycle approach, Well-to-Wheel), operating costs, and external costs. Operating costs include the direct vehicle running cost and the time cost for persons in vehicles (Akçelik, 2006). The total running costs to travel a particular distance are the sum of the cost of fuel/electricity needed to charge the battery or the cost of the diesel/gasoline used, plus the maintenance costs. The fuel/electricity consumption estimations are converted to direct vehicle running costs, based on an electricity charging cost of 0.20 euros/kWh, retail diesel price of 1.45 euros/L and gasoline price of 1.64 euros/L (Portuguese tariffs on February 2014). Maintenance costs (including depreciation) assume the price of 0.22 euros/km for buses and trucks, 0.15 euros/km for passenger cars and 0.13 euros/km for SEV (Melo et al., 2014). Without depreciation, the maintenance costs for buses and trucks are reduced to 0.14 euros/km, 0.05 euros/km for passenger cars and 0.03 euros/km for SEV (Gass, Schmidt, & Schmid, 2014; Lajunen, 2014; Palencia, Furubayashi, & Nakata, 2014). The time costs convert an average income to a value of time. Energy costs were based on the fuel (DGEG, 2014) and electricity statistics (Eurostat, 2014). External costs reflect the monetary quantification of CO_2 emissions provided by IMPACT (2008) and are updated with the current GDP per capita. Finally, the purchasing costs were based on market value of the considered (electric and conventional) vehicles, with an active lifetime in the fleet of 4 years or leasing time of 2 years.

RESULTS

Traffic Performance

Along previous research carried out on this topic, the authors evaluated the effects of replacing vans by SEV at different geographical levels (Melo et al., 2014). Within that study, it was shown that the best geographical context for the replacement to occur would be the street/block level. At the city level, the replacement should be lower than 10%, assuming a replacement rate of 1:1 for all scenarios. The study now further developed advances on that analysis as it uses varying replacement rates in order to assure the same delivery capacity (Table 2). The following description reveals how those replacement rates influence the traffic estimations.

The simulation of the impacts at the city level, referring to all vehicles (passenger cars, public transport – buses and taxis – and freight transport – trucks, vans, and SEVs) moving within the city network, indicate that the introduction of SEV replacing vans leads to a general increase in delay times on the traffic network (see Fig. 2).

The higher the replacement share of conventional commercial vehicles is, the worst are the results in terms of traffic performance. Along the 70% replacement share scenario, the overall increase in the delay time can even reach up to 10%. The average speed within the city network decreases and, thus, there are higher delay times and higher travel times. This context of congestion is also followed by lower distance traveled by vehicles. It is important to stress that the decrease in the distance traveled reflects the effect of a slower speed and higher delays, traveling within the same routes and paths as on the BAU scenario. When drivers face congestion, they keep to their initial planned routes. Under such context, the decrease in the distance traveled is not a positive indicator in terms of traffic performance.

The simulation of the impacts at the street level, referring to all vehicles moving within the street, follows the same patterns observed at the

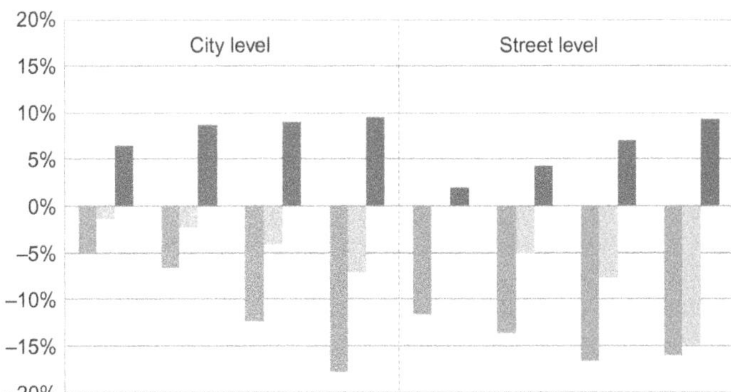

	Sc. 5%	Sc. 10%	Sc. 30%	Sc. 70%	Sc. 5%	Sc. 10%	Sc. 30%	Sc. 70%
Distance travelled (km)	–5%	–7%	–12%	–18%	–12%	–14%	–17%	–16%
Average speed (km/hour)	–1%	–2%	–4%	–7%	0%	–5%	–8%	–15%
Delay time (sec/km)	6%	9%	9%	10%	2%	4%	7%	9%

Fig. 2. Traffic Impacts of the Four Scenarios at City and Street Levels for All Road Users.

city level. Along the scenario of the 70% replacement share, the introduction of SEV to replace vans leads to a general increase in delay times up to 9%. The average speed within the street slightly decreases by 5 km/h (−15%) and, thus, there is a lower distance traveled by vehicles (−16%) and higher travel times (+1%), which are coherent with slower mobility levels.

This concordance of results between the city and the street level highlights the fact that the higher is the replacement share of conventional commercial vehicles; the worst are the results in terms of mobility. This situation is mostly based on the fact that SEV performance and its intrinsic characteristics impose lower speeds on other road users. Moreover, in order to assure the same delivery patterns, there is a need of a higher replacement rate of 2:1 for scenarios of 30% and 70% at the city level and for the scenario of 70% at the street level. Both factors together, as illustrated in Fig. 2, induce to the conclusion that SEV should be implemented at a replacement share lower than 5% if the goal is to promote a more sustainable mobility, considering all the road users.

Operational Performance and External Costs

The quantification of the impacts of each of the four scenarios by disaggregated group of stakeholders makes the different stakeholders interests more transparent and, thus, can help to optimize their process of decision making toward the promotion of more sustainable initiatives on city logistics. Fig. 3 illustrates the quantification of the operating and external costs, which are the main determinant issues for private and public stakeholders' decision making, respectively.

The operating and external costs of each scenario decrease with the replacement share of vans by SEV, both at the city and street levels, when all the road users are considered. Despite driving costs (time/person) increase due to the higher delay times, the running costs (maintenance plus fuel/electricity charging) decrease due to lower distance traveled and lower running energy costs per kilometer. The sum of those costs – operating costs – decreases with higher replacement shares, achieving reduction by −20% at the street level and by −11% at the city level.

The quantification of the environmental externalities approaches the monetary dimension of CO_2 emissions. As with the higher replacement share, there are lower fuel consumption levels derived from lower distances traveled and a lower number of vans running within the area at a lower speed, there is also a decreasing level of CO_2 emissions and consequently,

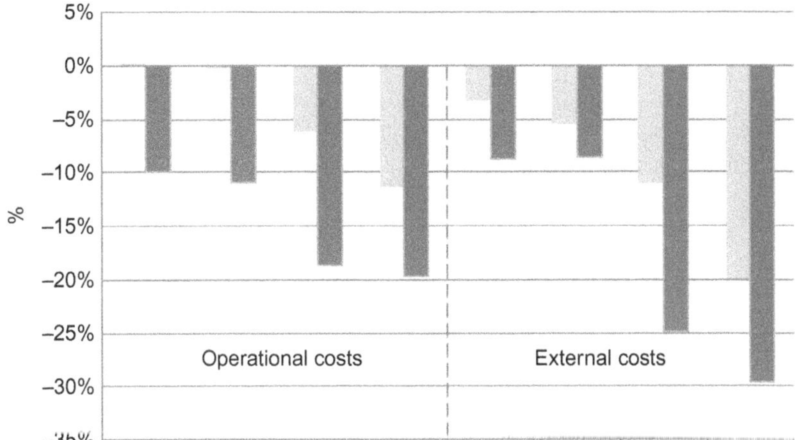

	Sc. 5%	Sc. 10%	Sc. 30%	Sc. 70%	Sc. 5%	Sc. 10%	Sc. 30%	Sc. 70%
City level	0.0%	–0.3%	–6.1%	–11.4%	–3%	–6%	–11%	–20%
Street level	–10%	–11%	–19%	–20%	–9%	–9%	–25%	–30%

Fig. 3. Operating and External Costs of the Four Scenarios for All Road Users.

lower environmental external costs. Altogether, these factors lead to general improvements on operating and external costs despite the unsatisfactory effects observed on their traffic performance (Figs. 1 and 2).

When considering the effects on conventional vans, similar effects are observed to the ones described for the sum of all road users, as shown in Fig. 4.

When these results are disaggregated by group of suppliers, it becomes noticeable that the effects are not homogeneous within the same group of suppliers. Table 3 presents the comparison of operating and external costs of suppliers' operations with light commercial vehicles, either running conventional vans or SEV.

The authors chose to present values based on the total effect caused by the two specific typologies of vehicle (vans and SEV) in the total network (city or street), in order to compare those impacts from the private and public perspectives, respectively. Trucks are not included on the analysis since SEVs do not replace them, due to their intrinsic delivery characteristics. The results from one of the vehicle typologies (either vans or SEVs) should be interpreted as reflecting the costs per hour of all vehicles from that typology operating within the city/street. Moreover, the results from the "suppliers" category refer to the sum of the results from all vans and SEVs operating within the network.

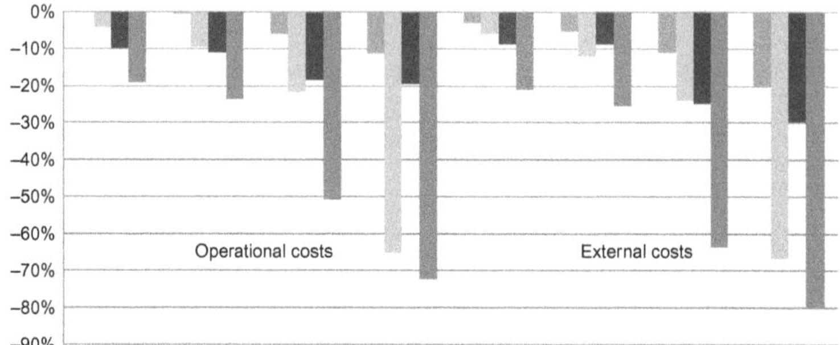

	Sc. 5%	Sc. 10%	Sc. 30%	Sc. 70%	Sc. 5%	Sc. 10%	Sc. 30%	Sc. 70%
Total–city level	0.0%	−0.3%	−6.1%	−11.4%	−3%	−6%	−11%	−20%
Vans–city level	−4%	−9%	−22%	−65%	−6%	−12%	−24%	−67%
Total–street level	−10%	−11%	−19%	−20%	−9%	−9%	−25%	−30%
Vans–street level	−19%	−24%	−51%	−72%	−21%	−26%	−64%	−80%

Fig. 4. Operational and External Costs of Vans versus All Road Users.

Table 3. Operating and External Costs for Suppliers Compared with BAU.

		Operational Costs/Hour (euro)				External Costs/Hour (euro)			
		Sc. 1 5%	Sc. 2 10%	Sc. 3 30%	Sc. 4 70%	Sc. 1 5%	Sc. 2 10%	Sc. 3 30%	Sc. 4 70%
City level	Suppliers	−44	−102	1,191	2,564	−7	−14	−28	−77
	Vans	−200	−424	−985	−2,953	−7	−14	−28	−77
	SEVs	156	321	2,176	5,517	0	0	0	0
Street level	Suppliers	−0.45	−0.53	−1.09	−0.91	−0.05	−0.06	−0.15	−0.19
	Vans	−0.47	−0.58	−1.27	−1.79	−0.05	−0.06	−0.15	−0.19
	SEVs	0.02	0.06	0.18	0.88	0.00	0.00	0.00	0.00

When increasing the replacement share of the scenario, operating costs on vans decrease and inversely increase on SEVs, while external cost on vans decreases. The explanation for this variation is that, higher the replacement share of vans by SEV, lower the number of vans operating within the area. Less vans running imply lower operational and external costs associated with this vehicle typology. Inversely, the higher the replacement share of vans by SEV is, higher the number of SEVs operating within the area. More SEVs imply more (total) operational costs on SEVs. As the purpose is to know the impact for all suppliers of replacing vans by SEVs, the quantification of those operational and external costs on a network total basis (rather than by kilometer) are determinant to evaluate if the initiative is indeed a good practice to be applied in the area.

On the second scenario, vans running within the city represent −424 euros/h on the operating costs than on the BAU (−9%), corresponding to having less than 720 vans/day distributing goods. On the same scenario, SEVs represent additional operating costs of 321 euros/h, corresponding to a replacement rate of 1 SEV:1 van to assure the same distribution capacity. This accounts for a total operational benefit for suppliers who are running light commercial vehicles of 102 euros/h (equivalent to 0.53 euros/vehicle/workday). With the 10% reduction of the number of vans running within the city, the external costs decrease by −6%. Private vehicles, taxis, and vans also present reductions in their external costs up to −5%, −7%, and −12%, respectively. Within the same scenario, the total external costs caused by suppliers (vans) would be reduced by 14 euros/h, equivalent to 0.07 euros/vehicle/workday, revealing the positive effects of such replacing share both for suppliers and the city.

The pattern observed on the street level is also positive, but with minor absolute effects. On the third scenario – the best theoretical one to this geographical scale of implementation – vans running within the street represent −1.27 euros/h on the operating costs than on the BAU (−51%), corresponding to having less than 18 vans/day distributing goods along the street. On the same scenario, SEVs would have additional operating costs of 0.18 euros/h, corresponding to a replacement rate of 2 SEVs:1 van. In what refers to external costs, the reduction of the number of vans circulating within the street leads to a reduction in the respective fuel consumption levels and consequent CO_2 emissions. As SEVs do not emit pollutants, the external costs of suppliers (vans and SEVs) decrease.

When public stakeholders promote the use of SEV as a sustainable policy on city logistics, the main argument to support their initiatives are: lower external costs and probable lower operating costs due to the lower costs of fuel. Private stakeholders, namely, suppliers, have reacted in a cautious way to this argument. The balance between private and public stakeholders' main interests and respective costs, when acquisition costs are not accounted, supports the replacement of vans by SEVs (Table 3), when the replacement rate is 1 SEV:1 van. The slight benefit in terms of costs may however not be robust enough to justify its implementation on a market share of 30% at the street level or of 10% at the city level. When the replacement rate is 2 SEVs:1 van, it is not advised to opt by this solution from an operational point of view. Considering the previous analysis, it is possible to conclude by now that SEVs should not have a market share higher than 5% and are only economically feasible (not considering purchasing and battery issues), when the replacement rate of 1:1 is sufficient to assure the same operational performance.

In the next section, the authors try to contribute to enlighten this argument, exploring the purchasing costs and battery issues.

(Un)Balancing Private and Public Stakeholders Perspectives: The Purchasing Costs and Battery Issues

When accounting the purchase costs, two different business models can be contemplated: the full acquisition and the leasing model. Considering the full acquisition of the vehicle, for a replacement rate of 1 SEV:1 van and comparing the cost of a conventional van with a SEV, equivalent to downsizing the vehicle capacity, up to 1.18 euros/h/vehicle can be saved. This value drops up to 0.82 euros/h in the leasing model option, as calculated by authors with the technical characteristics and vehicles conditions detailed, as described in the section "The Model: Base Conditions, Assumptions, and Outputs."

When considering the replacement rate of 2 SEVs:1 van, which is the case of scenarios equal and higher than 30% the city level and of the scenario 70% at the street level, authors estimated that replacing vans by SEVs can lead to savings of 0.19 euros/h and 0.13 euros/h in the purchase and leasing options, respectively. Therefore, for the specific case of downsizing the vehicle capacity, the replacement of vans by SEVs leads to lower purchasing costs.

Nonetheless, this is not what private stakeholders, namely suppliers, expect to compare. They are looking to compare vehicles with similar characteristics and technical performance. And in such conditions, if the replacement would occur with one electric duty vehicle with a similar weight and capacity, conventional vans would indeed have lower costs. In the full acquisition business model, conventional vans would represent a lower cost of 0.97 euros/h/vehicle than a similar EV. This value would be of 4.74 euros/h/vehicle in the leasing option, assuming a replacement rate of 1 EV:1 van. With a replacement rate of 2 SEVs:1 van, the replacement would imply additional costs of 4.10 euros/h/vehicle and 11 euros/h/vehicle in the acquisition and leasing options, respectively. These results confirm the high purchase cost of EVs as the major obstacle to spread its adoption (Kley et al., 2011; Vermie, 2002).

The other controversial issue when comparing conventional and EVs is the battery issue, namely, the charging cost and its acquisition/leasing cost. When considering the battery charging costs at a replacement rate of 1 SEV:1 van, the operating costs increase by 0.03 euros/km/SEV, which is lower than the operating costs of conventional vehicles. When considering the acquisition costs of a new replacement battery, the comparison benefits conventional vehicles over EVs. If the battery is obtained through leasing, the current values on the market vary between 80 and 160 euros/month, which is equivalent to additional operating SEV costs of 0.33 and 0.66 euros/h of operation, respectively. If the comparison is made with an EV with similar technical characteristics to a conventional van, the balance is more positive to the latter ones as EVs have, in general, higher purchasing and battery costs than SEVs.

CONCLUSIONS

The research conducted in this chapter confirms the insights from other studies and also provides an additional contribution to the topic. The innovative contribution to the topic is the simultaneous consideration of those who promote sustainable policies (public stakeholders) and those who try to follow those recommendations keeping the operational performance standards (private stakeholders). Moreover, the analysis enlightens transversal effects, namely, the impact on the mobility of the total area of specific users as well as the respective running and external costs. With such interdisciplinary approach, the authors were able to (a) highlight the importance of the replacement rate to make a decision on whether or not to use SEVs replacing vans; (b) confirm that SEVs are economically competitive with conventional vans if the replacement rate is 1:1; (c) confirm that SEVs have a better performance at the street level rather than at the city level; (d) quantify the maximum market share for SEV when used with normal traffic as a niche of market (lower than 5%), mostly dependent on its speed manufacturer limitation; and (e) explicitly quantify the comparison of SEVs with conventional vans revealing that although its benefits exist, they are not significant enough to drive suppliers for their adoption. The following paragraphs present these conclusions in more detail, referenced to the case study carried out in Porto.

The analysis revealed that the limited speed and the smaller capacity of SEV explain their negative impacts in terms of traffic effects. These two intrinsic characteristics of SEV worsen the traffic performance. On the one hand, speed decreases within the network. On the other hand, a more limited capacity implies an increasing number of vehicles in the traffic. In order to guarantee the same delivery patterns, for some of the tested scenarios, the replacement rate must be of 2 SEVs for 1 van that doubles the amount of vehicles. Despite, SEVs are economically competitive than EVs compared with conventional vehicles, the influence of the conversion rate might unbalance and predictably, making it economically unfeasible to choose SEV instead of conventional diesel vans. Under these conditions and considering a global perspective (all road users), the shift to SEV should be lower than 5% if the goal is to promote a more sustainable mobility. This share confirms the use of SEV as a niche market, with SEVs working in a complementary way to conventional vehicles. This conclusion is coherent with the current market tendency of the use of SEV on mail delivery, home delivery journeys, public institution services, and other specific applications.

Considering the cost analysis, from the public stakeholders' point of view, lower associated external costs and lower operating costs due to the lower costs of electricity have been the main arguments to support sustainable policies on city logistics. The case study described in this chapter confirms the validity of those arguments. However, the relevance of those benefits is overshadowed when purchasing and battery costs are considered and when traffic disturbance linked to speed limitations is accounted. From the private stakeholders' perspective, the use of SEV is not operationally or economically attractive enough to justify its widespread adoption. In terms of operational limitations, SEVs can only cope with parcels and not pallets and, consequently, SEVs can only cover specific types of business. Such context makes the economic feasibility analysis even more complex as it can require a conversion rate higher than 1:1 and such factor is determinant in the suppliers' decision of adopting SEVs. Likewise, it implies the existence and availability of a small consolidation center from where the vehicles depart and serve a high density of clients.

Regarding the battery issue, there still is a nebulous and complex business area that needs to be clarified. With the current market prices announced by producers, the battery cost can indeed be a relevant risk to the widespread usage of SEVs. Nonetheless, suppliers would be more confident if the replacement could assure the same operational and economic competitiveness. This cost risk can only be diminished through risk sharing with public stakeholders and subsidizing policies. The question is whether cities should compel themselves to actively developing this specific solution, as it is uncertain to spend a significant amount of time and money for such limited positive effects. However, if suppliers are willing to adopt them, cities should encourage the development and deployment of SEVs, creating the facilities to support their operation. The methodology presented in this chapter can be a reliable starting point to base the city's decision on whether to promote SEV as a sustainable city logistics policy, by defining the potential market share of SEV, the expected costs and mobility effects prior to its implementation.

ACKNOWLEDGMENTS

Thanks are due to Fundaçãopara a Ciência e Tecnologia for the Post-Doctoral financial supports (SFRH/BPD/79684/2011 and SFRH/BPD/74080/2010). The authors also acknowledge EU Project DOROTHY financed by the Programme Regions of Knowledge (REGIONS-2012-2013-1).

REFERENCES

Akçelik, R. (2006). Operating cost, fuel consumption and pollutant emission savings at a roundabout with metering signals. Seventh International Congress on Advances in Civil Engineering, 11–13 October 2006, Istanbul, Turkey.

Barredo, J., & Demicheli, L. (2003). Urban sustainability in developing countries' megacities: Modeling and predicting future urban growth in Lagos. *Cities*, *20*, 297–310.

Bartern, G., Reichmuth, D., Westbrook, J., Malczynski, L., West, T., & Manley, D. (2012). Parametric analysis of technology and policy tradeoffs for conventional and electric light-duty vehicles. *Energy Policy*, *46*, 473–488.

Birkner, M. (2012). Electromobility for heavy-duty vehicles (HDV). 26th Electric Vehicle Symposium (pp. 922–927).

Browne, M., Allen, J., & Leonardi, J. (2011). Evaluating the use of an urban consolidation centre and electric vehicles in central London. *IATSS Research*, *35*, 1–6.

Dablanc, L. (2007). Goods transport in large European cities: Difficult to organize, difficult to modernize. *Transportation Research Part A: Policy and Practice*, *41*(3), 280–285.

Dablanc, L. (2011). City distribution, a key element of the urban economy: Guidelines for practitioners. In C. Macharis & S. Melo (Eds.), *City distribution and urban freight transport* (pp. 13–36). Cheltenham: Edward Elgar.

Davis, B., & Figliozzi, M. (2013). A methodology to evaluate the competitiveness of electric delivery trucks. *Transportation Research Part E*, *49*, 8–23.

DGEG. (2014). Estatísticas e Preços. Direcção Geral de Energia e Geologia Online.

Duin, J., van Tavasszy, L., & Quak, H. (2013). Towards E(lectric)-urban freight: First promising steps in the electric vehicle revolution. *European Transport/Trasporti Europei*, *54*(9), 2. ISSN 1825-3997.

Etienne, N. (2012). Clean vehicles deliveries in Brussels. BESTFACT Urban Freight Workshop: "Public and Private Partnerships for Zero Emission Urban Freight in Europe", Belgium, Brussels, 6–7 December 2012.

Eurostat. (2014). *Electricity statistics*. Retrieved from http://epp.eurostat.ec.europa.eu/statistics_explained/index.php/Electricity_market_indicators. Accessed on the July 13, 2014.

Feng, W., & Figliozzi, M. (2013). An economic and technological analysis of the key factors affecting the competitiveness of electric commercial vehicles: A case study from the USA market. *Transportation Research Part C: Emerging Technologies*, *26*, 135–145.

Gass, V., Schmidt, J., & Schmid, E. (2014). Analysis of alternative policy instruments to promote electric vehicles in Austria. *Renewable Energy*, *61*, 96–101.

Giuliano, G., O'Brien, T., Dablanc, L., & Holliday, K. (2013). *Synthesis of freight research in urban transportation planning*. Washington, DC: National Academy of Sciences.

IMPACT. (2008) *Handbook on estimation of external costs in the transport sector*. Delft: European Commission.

Kley, F., Lerch, C., & Dallinger, D. (2011). New business models for electric cars – A holistic approach. *Energy Policy*, *39*, 3392–3403.

Lajunen, A. (2014). Energy consumption and cost–benefit analysis of hybrid and electric city buses. *Transportation Research Part C: Emerging Technologies*, *38*, 1–15.

Leonardi, J., Browne, M., & Allen, J. (2012). Before-after assessment of a logistics trial with clean urban freight vehicles: A case study in London. *Procedia – Social and Behavioral Sciences*, *39*, 146–157.

Melo, S. (2010). *Evaluation of urban goods distribution initiatives towards mobility and sustainability: Indicators, stakeholders and assessment tools*. Faculty of Engineering of Porto, Portugal: University of Porto.

Melo, S. (2011). Evaluation of urban goods distribution initiatives: An empirical overview in the Portuguese context. In C. Macharis & S. Melo (Eds.), *City distribution and urban freight transport* (pp. 234–260). Cheltenham: Edward Elgar.

Melo, S., Baptista, P., & Costa, Á. (2014). Comparing the use of small sized electric vehicles with diesel vans on city logistics. *Procedia – Social and Behavioral Sciences*, *111*, 350–359.

Palencia, J., Furubayashi, T., & Nakata, T. (2014). Techno-economic assessment of light-weight and zero emission vehicles deployment in the passenger car fleet of developing countries. *Applied Energy*, *123*, 129–142.

Ruesch, M., & Petz, C. (2008). Best practice update: E-commerce and urban freight distribution (home shopping). *BESTUFS II, Best Urban Freight Solutions II*, 101. Deliverable 2.4 Part I.

Russo, F., & Comi, A. (2012). City characteristics and urban goods movements: A way to environmental transportation system in a sustainable. *Procedia – Social and Behavioral Sciences*, *39*, 61–73.

SEStran. (2010). *Sustainable urban distribution*. Edinburgh: Colin Buchanan and Partners.

Thiel, C., Perujo, A., & Mercier, A. (2010). Cost and CO_2 aspects of future vehicle options in Europe under new energy policy scenarios. *Energy Policy*, *38*(11), 7142–7151.

van Rooijen, T., & Quak, H. (2010). Local impacts of a new urban consolidation centre – The case of Binnenstadservice.nl. *Procedia – Social and Behavioral Sciences*, *2*(3), 5967–5979.

Vermie, T. (2002). ELCIDIS electric vehicle city distribution. Final report, Rotterdam.

Wolpert, S., & Reuter, C. (2012). Status quo of city logistics in scientific literature systematic review. *Transportation Research Record: Journal of the Transportation Research Board*, 110–116.

ABOUT THE EDITORS

Cathy Macharis is a professor at the Vrije Universiteit Brussel and visiting professor at University of Gothenburg. Her research group MOBI – (Mobility, Logistics and Automotive Technology) is an interdisciplinary group focusing on sustainable logistics, electric and hybrid vehicles and travel behaviour. She has been involved in several regional, national and European research projects dealing with topics such as the location of intermodal terminals, assessment of policy measures in the field of logistics and sustainable mobility, electric and hybrid vehicles, etc.

Sandra Melo has been focused on urban freight transport and green logistics. She completed her Ph.D. in 2010 at FEUP, Porto, Portugal, investigating the influence of the power of the local context on the performance of last mile 'best practices' towards a more sustainable mobility, considering public and private (freight and passenger) transport. She is currently Post-doc researcher at IDMEC-IST, consultant on Transport Policy and Planning and Senior Member of the Portuguese Civil Engineers Chamber. Her current research focuses on urban logistics, sustainable freight transport, transport policy evaluation and transport behaviour decision-making.

Johan Woxenius is professor of Maritime Transport Management and Logistics at University of Gothenburg since 2008. He received the degrees of M.Sc. (Industrial engineering), Ph.D. and associate professor/docent at Chalmers University of Technology. His main research field is maritime and intermodal freight transport and the research covers sustainability, industrial organisation, production systems, traffic designs, information systems and technology surveys and evaluations. He is member of the board of the School of Business, Economics and Law at University of Gothenburg and leads the university's part of the maritime research and education programme Lighthouse, the logistics research centre Northern LEAD and the Area of Advance Transport, all run together with Chalmers. He is member of the Royal Swedish Academy of Engineering Sciences, on the scientific advisory boards of several research funding organisations, editorial boards of journals and he is member of the Scientific Committee of the World Conference on Transport Research Society.

Tom van Lier is a postdoctoral researcher in the research group MOBI (Vrije Universiteit Brussel) led by Professor Cathy Macharis. His work focuses on evaluating the sustainability of transport solutions by means of external transport cost calculations. He has been involved in several research projects dealing with topics such as social cost-benefit analysis of transport options, external cost savings of freight bundling, carbon footprint calculations and life-cycle assessment of transport services. He is also involved in the Lean & Green project in Flanders (led by the Flemish Institute for Logistics) as a neutral assessor for the plans of actions of companies to reduce their carbon footprint.

ABOUT THE AUTHORS

Patrícia Baptista received a Chemistry degree in 2006 and her PhD degree in Sustainable Energy Systems within the MIT Portugal Program in 2011 from Instituto Superior Técnico, University of Lisbon, Portugal. As a postdoctoral researcher, her main research topics have been on the quantification of energy and environmental impacts of alternative options for the transportation sector and on how information and communication technologies can help characterizing vehicle use and influencing driver behavior to promote energy efficiency and road safety. She has also managed and participated in R&D projects, published several journal papers, supervised MSc students thesis and had teaching experiences as invited lecturer.

Maria Björklund is an Associate Professor of Logistics Management at the Department of Management and Engineering, Linköping University, Sweden. She received her PhD in 2005 from Lund University, Sweden (the dissertation was on environmental considerations when purchasing transport services). Dr Björklund has published journal articles in e.g. International Journal of Physical Distribution & Logistics Management, Journal of Business Research, Journal of Purchasing and Supply Management, Benchmarking: an International Journal, International Journal of Supply Chain Management, Journal of Social Responsibility, and International Journal of Productivity & Performance Management. She is in the Editorial Advisory Board of International Journal of Physical Distribution & Logistics Management. Her areas of speciality include sustainable supply chain management and logistics, with a focus on environmental and social considerations in logistics and purchasing, green performance measurement, city logistics as well as product development and innovation to support a more sustainable development of logistics.

Ming Chen is senior consultant for the TNO groups 'Sustainable Transport and Logistics' and 'Strategy and Policy' and EU account manager for logistics. Expert in the field of economics and freight transport modeling and forecasting, data- and information systems, sustainability and impact assessment. Broad international experience and he has been coordinator of several large research projects for the European Commission. Currently he is project manager in the project Desire on the development of a system of Indicators for a Resource efficient Europe and the project CarbonCap on Carbon emission mitigation by Consumption-based Accounting and Policy. Further he is currently involved in several other EU projects including project involving Forward looking scenarios (Flagship), Transport modeling (High-Tool), market and impact assessment (Spectrum and COFRET). He has performed several internal studies on sustainable logistics and energy related aspects amongst which the impact of oil scarcity on the potential for economic growth.

Álvaro Costa is Professor of Transports at University of Porto, FEUP, and CEO of TRENMO (Transportes, Engenharia e Modelação). He holds a PhD in Transport Economics and has broad experience as project manager on R&D projects on city logistics, transport networks, urban mobility and transport operations management.

Sharon Cullinane has worked in transport research for more than 20 years. She completed her PhD at Plymouth University on the effects of heavier vehicles in 1987. Since then she has worked in universities around the world including Oxford University, the University of Hong Kong, the Ecole Superieur de Commerce in Rennes, the Egyptian National Institute of Transport in Cairo and most recently, Heriot-Watt University in Scotland. She is now a Visiting Fellow at the University of Gothenburg. Her work throughout her career has focussed on the environmental and behavioural aspects of transport; both freight and passenger.

Jonas Flodén has a PhD in Business Administration focusing on logistics. He is a senior lecturer at the Department of Business Administration, School of Business, Economics and Law at University of Gothenburg, Sweden and he is Head of section for Industrial and Financial Management & Logistics at the department. Dr Flodén received his PhD in 2007 on a thesis on modelling of intermodal transport systems. His main research areas are intermodal freight transport and business models. Teaching areas include intermodal transport, supply chain management and information systems.

Helena Forslund is a Professor of Logistics at Linnaeus University, Sweden. She received her PhD from Institute of Technology at Linköping University, Sweden. Dr Forslund has published journal articles in e.g. International Journal of Physical Distribution & Logistics Management, International Journal of Operations & Production Management, International Journal of Retail & Distribution Management, International Journal of Productivity & Performance Management and Management of Environmental Quality;

an International Journal. She is in the Editorial Advisory Boards of International Journal of Physical Distribution & Logistics Management and of Industrial Management & Data Systems. Her research interests are in the area of supply chain management, performance management and process integration. She is also director for the master's program in Business Process and Supply Chain Management at Linnaeus University.

Jesus Gonzalez-Feliu is research engineer in data production and analysis the French Centre National de la Recherche Scientifique. Graduated as an engineer from the INSA-Lyon in civil engineering and town planning in 2003, he obtained a MSc in sustainable transport and mobility and then a PhD in computer and systems engineering at the Politecnico of Turin (Italy). He joined the LET in 2008 as a post-doc before taking on his present position. His main research interests are city logistics demand modelling, multi-stage systems, shopping trip generation and simulation, group decision making, logistics sharing and scenario assessment, among others. He is author or co-author of more than 30 papers in international peer-review journals or books, and more than 20 communications in international conferences. He also coordinated an international scientific book on sustainable urban logistics (Springer) and two special editions of international peer-reviewed scientific journals.

Anne Goodchild is the Allan and Inger Osberg Associate Professor of Civil and Environmental Engineering at the University of Washington. Her research interests lie in the analysis of logistic systems, with an emphasis on freight transportation and sustainability. Recent research has evaluated CO_2 emissions in strategic routing and schedule planning in urban pick-up and delivery systems, replacement of personal travel by freight travel in delivery systems, and the relationship between truck activity and urban form. She holds a PhD and MS in Civil and Environmental Engineering from the UC at Berkeley, and a BS (with High Honors) in Mathematics from the UC at Davis. Before attending Berkeley she worked in consulting for 5 years in Europe and North America, for PricewaterhouseCoopers LLP and Applied Decision Analysis Inc., modeling business problems such as airline fleet maintenance scheduling. She serves as Chair of TRB's Intermodal Freight Transportation committee.

Egbert Guis is specialized in (re)organizing and especially renewing supply chains in changing circumstances. He applied this skill throughout the years as manager or consultant in several sectors, both national and international. He was frequently involved in green initiatives regarding city distribution, both from business and research perspective. Crucial to him is to promote collaboration between all supply chain partners using fact-based evidence. His contribution to TNO is mainly in the support of the private sector with available knowledge within TNO to solve their business issues. But he was also involved in some projects for the European Commission like Reorient, Promit and Retrack. He had a main contribution to the development of the ReponseTM simulation model and also placed Risk Management on the research agenda of TNO. At this moment he is a supply chain consultant, running projects besides TNO to support companies and government implementing Urban Consolidation Centers.

Hyun-Chan Kim is a PhD candidate in the Department of Civil and Natural Resources Engineering of the University of Canterbury. Mr Kim holds degrees in Transport Engineering from the University of Utah, USA, and Logistics Engineering from the Korea Maritime University. He is currently finalizing his PhD on developing a mode choice model for NZ freight transport. He worked in the petroleum industry as a logistics manager for five years in the United States before starting postgraduate study. His research has involved mainly two areas, mode choice and trip distribution. He has been involved with several travel behaviour researches including dynamic traveller response modelling for degraded transport network, BRT ridership forecasting and service planning, and modelling freight transport mode choice and modal shift. His research and professional interests are sustainable transport, logistics and freight transport, modelling travel behaviour, discrete choice models, and transport policy for modal shift.

Diana Kusumastuti holds a PhD in Transportation Sciences from Hasselt University, Belgium and degrees in Architecture Engineering and Urban Management and Development. She has been working as a Postdoctoral Research Fellow in Transportation at the Department of Civil and Natural Resources Engineering of the University of Canterbury since 2013. Earlier, she worked as a Researcher at the Centre of Transport Studies of the University of Twente in the Netherlands. To date, she has been involved in numerous research projects, including 'SUNSET', an EU funded research project under FP7 about promoting sustainable mobility through a mobile application. Furthermore, she has developed a method to elicit people's travel-related decision-making processes and has used the outcomes to generate mental-level models to predict people's travel behaviour. Her research interests include, amongst others, choice modelling, stated preference survey, transport planning and policy, and transport behaviour decision-making.

Dries Meers holds a Master of Science in Urban Geography and currently works as a research associate at the MOBI – Mobility, Logistics and Automotive technology research group of Prof. Dr Cathy Macharis at the Vrije Universiteit Brussel. His PhD work focuses on intermodal transport and the further development of a Location Analysis Model for Belgian Intermodal Terminals. His main interests are in modal choice and the analysis of policy measures and competition in the intermodal transport sector.

Katrin Molina-Besch is a PhD student at the division of Packaging Logistics at Lund University. Katrin holds an Environmental Engineering degree from RheinMain University of Applied Sciences in Germany and a Master's degree in Environmental Management and Policy from the International Institute of Industrial Environmental Economics (IIIEE) at Lund University. Prior to her PhD studies, Katrin worked for seven years in the packaging industry with life-cycle assessment. Her research focuses on environmental issues in packaging logistics, such as how to develop packaging systems so that environmental impacts along the supply chain are minimized. Katrin is course coordinator and teacher in a master course about packaging technology and development at Lund University.

Joëlle Morana is an assistant professor with capacitation to direct PhD theses at Université Lumière Lyon 2. After working for 10 years as a cost controller with the multinational STMicroelectronics, she returned to her studies to prepare for an Advanced Studies Degree Strategy option in 1999, followed by her defence of her PhD thesis in October 2002. Her thesis, on the synergy between Supply Chain Management and Dashboards, was rewarded by the 2002 Prize for a cross-disciplinary thesis in accounting and management. Her main research subjects are sustainable supply chain management, synergy between supply chain management and dashboards, reverse logistics and tracking, among others. She is author or co-author of more than 20 papers in peer-review journals or international books, 3 books (two of them on Sustainable Supply Chain Management, published respectively by Hermès-Lavoissier and Wiley-ISTE) and more than 10 communications in international conferences.

Nina Nesterova is a research scientist at TNO. She holds two Master's Degrees: in International Economics (Russia) and Logistics and Organisation (France) and a PhD degree in International Transport (France). She has more than 10 years of professional experience in research and consulting projects in such topics as: transport economics, intermodal transport, transport planning and policy, cross-border issues, logistics, sustainable transport. Within TNO Nina is specialized on freight transport research and consultancy, with a particular focus on sustainable transport policy.

Alan Nicholson holds degrees in Civil Engineering and in Transportation and Traffic Planning. He has been teaching and researching various aspects of transportation planning and engineering at the University of Canterbury since 1981. He has been the Director of the Transportation programme since 2002 and was the Head of Civil and Natural Resources Engineering from 2005–2009. He has been a Fellow of the NZ Institution of Engineers since 2000, and was National Chairman of the Institution's Transportation Group from 2004 to 2007. From 1981 to 2009, he has advised the NZ Government's highways agencies on transportation research needs and priorities. He is an expert advisor on transportation research for the Australian Research Council and the Natural Sciences and Engineering Research Council of Canada, and is a member of the International Scientific Committees for the International Symposia on Transport Network Reliability and on Transport Simulation.

Henrik Pålsson is an Associate Professor in packaging logistics at Lund University, where he also was awarded his PhD in 2009 and MSc in Mechanical Engineering 2000. His research and teaching focuses on packaging logistics, sustainable freight transport and sustainable supply chain management. Dr Pålsson has published books, academic reports and scientific articles on these topics in international journals, for instance, in International Journal of Physical Distribution and Logistics Management, Packaging Technology and Science and Benchmarking: an International Journal. Prior to his research career, Dr Pålsson worked as a mechanical engineer and as a consultant with logistics management for five years.

Hans Quak is a senior research scientist at TNO. He mainly focuses on urban freight transport and sustainable logistics. He worked in EU projects FREVUE, Straightsol (both WP leader), and CityLog. Next, he was project manager of the Transumo-project 'Transition towards sustainable urban freight transport' and (co)authored several research reports and papers on freight transport, logistics and (local) policy in these areas. Hans is TNO's project leader in VREF Center of Excellence Sustainable Urban Freight Systems. Before, from 1997 to 2002 Hans studied Business Administration in Rotterdam (master in Logistics Management) after which he started as a PhD candidate at the Rotterdam School of Management in January 2003. Hans successfully defended his PhD thesis 'Sustainability of Urban Freight

Transport' in March 2008. Hans published papers in international top journals, such as Journal of Operations Management (best paper 2007 finalist award) and Transportation Science. Hans was also appointed as assistant professor for one day a week at the Delft University of Technology.

Richard Smokers holds a PhD in experimental physics, but since then has gained 22 years of experience in technology assessment and policy-oriented studies in the field of transport, energy and environment. He works for TNO as strategic consultant sustainable mobility, carrying out projects for governments and a range of private stakeholders. During his entire career Richard has worked in interdisciplinary projects combining knowledge on vehicle and propulsion system technologies with e.g. well-to-wheel analyses, traffic and mobility research, air quality modelling and transition and innovation theory. From 2005 onwards Richard has worked as consultant for DG CLIMA in a series of projects on regulation of the CO_2 emission of passenger cars and light commercial vehicles. More recently Richard has expanded his field of work to include the assessment of CO_2 reduction options for HD vehicles and wider aspects of sustainable logistics, including the development and application of carbon footprinting methodologies.

Edith Sorkina has a Licentiate Degree in Business Administration and is currently a PhD student at Department of Business Administration, School of Business, Economics and Law at University of Gothenburg, Sweden. The title of the licentiate thesis defended in 2013 was Customer Managed Intermodal Transport Chains. Her main research area is intermodal freight transport, focusing on organizational aspects, development process, business models, risk, and shippers' role in the intermodal system.

Mathieu Strale is a researcher at the Institute for Environmental Management and Land-use Planning (Institut de Gestion de l'Environnement et d'Aménagement du Territoire – IGEAT), Free University of Brussels (ULB), Belgium. Having finished his PhD in geography in 2013 on the location of logistics activities and their territorial impacts, Mathieu Strale has developed an expertise on freight transport topics and related issues. He also studies urban supply chains and their socio-economic, environmental or political ties. His research covers Northwest Europe as well as the Brussels and surrounding spaces. By developing an empirical approach based on the location and organisation of stakeholders and on the analysis of their decisions and actions, Mathieu Strale develops an original and often neglected point of view on these topics.

Lóránt Tavasszy is principal scientist at the national research institute TNO in Delft, the Netherlands and holds a chair in Freight Transport and Logistics at the Delft University of Technology. His research is on freight transport and logistics modelling, transport policy evaluation and sustainable transport systems. Recent work included the modelling of global maritime container networks, impact assessment of the internalization of external costs and new concepts for sustainable city logistics. Prof. Tavasszy has produced over 100 publications for international journals, books and conferences and holds regular guest lectures in the Netherlands and abroad. He is chair and member of various committees in the international research community (TRB, WCTR, AET) including the European Technology Platforms ERTRAC and ALICE, and is core partner in the Volvo Research Foundation's Center of Excellence on Sustainable Urban Freight Systems.

Tom Vermeiren is a senior management consultant specialized in strategic and financial advice to the port sector. Tom has more than 17 years of expertise working in a dedicated team of port consultants at PricewaterhouseCoopers. He has conducted various studies for public and private parties in the field of port economics, strategic planning and corporate finance. He has worked on assignments in Europe, North America, Middle East, Caribbean, China and India. Building on his expertise, Tom completed his PhD at the Vrije Universiteit Brussels where he carried out fundamental research about competing intermodal door-to-door chains through Europe's main ports. Tom's main areas of specialization include: feasibility studies & business cases, market assessments & competitive positioning, shareholder value analysis, privatisation & transactions, concession & pricing strategies, business modelling, location strategy & site selection, regional economic development policy.

Erica Wygonik's practice examines various transportation modes to identify opportunities to reduce system-wide impacts. Her work focuses on strategic use of home delivery and other mixed-mode transportation strategies. She has explored the state of freight modeling and data sharing, considered ways to compare freight modes, modelled bike-share, and evaluated delivery services across land densities. She holds Phd and MS degrees in Transportation Engineering from the department of Civil and Environmental Engineering at the University of Washington, a Bachelor of Engineering degree from the Thayer School of Engineering at Dartmouth College, and a Bachelor of Arts degree in Cognitive Science from Dartmouth College. Before matriculating at the University of Washington, she was a senior associate at RSG, where she led the microsimulation and traffic operations practice areas.

INDEX